A Handbook of Transcription Factors

SUBCELLULAR BIOCHEMISTRY

Recent Volumes in this Series

For further volumes:
http://www.springer.com/series/6515

T.R. Hughes
Editor

A Handbook of Transcription Factors

Springer

Editor
T.R. Hughes
University of Toronto
Centre for Cellular and Biomolecular
Research
160 College St.
M5S 3E1 Toronto
Canada
t.hughes@utoronto.ca

ISSN 0306-0225
ISBN 978-90-481-9068-3 e-ISBN 978-90-481-9069-0
DOI 10.1007/978-90-481-9069-0
Springer Dordrecht Heidelberg London New York

Library of Congress Control Number: 2011926500

Printed on acid-free paper

Springer is part of Springer Science+Business Media (www.springer.com)

Preface

The major goal of "A Handbook of Transcription Factors" is to provide a resource encompassing major facets of the molecular biology of TFs. As a Handbook, this volume is intended to provide a broad overview of this increasingly complex field, and is aimed at providing general context rather than fine details of specific examples. After decades of study of TFs at the molecular level, over 100,000 TF-related publications on Medline, hundreds of genome sequences, and continuous technological advances, a comprehensive review on TFs is not possible, even in book format. And since most reviews focus on specific topics, it can be difficult to get a perspective on what is known, what is not known, and what the global problems are. Topics in this book include the TF repertoire in both prokaryotes and eukaryotes, TF targeting and specificity, the properties of regulatory sequence, the interaction of TFs with chromatin, and mechanisms of TF action. The chapters are written by a team of experts, and highlight the current state of knowledge and research, as well as numerous challenges. I hope that this book will serve as a guide and reference for readers of all levels.

Toronto, ON T.R. Hughes

Contents

Contributors

Asifa Akhtar Laboratory of Chromatin Regulation, Max Planck Institute of Immunobiology, 79108 Freiburg, Germany, akhtar@immunbio.mpg.de

Harm van Bakel Banting and Best Department of Medical Research, University of Toronto, Toronto, ON, Canada M5S 3E1, hvbakel@gmail.com

Thomas R. Bürglin Department of Biosciences and Nutrition, and Center for Biosciences, Karolinska Institutet, Hälsovägen 7, Novum, SE 141 83 Huddinge, Sweden, Thomas.burglin@ki.se

Derek Caetano-Anolles Department of Cell and Developmental Biology, Institute for Genomic Biology, University of Illinois, Urbana, IL 61801, USA, dcaetan2@illinois.edu

Raghunath Chatterjee Laboratory of Metabolism, NCI, NIH, Bethesda, MD 20892, USA, chatterjeer2@mail.nih.gov

Peggy J. Farnham Department of Biochemistry and Molecular Biology, Norris Comprehensive Cancer Center, University of Southern California, Los Angeles, CA 90033, USA, pfarnham@usc.edu

Yair Field Department of Computer Science and Applied Mathematics, Weizmann Institute of Science, Rehovot 76100, Israel, yair.field@weizmann.ac.il

Peter Fitzgerald Genome Analysis Unit, NCI, NIH, Bethesda, MD 20892, USA, pcf@helix.nih.gov

Seth Frietze Department of Biochemistry and Molecular Biology, University of Southern California, Los Angeles, CA 90033, USA, frietze@usc.edu

T.R. Hughes Banting and Best Department of Medical Research, Donnelly Centre for Cellular and Biomolecular Research, University of Toronto, Toronto, ON, Canada M5S 3E1, t.hughes@utoronto.ca

Arttu Jolma Department of Biosciences and Nutrition, SE-171 77, Stockholm, Sweden, arttu.jolma@ki.se

Henry Krause Banting and Best Department of Medical Research, The Department of Molecular Genetics, University of Toronto, Toronto, ON, Canada M5S 3E1; The Donnelly Centre for Cellular and Biomolecular Research, University of Toronto, Toronto, ON, Canada, M5S 3E1, h.krause@utoronto.ca

Nicholas M. Luscombe EMBL-European Bioinformatics Institute, CB10 1SD, Cambridge, UK; Genome Biology Unit, European Molecular Biology Laboratory, 69117 Heidelberg, Germany; EMBL-Heidelberg Genome Biology Unit, D-69117 Heidelberg, Germany (Joint appointment), luscombe@ebi.ac.uk

Aleksandar S. Necakov European Molecular Biology Laboratory, 69117 Heidelberg, Germany, necakov@embl.de

Duncan T. Odom Cancer Research UK, Li Ka Shing Centre, University of Cambridge, Cambridge, UK, duncan.odom@cancer.org.uk

Keith Pardee Department of Biomedical Engineering, Boston University, Boston, MA 02215, USA; Wyss Institute for Biologically Inspired Engineering, Harvard University, Boston, MA 02215, USA, keith.pardee@wyss.harvard.edu

Eran Segal Department of Computer Science and Applied Mathematics, Weizmann Institute of Science, Rehovot 76100, Israel, eran.segal@weizmann.ac.il

Aswin Sai Narain Seshasayee EMBL-European Bioinformatics Institute, Wellcome Trust Genome Campus, CB10 1SD Cambridge, UK, aswin@ncbs.res.in

Eilon Sharon Department of Computer Science and Applied Mathematics, Weizmann Institute of Science, Rehovot 76100, Israel, eilon.sharon@weizmann.ac.il

Karthikeyan Sivaraman EMBL-European Bioinformatics Institute, Wellcome Trust Genome Campus, CB10 1SD Cambridge, UK, ksivan@ebi.ac.uk

Lisa Stubbs Department of Cell and Developmental Biology, Institute for Genomic Biology, University of Illinois, Urbana, IL 61801, USA, ljstubbs@illinois.edu

Younguk Sun Department of Cell and Developmental Biology, Institute for Genomic Biology, University of Illinois, Urbana, IL 61801, USA, sun29@illinois.edu

Jussi Taipale Department of Biosciences and Nutrition, SE-171 77, Stockholm, Sweden; Genome-Scale Biology Research Program, University of Helsinki, FI-00014, Helsingin Yliopisto, Finland, jussi.taipale@ki.se

Juan M. Vaquerizas EMBL-European Bioinformatics Institute, CB10 1SD Cambridge, UK, jvaquerizas@ebi.ac.uk

Charles Vinson Laboratory of Metabolism, NCI, NIH, Bethesda, MD 20892, USA, Vinsonc@mail.nih.gov

Matthew T. Weirauch Banting and Best Department of Medical Research, Donnelly Centre for Cellular and Biomolecular Research, University of Toronto, Toronto, ON, Canada M5S 3E1, matt.weirauch@utoronto.ca

Chapter 1
Introduction to "A Handbook of Transcription Factors"

T.R. Hughes

Abstract This chapter briefly summarizes the topics in this volume.

1.1 Overview

Transcription factors (TFs) are conventionally defined by their ability to bind specific DNA sequences and regulate transcription. They can perform these functions in many ways. The repertory of known DNA-binding domains, interactions with other TFs and chromatin proteins, and means of influencing transcription continue to grow. TFs have long fascinated molecular biologists, as they are often identified as key metabolic or developmental regulators, and at least initially provided an easily understood scheme for the orchestration of gene expression, cell differentiation, and homeostasis. Genome sequencing and genomic analyses, however, have abundantly confirmed early suspicions that cells interpret genomic information by mechanisms that are both complicated and varied. The ever-increasing amount of data presents new challenges and opens new horizons in the analysis of TFs and their functions. This book is intended to provide a broad overview of this increasingly complex field. This introductory chapter gives an overview of the material contained in the book, which includes catalogues of TFs, reviews of specific prominent families, methods for analysis of DNA-binding, interactions with chromatin, and modes in which transcriptional output is controlled.

1.2 Families of DNA-Binding Domains

Transcription factors are typically defined on the basis of containing one or more DNA-binding domains (DBDs) which encode a sequence-specific DNA-binding module, and TFs are often classified by the type of DNA-binding domain they

T.R. Hughes (✉)
Banting and Best Department of Medical Research, Donnelly Centre for Cellular and Biomolecular Research, University of Toronto, Toronto, ON, Canada M5S 3E1
e-mail: t.hughes@utoronto.ca

T.R. Hughes (ed.), *A Handbook of Transcription Factors*, Subcellular Biochemistry 52,
DOI 10.1007/978-90-481-9069-0_1,

contain. Other parts of the protein can contribute to and influence the intrinsic DNA-binding activity, including sequences that flank the DBD and that mediate dimerization (e.g. [1]). There are also proteins that possess sequence-specific DNA-binding activity but no known DBD [2, 3], and there are undoubtedly additional DBDs remaining to be discovered among the many orphan domain types. Nonetheless, the known DBDs are already prevalent in most genomes, so it is reasonable to use currently-known DBDs as a baseline to catalogue TFs in sequenced genomes. Chapter 2 (Seshasayee, Sivaraman, and Luscombe) and Chapter 3 (Weirauch and Hughes) take this approach to survey TFs in prokaryotes and eukaryotes, respectively, and briefly describe the character and distribution of prominent classes of TFs. There are remarkable differences between these two catalogues: for example, there are very few DBD classes shared between prokaryotes and eukaryotes. Chapter 2 – the only chapter in this book about prokaryotic TFs – also outlines the global properties and architectures of transcriptional networks in model bacteria.

Most DBD types are found in many species and lineages, and in multiple proteins within the same genome. Expansion of TF families is thought to occur by evolutionary mechanisms including duplication and divergence, domain accretion, and in some cases, diversifying selection on the DNA-contacting residues, which presumably indicates a selection for novel DNA-binding activities. Every DBD (and indeed every TF) has its own story, but a handful of DBD classes are particularly prevalent in both numbers and in the biomedical literature. Chapter 4 (Stubbs, Sun, and Caetano-Anolles), Chapter 5 (Bürglin), and Chapter 6 (Pardee, Necakov, and Krause) address three that are of particular interest due to their recent expansion in metazoans and large numbers in human. The C2H2 zinc finger (Chapter 4) is the largest class of putative TFs in human, and the homeodomain (Chapter 5) is the second-largest. Both classes are found in virtually all eukaryotes and presumably date to the origin of eukarya or earlier, but have undergone extensive expansion and diversification; Chapters 4 and 5 describe both the evolutionary mechanisms and likely functional consequences. Nuclear receptors (Chapter 6) have a less certain origin, and are found almost entirely in metazoans. These proteins are of specific interest because they are characterized by the presence of both a DNA-binding domain and a ligand-binding domain which controls the activity of the TF. The ligands include hormones, metabolites, and other physiological regulators, making them especially relevant to human health.

1.3 DNA-Binding Activity, TF Targeting, and the Influence of Chromatin

Since the defining feature of TFs is their sequence-specific DNA-binding activity, it is of interest to characterize their intrinsic sequence specificity. Chapter 7 (Jolma and Taipale) reviews methods for determining the sequence preferences of TFs in vitro. The data from these experiments are useful because many lines of investigation involve scanning sequences for potential binding sites. Given the central importance of this task in building models of transcriptional regulation, it is remarkable that there is currently no generally-accepted gold standard for either models or scanning. DNA-binding specificities are represented variously by consensus sequences (e.g. using IUPAC codes), Position Weight Matrices (or PWMs) (which are visualized as "sequence logos"), or by a table of affinities (or relative affinities) to individual sequences. There is ongoing controversy about how well conventional motif models represent the actual sequence preferences of the TFs [4–6], how motifs should be derived [7], and what is the best approach for scoring genomic sequences using motif models [8]. The newer methods described in Chapter 7 should enable the relative merits of different models to be examined in greater detail than has previously been possible.

The intrinsic DNA-binding activity of an individual TF is only one of several parameters that can determine the sites it binds in the genome. It has long been recognized that site occupancy can also be influenced by accessibility of sites within chromatin, by cooperation or competition with other sequence-specific DNA-binding proteins, and by interactions with chromatin proteins and chromatin modifications. In addition, while the in vivo binding landscape is sparse, it is not binary; some binding sites will be bound more strongly/frequently than others. Chapter 8 (Odom) gives an overview of methods for studying DNA–protein interactions in vivo.

Chapter 8 also raises intriguing questions regarding the function and evolution of in vivo-bound sites. Indeed, many of the outstanding questions in gene regulation and TF function in eukaryotes relate to our incomplete understanding of how regulatory sites (i.e. TF binding sites) are specified, particularly in large genomes, and how the regulatory output of these sites is determined. A variety of models have been proposed (e.g. [9–13]), with a common assumption being that TFs or ensembles of TFs must compete with other factors, and that the specific arrangement of the DNA sequences must somehow specify both the regulatory site locations and the regulatory consequences of TF binding. Elucidating the full details of how these processes are enacted in living cells is a very active research area. Understanding how the chromatin state is established, what different chromatin configurations signify, and how combinations of factors influence the recruitment and productive elongation of RNA polymerase is critical to understanding how TFs work, how transcription is regulated, and to a significant extent the constraints under which genomes evolve. These questions are also critical to the burgeoning field of epigenetics: TFs are not only affected by chromatin; they are also the prime candidate as the mechanism that underlies the establishment of the chromatin landscape, since by definition they are the major class of molecules that can distinguish among different DNA sequences.

Given the importance of these topics, several chapters are dedicated to presenting different perspectives on the interactions of TFs with chromatin and other nuclear factors, and how combinations of factors can determine regulatory sites and activities. Chapter 9 (Field, Sharon, and Segal) considers how regulatory sites are specified, with an emphasis on the accessibility of sites in chromatin, the role of DNA sequence in facilitating accessibility, and competition and cooperation among factors for DNA-binding. Chapter 10 (Vinson, Chatterjee, and Fitzgerald) specifically examines the occurrence of TF binding sites at specific positions in metazoan promoters, and presents evidence that the fundamental architecture of promoters is remarkably flexible over evolutionary time. Chapter 11 (van Bakel) gives an overview of the interactions of TFs with chromatin. Together, these chapters highlight the complexity of TF targeting and regulatory sequence definition, but also underscore the fact that there are many experimental and computational approaches that can be applied, and that a remarkable amount of progress has been made.

1.4 Transcriptional Regulatory Activity

The name "Transcription Factor" comes from the ability of TFs to modulate transcription. Traditionally, TFs have been classified as activators or repressors, often depending on the response of heterologous reporter constructs and/or specific genes to perturbation of the TF activity [3]. However, this classification may be an oversimplification; as noted above, it is a long-standing observation that the specific influence of a TF on transcription depends on where it is, and what other factors it is working with. Chapter 12 (Frietze and Farnham) reviews effector domains in eukaryotic TFs, and how they control transcription. Consistent with the growing importance of chromatin in gene regulation, many of the mechanisms are based on influencing local chromatin, rather than directly interacting with RNA polymerase. For example, many TFs interact with coactivators and/or corepressors, which in turn influence local chromatin state, often by catalyzing or otherwise controlling histone modifications, which can in turn recruit additional factors, regulate DNA accessibility, and promote or prevent transcription. In addition, while DNA-binding domains typically have easily-recognized conserved structural motifs with ancient origins, at least some effector domains appear to be inherently physically unstructured. Moreover, individual TF effector domains can interact with a multitude of different partners, providing a mechanistic basis for context specific regulatory activity.

Finally, Chapter 13 (Vaquerizas, Akhtar, and Luscombe) discusses one role of TFs that is rapidly gaining increasing attention: their influence on the large-scale 3-dimensional arrangement of DNA. Such a role is widely accepted as a function of for prokaryotic Nucleoid-Associated Proteins (as described in Chapter 2), and is also implicit in traditional models of eukaryotic regulatory site action. The action of enhancers, repressors, and insulators – three major classes of regulatory sequences – must depend in some way upon the coordination of DNA topology and/or communication between loci along the DNA, since they are all capable of acting at a distance. Until recently, it has been difficult to make measurements of large-scale DNA topology, and global approaches are still limited in resolution; however, technologies for both single-locus measurements and genome-wide surveys are developing rapidly and will likely represent a corner piece in solving some of the more puzzling aspects of gene regulation and TF activity.

1.5 Conclusion and Perspective

The molecular biology of TFs encompasses a vast literature, and is intertwined with many disciplines. As a handbook, this volume is intended to provide a starting point, overall picture, and ready reference. Relevant topics that are treated here only peripherally include the evolution of genomes, the biochemistry of protein–DNA interactions and protein-protein interactions, the myriad chromatin and histone modifications, functional genomics and systems biology, biological networks, and the biochemistry of RNA polymerases and their many accessory factors. The roles of noncoding RNA in chromatin and transcriptional regulation, the entire branch of archaebacteria, and the signalling pathways that regulate TFs are also, in general, not considered here. Many excellent texts and review papers are available on these subjects (e.g. [14–23]). Readers may also wish to consult other texts that specifically consider TFs (e.g. [24–26]).

An additional highly-relevant topic is the recognition "codes" that relate DBD amino-acid sequence to DNA sequence preferences of TFs [27]. Despite difficulty in deriving such codes, the concept has endured (and been extended to other aspects of TF function, such as the encoding of regulatory information in genome sequence [28]). Indeed, there is evidence that with sufficient processing power these types of problems are computationally approachable [29], although perhaps not in exactly the ways originally intended – for both TF-DNA recognition codes and regulatory codes, it is possible that biophysical models may fare better than logical models. On the whole, this Handbook contains relatively little on computational biology, despite the fact that research in transcriptional regulation is becoming increasingly an enterprise in data analysis. Indeed, a long-standing objective is to create a computer program that can predict the transcriptional output of a cell, given the biochemical activities of its myriad components (e.g. the sequence specificities of TFs [30, 31]). I hope this book will provide both information and motivation for readers who are exploring these and other problems in the biology of TFs.

References

1. Joshi R, et al. (2007) Functional specificity of a Hox protein mediated by the recognition of minor groove structure. Cell 131:530–543
2. Hu S, et al. (2009) Profiling the human protein-DNA interactome reveals ERK2 as a transcriptional repressor of interferon signaling. Cell 139:610–622
3. Fulton DL, et al. (2009) TFCat: the curated catalog of mouse and human transcription factors. Genome Biol 10:R29
4. Benos PV, Bulyk ML, Stormo GD (2002) Additivity in protein–DNA interactions: how good an approximation is it? Nucleic Acids Res 30:4442–4451
5. Maerkl SJ, Quake SR (2007) A systems approach to measuring the binding energy landscapes of transcription factors. Science 315:233–237
6. Badis G, et al. (2009) Diversity and complexity in DNA recognition by transcription factors. Science 324:1720–1723
7. Tompa M, et al. (2005) Assessing computational tools for the discovery of transcription factor binding sites. Nat Biotechnol 23:137–144

8. Granek JA, Clarke ND (2005) Explicit equilibrium modeling of transcription-factor binding and gene regulation. Genome Biol 6:R87
9. Arnosti DN, Kulkarni MM (2005) Transcriptional enhancers: Intelligent enhanceosomes or flexible billboards? J Cell Biochem 94:890–898
10. Berman BP, et al. (2002) Exploiting transcription factor binding site clustering to identify cis-regulatory modules involved in pattern formation in the *Drosophila* genome. Proc Natl Acad Sci U S A 99:757–762
11. Wunderlich Z, Mirny LA (2009) Different gene regulation strategies revealed by analysis of binding motifs. Trends Genet 25:434–440
12. Segal E, Widom, J (2009) DNA sequence to transcriptional behaviour: a quantitative approach. Nat Rev Genet 10:443–456
13. Wasson T, Hartemink AJ (2009) An ensemble model of competitive multi-factor binding of the genome. Genome Res 19:2101–2112
14. Davidson EH (2006) The Regulatory Genome. Academic Press, London
15. Rice PA, Correll CC (2008) Protein-nucleic acid interactions: structural biology. Royal Society of Chemistry, Cambridge, UK
16. Kouzarides T (2007) Chromatin modifications and their function. Cell 128:693–705
17. Berger SL (2007) The complex language of chromatin regulation during transcription. Nature 447:407–412
18. Bernstein BE, Meissner A, Lander ES (2007) The mammalian epigenome. Cell 128:669–681
19. Bonneau R (2008) Learning biological networks: from modules to dynamics. Nat Chem Biol 4:658–664
20. Alon U (2007) An Introduction to Systems Biology: Design Principles of Biological Circuits. Chapman & Hall/CRC, Boca Raton, FL
21. Junker BH, Schreiber F (2008) Analysis of Biological Networks, Wiley Series on Bioinformatics. Wiley, Hoboken, NJ
22. Moazed D (2009) RNAs in transcriptional gene silencing and genome defence. Nature 457:413–420
23. Nagano T, Fraser P (2009) Emerging similarities in epigenetic gene silencing by long noncoding RNAs. Mamm Genome 20:557–562
24. Latchman DS (2008) Eukaryotic Transcription Factors, Fifth Edition. Academic Press, London
25. Locker J (2000) Transcription Factors, Human Molecular Genetics Series. BIOS Scientific Publication, Oxford
26. Ghosh D, Locker J (1996) Transcription Factors: Essential Data. Wiley, New York, NY
27. Pabo CO, Sauer RT (1984) Protein-DNA recognition. Annu Rev Biochem 53:293–321
28. Michelson AM, Bulyk ML (2006) Biological code breaking in the 21st century. Mol Syst Biol 2:2006 0018
29. Liu LA, Bader JS (2007) Ab initio prediction of transcription factor binding sites. Pac Symp Biocomput 2007:484–495
30. Beer MA, Tavazoie S (2004) Predicting gene expression from sequence. Cell 117:185–198
31. Middendorf M, Kundaje A, Wiggins C, Freund Y, Leslie C (2004) Predicting genetic regulatory response using classification. Bioinformatics 20(Suppl 1):i232–40

Chapter 2
An Overview of Prokaryotic Transcription Factors

A Summary of Function and Occurrence in Bacterial Genomes

Aswin Sai Narain Seshasayee, Karthikeyan Sivaraman, and Nicholas M. Luscombe

Abstract Transcriptional initiation is arguably the most important control point for gene expression. It is regulated by a combination of factors, including DNA sequence and its three-dimensional topology, proteins and small molecules. In this chapter, we focus on the trans-acting factors of bacterial regulation. Initiation begins with the recruitment of the RNA polymerase holoenzyme to a specific locus upstream of the gene known as its promoter. The sigma factor, which is a component of the holoenzyme, provides the most fundamental mechanisms for orchestrating broad changes in gene expression state. It is responsible for promoter recognition as well as recruiting the holoenzyme to the promoter. Distinct sigma factors compete with for binding to a common pool of RNA polymerases, thus achieving condition-dependent differential expression. Another important class of bacterial regulators is transcription factors, which activate or repress transcription of target genes typically in response to an environmental or cellular trigger. These factors may be global or local depending on the number of genes and range of cellular functions that they target. The activities of both global and local transcription factors may be regulated either at a post-transcriptional level via signal-sensing protein domains or at the level of their own expression. In addition to modulating polymerase recruitment to promoters, several global factors are considered as "nucleoid-associated proteins" that impose structural constraints on the chromosome by altering the conformation of the bound DNA, thus influencing other processes involving DNA such as replication and recombination. This chapter concludes with a discussion of how regulatory interactions between transcription factors and their target genes can be represented as a network.

N.M. Luscombe (✉)
EMBL-European Bioinformatics Institute, CB10 1SD, Cambridge, UK; Genome Biology Unit, European Molecular Biology Laboratory, 69117 Heidelberg, Germany; EMBL-Heidelberg Genome Biology Unit, D-69117 Heidelberg, Germany (Joint appointment)
e-mail: luscombe@ebi.ac.uk

T.R. Hughes (ed.), *A Handbook of Transcription Factors*, Subcellular Biochemistry 52,
DOI 10.1007/978-90-481-9069-0_2,

2.1 Introduction: Regulation of Transcription Initiation in Bacteria

Flow of genetic information from DNA to proteins via transcription and translation is a tightly regulated process in bacteria, enabling optimal use of valuable nutritional resources and ensuring survival in rapidly changing environments. The initiation of transcription is arguably the most important control point for regulating gene expression. It is controlled by a wide range of molecule types: cis-acting DNA sequence and structural elements, and trans-acting proteins and small molecules.

Transcription initiation begins with the recruitment of the RNA polymerase (RNAP) holoenzyme – a complex of the catalytically capable RNAP apoenzyme and a "σ-factor" – to a specific locus upstream of the gene known as its "promoter". The σ-factor is responsible for promoter recognition as well as recruiting the holoenzyme to the promoter. The complex of RNAP holoenzyme and DNA (promoter) thus formed is called the "closed complex" [1]. In many cases, the σ-factor also facilitates the formation of the transcription bubble, i.e. the "open complex", by stabilising the unwound DNA around 10 bp upstream of the transcription start site. Amidst extensive abortive initiation events, where the RNAP holoenzyme dissociates from the DNA after synthesising <15nt of RNA [2], processive elongation ensues followed by termination.

Successful transcription initiation requires several key components such as (a) DNA sequence and topology that permit promoter recognition, (b) σ-factors that can recognise promoters, (c) free RNAP for recruitment to the promoter concerned, and (d) trans-acting transcriptional regulators and their small molecule modulators, that enable condition-dependent differential gene expression.

In this chapter, we primarily discuss trans-acting protein factors that determine RNAP recruitment to promoters: namely σ-factors and transcription factors. The different categories of trans-acting protein factors are illustrated in Fig. 2.1. Other determinants like promoter architecture and the activity of RNAP apoenzyme have been extensively reviewed elsewhere, and will not be discussed here. First, we introduce the different families of σ-factors and highlight certain genome-scale investigations of their function. We then discuss transcription factors by focusing on their computational identification and occurrence in bacterial genomes. We will also discuss functional examples of transcription factors that regulate gene expression. Third, we highlight examples of functional interpretations derived from genome-scale analyses of transcriptional regulatory network structure. Fourth, we briefly discuss the architecture and evolution of transcription regulatory networks in *Escherichia coli*. Finally, we conclude the chapter with specific open questions that need to be addressed. Most of our discussion will pertain to the bacterium *E. coli*, for which there is extensive genomic scale experimental data.

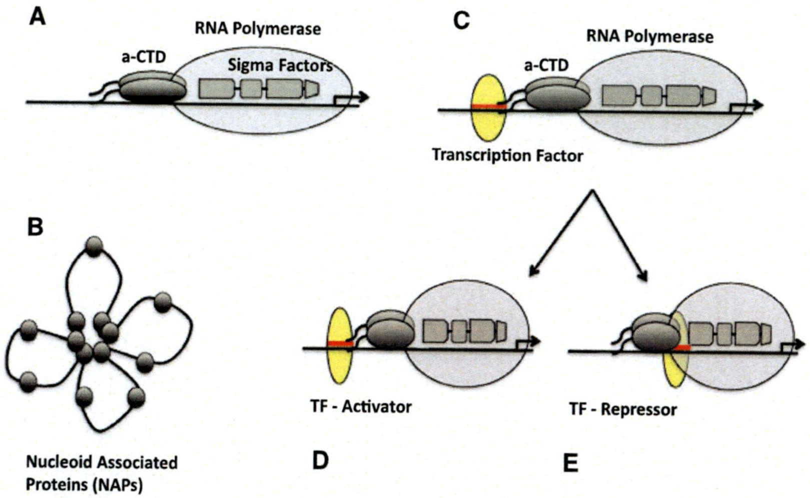

Fig. 2.1 Different groups of transcription factors based on their activity. Here we illustrate the different groups of transcription factors that are discussed in the text. *Panel A* depicts sigma factors that are an integral part of the transcription machinery (RNA polymerase holoenzyme). *Panel B* shows Nucleoid Associated Proteins (NAPs), which bind to chromosomal DNA and assist the formation of 3-dimensional nucleoid structure. *Panel C* depicts the classical TFs that aid the RNA polymerase holoenzyme in regulating transcription. These can be functionally divided into activators (*panel D*) or repressors (*panel E*) depending on their binding site relative to the transcriptional start site

2.2 Core Regulatory Members of the RNA Polymerase: The σ-Factors

σ-factors determine promoter specificity and are an integral part of the transcriptional machinery and the closed complex. These proteins provide most, if not all, of the determinants for promoter recognition and open complex formation, but only in complex with the rest of the RNAP [3].

There are two evolutionarily distinct families of σ-factors: σ^{70} and σ^{54}. Typically, most transcription in rapidly growing cells is mediated by what is called the major σ-factor, which belongs to the σ^{70} family. Many bacterial genomes also code for several alternative σ-factors, which regulate specific sets of genes under different stresses and growth transformations, thus representing the most fundamental means of achieving major changes in transcription. Most alternative σ-factors also belong to different subgroups of the σ^{70} family. Whereas members of this family carry out open complex stabilisation on their own (as part of the RNAP holoenzyme), members of the second family, named σ^{54}, require additional activators belonging to the AAA+ ATPase family to unwind the DNA. The σ^{70} family is almost ubiquitous in bacteria, and is mostly represented by multiple members. On the other hand, the σ^{54} family is found only in ~65% of sequenced bacterial genomes, and where present comprises a single member [3, 4]. For example, *E. coli* K12 encodes six members of the σ^{70} family (the major sigma factor RpoD, RpoH, RpoS, RpoE, FliA and FecI) but only one σ^{54} protein (RpoN).

Different σ-factors in bacterial cell compete for a limited number of RNAP apoenzyme molecules, and the outcome of this competition determines the cellular gene expression state. The dynamics of this competition depend on (i) relative concentrations of various sigma factors [5–8], (ii) presence of σ-factor sequestering anti-σ-factor proteins (Rsd in *E. coli*) [9], (iii) presence of modulating small molecule second-messengers such as (p)ppGpp [10], (iv) small non-coding RNA such as 6S RNA [11], (v) presence of other players such as H-NS [12], and (vi) finally, the ability of the sigma factor to recognize evolutionarily divergent promoter sites [13].

All these factors play a role in determining the outcome of stress σ-factor (RpoS) regulation in *E. coli*. During the stationary phase, RpoS is highly expressed, albeit at a third of the RpoD expression levels. However, the major σ-factor itself is sequestered by its anti-σ-factor Rsd. Also, the presence of (p)ppGpp during starvation conditions reduces transcription from the RpoD promoters, as does the 6S RNA. The presence of H-NS on chromosomal DNA also negatively impacts transcriptional initiation by RpoD. Also, molecular level studies have shown that RpoS is more tolerant to mutations in its promoters, and hence is more robust at initiating transcription from mutant promoters. All these factors facilitate transcription by RpoS at the promoters. Further, the activity of RpoS is also enhanced by the presence of A/T rich tracts upstream, and sometimes downstream, of the promoter [14].

Thus, a combination of dynamic (small molecules/proteins) and static properties (promoter sequence/architecture) determines the condition specific dominance of various sigma factors. However, it is not known how much the target gene repertoires (regulons) of different σ-factors in an organism overlap with one another. Even though a recent study reports a significant overlap between the regulons of two distinct σ-factors (RpoD and RpoH) in *E. coli* [15], these conclusions are controversial and await further clarification [16].

The role of σ-factors in initiating transcription, coupled with results of earlier molecular studies [17–19], suggested that the σ-factor dissociates after a successful initiation. However, later studies have shown that as much as 90% of early elongation complexes contain the σ-factor [20, 21] and provide evidence for some σ-factor retention well inside gene bodies [22]. These studies, in concert with earlier results, suggest that σ-factors play a complex role by regulating expression during initiation and controlling RNAP pausing in the elongation phases [23, 24].

2.3 Transcription Factors

Transcription factors (TFs) are proteins that bind to specific sequences on the DNA near their target genes, thus modulating transcription initiation. TFs can activate or repress transcription depending where they bind relative to the transcription start site of the target gene [1]. Each TF regulates a set of genes, in response to specific environmental and/or intracellular triggers. A complete transcriptional regulatory interaction between a TF and its target gene-(s) encompasses (1) signal sensing, (2) signal transduction, (3) the TF; and (4) the target gene-(s) [25]. In the following sections, we will focus primarily on identification of the TFs and transcriptional regulation by these TFs.

2.3.1 Identification and Genomic Distribution of Transcription Factors

Both prokaryotic and eukaryotic TFs are generally identified by the presence of a DNA-binding domain using sequence searches against protein family databases such as PFAM [26], and by BLAST-based [27] detection of homologs of experimentally-verified TFs. Several databases of computationally identified transcription factors are publicly available; most are specific to certain phylogenetic groups such as the FlyTF [28], and RegulonDB [29]. On the other hand, DBD (which in this chapter refers to "DNA-Binding Domain Database") includes many completely sequenced genomes [30]. This database contains TF predictions for about 480 of >1,000 bacterial genomes that have been completely sequenced.

Transcription factors in the above-mentioned DBD contain one of 131 distinct protein families or domains, of which 61 are found in bacteria. Such studies showed that the number of TFs scales in a nearly quadratic fashion with genome size [31–33]. For bacteria with comparatively large genomes such as *E. coli* and *Bacillus subtilis*, TFs account for ∼6% of their total gene count. These organisms may require a large proportion of transcription factors in order to regulate functionally specialised groups of genes or they might make use of more complex, and longer cascades of regulatory interactions [34]. On the other hand, organisms in host-associated symbiosis or parasitism have an extremely poor TF gene content consistent with their lack of need for sensing and responding to changing environments. Examples include *Mycobacterium leprae* [35] which encodes only 42 TFs (2.4% of gene count), and *Rickettsia prowazekii* [36] which has only nine TFs (<1%).

The *E. coli* genome is predicted to code for around 270 TFs, which accounts for 6% of protein-coding genes in this organism [33]. Based on the hierarchical classification of protein structures in the SCOP database, it was found that these TFs all belong to one of 11 different families, of which 10 contain the helix-turn-helix structural motif. Over 75% of all predicted TFs in *E. coli* contain an additional domain, belonging to a wider range of 46 different protein families. These domains are largely involved in sensing signals. Significantly, 40–50% of all TFs contain a second domain that can potentially bind to small-molecules [33, 37] and more than a third of these have been experimentally verified according to the Ecocyc database [38]. Such a high percentage of TFs with small-molecule-binding capability is not known in eukaryotes [39]. Another 10% of TFs are part of two-component signalling cascades where they are phosphorylated by an upstream histidine kinase, which in almost every case is the top-level signal sensor. Overall, these patterns of domain coupling suggest extensive and immediate interactions between signals and the transcriptional machinery, which in eukaryotes takes place through longer cascades of signal-transduction events.

2.3.2 Classification of Transcription Factors Based on Their Regulatory Scope: Global and Local Regulators

TFs in bacteria can have either a broad or a narrow regulatory scope. The scope of regulation of various TFs can be studied for the *E. coli* genome using the RegulonDB database. This is a collection of experimentally validated and computationally predicted TF–target interactions for majority of TFs in *E. coli* genome. Despite not representing many TFs, this database is useful for analyzing trends of TF–target interactions in the genome.

A cursory analysis of RegulonDB reveals that ten TFs in *E. coli* are responsible for more than 61% of regulatory interactions in this bacterium. Thus, a small proportion of TFs in *E. coli* have a global scope (global TFs), while most others target specific gene (s) and/or operon (s) (local TFs). This leaves an open question of classifying a TF as "global" or "local", which was addressed by Martinez-Antonio and Collado-Vides [40].

Martinez-Antonio and Collado-Vides have defined a set of characteristics that distinguish global TFs from "local" players that go beyond the number of genes it regulates [40]. These characteristics include (1) number and nature of co-regulating TFs, (2) ability to regulate genes which belong to target-groups of different σ-factors, (3) capacity to regulate genes belonging to diverse functional categories, and (4) potential to respond to a wide range of environmental conditions. Besides these characteristics, global TFs have been recently shown to bind extensively to the chromosomal DNA, not necessarily causing expression changes in proximal genes [41]. Only seven TFs in *E. coli* satisfy all the above criteria to be a global TF: the catabolite-responsive CRP, anaerobiosis regulators FNR and ArcA, the feast or famine LRP, and three other DNA structuring proteins FIS, IHF and H-NS. Based on an analysis of target genes involved in small molecule metabolism, we have shown that six of the seven above TFs regulate multiple functional categories, but show a statistical enrichment for targeting a single function. On the other hand, most of the remaining TFs regulate genes from a single metabolic pathway or a broader functional grouping of pathways [42].

Moreover, at least five of the above seven global TFs have been classified as "nucleoid-associated proteins" (NAP) (Fig. 2.1b), primarily based on their ability to bind extensively to the DNA and to alter the topology of the bound DNA by bending, bridging or wrapping it. However, such classification is unlikely to be definite in the absence of further data; for example, there is evidence that one of the global TFs not usually considered as a NAP – FNR – can bend DNA. Finally, some global

TFs have signal sensing or phosphorylation-receiving domains, which regulate their DNA binding activity; the activities of other global TFs may be regulated primarily at the level of their expression levels and/or competition or interaction with other proteins. Different NAPs show distinct patterns of gene expression during batch growth and also differ from each other in the degree of sequence specificity (see below); for instance H-NS displays preferential binding to A/T-rich sequences, and the [A/G]ATA[A/T][T/A] motif in particular, whereas others such as Hu have not been associated with any motifs so far. The properties of global TFs are illustrated with examples below.

2.3.3 *Signal Dependent Activity of Global Regulators: CRP and LRP*

2.3.3.1 Lrp: The Feast or Famine Global Transcription Factor

Lrp was first identified as a regulator of branched amino acid transport [43]. It was also observed in many cases that in turn its own activity is modulated by the amino acid leucine, which acts as a nutritional indicator [44, 45]. In *E. coli*, the TF regulates genes involved in amino acid metabolism and transport, and non-metabolic functions such as pili biosynthesis. A recent study interrogating the genome-wide binding of Lrp to the DNA identified sequence-specific interactions with ~140 chromosomal sites with an identifiable sequence motif, thus expanding the catalogue of known Lrp targets by a factor of five [46, 47]. The authors showed that absence of leucine and stationary phase increase the number of Lrp-binding regions by 3 to 4-fold, the latter effect in agreement with the inverse relationship between Lrp expression and growth rate.

Lrp and its signal, leucine, can interact in three distinct ways: (a) independent response where leucine has no effect on Lrp action; (b) concerted response in which leucine enhances the effect of Lrp; and (c) reciprocal response in which leucine antagonises the effect of Lrp. Lrp exists largely in two forms: octameric (Lrp8) and hexadecameric (Lrp16). Leucine binding favours the dissociation of Lrp to the octameric form (Lrp8-leu) [48]. Differences among promoters in their affinities to the different oligomeric forms of Lrp might explain the manner in which they are regulated by leucine [48].

Lrp can also bend and wrap the DNA [49], and its ortholog in *Bacillus subtilis* can, in addition, help form DNA bridges [50, 51]. These results, combined with its global scope of binding, imply that Lrp can influence the 3D topology of the chromosome. For these reasons, Lrp is considered as a NAP.

2.3.3.2 Crp and Transcriptional Responses to Carbon-Source Nutrition

Crp is the most prolific global transcription factor in *E. coli*, based on the information available in RegulonDB [29]. It is activated by the binding of the second messenger cyclic-AMP (cAMP) in response to glucose starvation and other stresses.

Though commonly described in the context of catabolite repression (utilization of an alternative carbon source in the absence of glucose), a microarray study investigating gene expression changes in a *Δcrp* strain revealed a much broader regulatory scope for CRP [52], including regulation of motility in *E. coli* [53]. Another study investigating differential expression of genes following a change of carbon source from glucose to another (of poorer quality) highlighted that most targets of CRP are likely to be regulated indirectly [54]. Genome-wide binding studies on Crp in *E. coli* revealed fewer strong binding sites (~70) than expected, with a relative high background generated by many weak binding events at low-affinity sites [55]. The study also noted that only a minority of binding events directly affected target gene transcription. Based on these results and the ability of CRP to bend DNA [56, 57], the authors of this study [55] propose that CRP is too a NAP.

2.3.4 *Expression and Protein–Protein Interaction Dependent Activity of Global Regulators: FIS and H-NS*

2.3.4.1 Fis: An Enigmatic Transcriptional Regulator

Fis is a versatile DNA binding protein that can affect multiple processes including transcription. In *E. coli*, it is thought to be a major regulator of growth transitions [58]. Fis is expressed in a growth phase dependent fashion, showing high expression during logarithmic growth [59]. It activates more genes than it represses [41], though it represses several non-essential genes during exponential growth [60–62]. At least two independent genomic studies in *E. coli* have demonstrated that Fis mediates global changes in gene expression with over 20% of all genes being affected by Fis [41, 63, 64]. Δfis mutants of *E. coli* show unnaturally high negative supercoiling during stationary phase growth [58], which might lead to a general increase in transcription during this phase of growth.

Though certain FIS-binding characteristics such as localisation to gene-upstream regions may be associated with gene expression, it is being realised that, as with CRP [55], a majority of Fis binding events do not lead to proximal gene expression changes [41]. This might be because Fis has complex effects on the 3D topology of chromosomal DNA [65, 66] that go beyond just proximity binding effects.

2.3.4.2 H-NS: "The Genome Sentinel"

H-NS is a global repressor of gene expression in enterobacteria and is one of the best-studied NAPs. It is expressed throughout all the growth phases in *E. coli* and simultaneously affects DNA structure and transcription by forming DNA–H-NS–DNA bridges and reinforcing plectonemically supercoiled structures [67–71]. Genome-scale analysis [41, 72] showed that H-NS binds to tracts of DNA [72] and it spreads linearly from high affinity sites to flanking lower affinity regions [41]. This analysis further provided genome-scale evidence for the existence of two modes of H-NS-mediated gene regulation. Short binding regions provide mild modulation, typically repression, of the expression of proximal genes whereas long binding tracts lead to total transcriptional silencing [41].

Genome-scale investigations of H-NS-binding in *Salmonella* revealed a surprising mechanism for bacterial defence against foreign DNA: the protein selectively silences the transcription of large numbers of horizontally acquired genes, including those within its major pathogenicity islands [73, 74]. This arises because the protein preferentially binds A/T-rich DNA, and these acquired genomic regions tend to display high AT-content. Removal of H-NS leads to uncontrolled expression of several pathogenicity islands, which has deleterious consequences for bacterial fitness. The mechanism appears to be general for other enterobacteria, since introduction of non-native plasmids into Δ*hns* cells can cause severe growth and infectivity defects [74–76]. Although the acquired genes are silenced during log growth, the combination of H-NS interactions with other regulatory factors and promoter-binding by the stress-associated RpoS σ-factor enables expression under stress conditions [77–79]. Thus, H-NS enables DNA to be acquired from exogenous sources, while avoiding their unregulated expression.

Thus, global regulators such as Lrp, CRP, Fis and H-NS modulate gene expression on a genome wide scale, in response to various stresses. Their responses are characterized by a global scope combined with a specific focus, such as repression of horizontally acquired genes by H-NS.

2.3.5 Local Transcription Factors and Specific Responses

The global TFs set the generic response mode such as stress, starvation and utilization of alternative carbon sources. However, in many cases, they are aided by many other TFs that make up the bulk of TF repertoire in the bacterial genome. These specific TFs, also known as local TFs, usually have a restricted regulatory scope comprising a few genes or operons. These are nonetheless responsible and necessary for regulation of their respective targets. In many known cases these TFs also act as signal sensing modules by sensing the environmental concentration of their small molecule “trigger”. We will discuss two specific examples of local TFs, both of which bind to a small molecule metabolite that modulates their activity.

LacI is a canonical local TF, which regulates the expression of the *lac* operon, in response to a combination of glucose starvation (CRP/c-AMP) and presence of allolactose inside the cell (LacI). The regulation of the *lac* operon also presents a classic case of combinatorial regulation by CRP. When the cell senses the absence of glucose, and the presence of alternative carbon source in the form of lactose/allolactose, the *lac* operon is activated and lactose catabolism ensues. So far, the only known target of LacI in *E. coli* is the *lac* operon.

Another example of specific local regulation involves the tryptophan synthesis operon (*trp*), which is regulated by the TrpR (trp Repressor). TrpR senses the levels of free tryptophan, which is the end-product of the *trp* operon, inside the cell by binding it. When levels of tryptophan increase inside the cell, the repressor binds to the amino acid, which stabilizes its active conformation [80], allowing it to bind upstream of the *trp* operon. Upon depletion of intracellular tryptophan, this process is reversed and the repression is relieved.

There are many such examples of specific repression/activation of genes and pathways by local TFs in bacteria.

2.4 Structure and Evolution of Bacterial Transcriptional Regulatory Networks

The ensemble of TF-target gene interactions in a bacterium determines its gene expression profile, and subsequently, its temporary phenotype. Such interactions can be analyzed in the form of networks, in order to gain a deeper understanding of bacterial biology. In this section, we will introduce bacterial gene regulatory networks and discuss their implications.

2.4.1 Modular Architecture of the Transcriptional Regulatory Network

A functional module is defined as a discrete entity whose function is separable from those of other modules [81]. Although there are numerous algorithms for identifying modules based on network topologies [82–85], perhaps the best characterised types of modules are network motifs that were originally described by Alon and colleagues [86]. Network motifs can be thought of as recurring circuits of regulatory interactions between TFs and target genes. Such motifs were originally defined in *E. coli*, in which they were detected as patterns of connections that occurred in the transcriptional network more often than would be expected in random networks.

One of the most important motifs is called the Feed Forward Loop (FFL), in which TF A regulates TF B and both A and B regulate a target gene C (Fig. 2.2a). The top-level TF in many FFLs is a global regulator: this is particularly exemplified by the classical catabolite repression which involves CRP as the top-level regulator and one of various sugar-responsive local TFs as the second regulator. Removal of global TFs from the dataset led to loss of many FFLs within the network [82, 84, 86], highlighting their importance in establishing this motif.

In addition to describing topological relationships between TFs and targets, different types of network motifs have been shown to carry out specific information-processing functions that are particularly suited to the biological requirements of the involved genes. For instance, FFLs filter out transient or rapidly varying input signals, thus enforcing the requirement of persistent signals for activation [86]. Thus an interesting question that can be addressed using network-based approaches is whether different types of cellular functions are regulated by distinct network architectures. For instance, the use of FFLs in controlling sugar metabolism ensures that catabolic enzymes are not expressed unless there are steady levels of the correct nutrients in the environment.

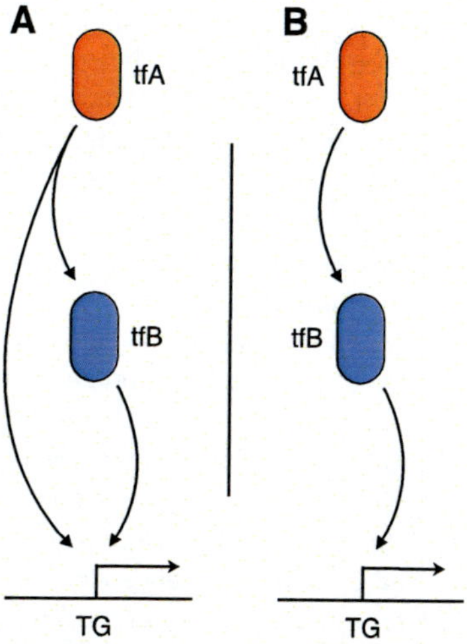

Fig. 2.2 Two prominent architectures found in *E. coli* gene regulatory networks. Two sub-network architectures are prominently found in *E. coli* gene regulatory networks. They are the feed forward loop (FFL – *Panel A*), and the cascade (*Panel B*). In a FFL, a primary TF (tfA in figure) regulates a secondary TF (tfB) and both tfA and tfB regulate the expression of the target gene (TG in figure). Such combinatorial regulation is observed in the regulation of catabolic genes by the global regulator CRP. The second architecture, the cascade, occurs in regulation of developmental processes such as flagellar biosynthesis. In this mode of regulation, a primary TF (tfA) regulates a secondary TF (tfB), which in turn regulates the target genes

2.4.2 Subnetwork Architectures for Different Gene Functions

An important question is how these network motifs combine to form the whole regulatory system. Using symbols for different types of motifs can help depict an entire regulatory system in a compact way. In *E. coli*, it becomes immediately clear that FFLs feed into a layer of densely interconnected TFs, an arrangement commonly known as multi-input motifs (MIMs). Here, each TF regulates many target genes, and in turn each target is controlled by many TFs; thus a MIM can be conceptualised as a gate-array that translates multiple inputs into multiple outputs. *E. coli* has several discrete MIMs with hundreds of output genes, each responsible for a broad biological function, such as anaerobic growth and stress response.

Long regulatory cascades are rare in *E. coli*: thus most FFLs connect directly into a MIM, and in most cases, each MIM produces a final output. A possible reason for this shallow architecture is that single-celled organisms need to respond rapidly to changing environmental conditions. An exception is the relatively long cascade controlling flagella assembly: the temporal ordering afforded by multiple TFs is thought to be useful in processes requiring several stages to complete. This type of mechanism also helps explain the experimentally observed temporal programme in the expression of flagella biosynthesis genes [87].

Despite the discrete network organisation of different cellular functions (such as sugar metabolism and flagella assembly above), there is also a great deal of interconnection between them. In particular, glucose is a positive regulator of biofilm formation [88], thus linking sugar metabolism/carbon nutrition with long-term cellular decisions. This is potentially due to CRP, which is indirectly controlled by glucose availability and is a top-level regulator of both sugar metabolism and these developmental processes. A second control point integrating these two functions operates at a post-transcriptional level [89].

Architectural features of regulatory sub-networks can vary even within a single functional group. For instance, the three broad functions within metabolism, viz. catabolism, anabolism and central metabolism, differ from each other in the number and types of their regulators [42]. The genes involved in catabolism undergo combinatorial regulation, with a global regulator such as CRP and a local TF. On the other hand, anabolic pathways are often regulated by a single specific TF, and the central metabolism is regulated by multiple global TFs [42]. Further, despite the similarity in network architectures of catabolic genes, different sugar operons display distinct output patterns in response to input signals [90].

2.4.3 Evolution of Transcription Networks: Implications for Regulatory Networks

TFs and their networks are dynamic evolving entities. In fact, TFs are less conserved that other protein types such as enzymes [34, 91]. Such evolution is often directed by the environment of the bacterium and, in some cases, its interaction with a higher eukaryotic host. Interaction of bacteria with higher eukaryotes, often as pathogens, means that certain transcriptional response networks in phylogenetically distinct organisms may undergo convergent evolution. The outcome of such evolution is that phylogenetically unrelated networks might assume similar functional architectures, where related ones will differ. The evolution of transcriptional regulatory networks between phylogenetically related organisms, and its driving forces, pose some of the important questions to be addressed in this field.

2.5 Conclusions

Transcriptional regulation is essential for ensuring that the correct genes are expressed at the right amounts at the appropriate time. It is controlled by a combination of cis-effects such as DNA sequence and topology, and trans-acting factors, the focus of this chapter. Sigma factors, a component of the RNA polymerase holoenzyme, are responsible for promoter-recognition and recruitment of the holoenzyme to specific promoters; therefore they provide the most fundamental level of control for the expression of large numbers of genes. Among DNA-binding TFs, global regulators target a disproportionately large numbers of genes, and exert their control over diverse functional categories. In *E. coli*, five out of seven global TFs are also nucleoid-associated proteins, "histone-like" proteins that bind extensively to the genome, and alter the topology of the bound DNA. The role of such proteins appear to extend well beyond the traditional confines of transcriptional regulation, since a large proportion of binding sites do not appear to cause expression changes in proximal genes. Finally, local TFs comprise most of the regulatory repertoire in bacterial genomes, and usually have a narrow regulatory scope restricted to specific gene functions.

A crucial point to consider in bacterial gene regulation is that RNA polymerase is in very short supply: in *E. coli* there are estimated ~1,500 to ~11,500 polymerase molecules per cell depending on growth condition. In combination, the above factors ensure that the RNA polymerase holoenzyme is correctly distributed among the 2,000 or so competing promoters in the genome. Molecular and biophysical studies over the past 50 years have elucidated distinct mechanisms for modulating the expression of individual genes: some mechanisms allow for fine tuning of expression levels, whereas others define much sharper transitions between active and inactive transcriptional states. In contrast, genome-scale studies during the last decade have generated unprecedented quantities of information describing the location of binding sites; however, our understanding of how all these binding events lead to transcriptional regulation is still very preliminary. A major challenge over the next decade will be to bridge the gap between the detailed molecular descriptions and genome-scale overviews so that we can understand how every gene in a bacterial genome is transcriptionally regulated.

References

1. Browning DF, Busby SJ (2004) The regulation of bacterial transcription initiation. Nat Rev Microbiol 2:57–65
2. Goldman SR, Ebright RH, Nickels BE (2009) Direct detection of abortive RNA transcripts in vivo. Science 324:927–928
3. Gruber TM, Gross CA (2003) Multiple sigma subunits and the partitioning of bacterial transcription space. Annu Rev Microbiol 57:441–466
4. Pérez-Rueda E, Janga SC, Martínez-Antonio A (2009) Scaling relationship in the gene content of transcriptional machinery in bacteria. Mol Biosyst 5:1494–1501
5. Holland AM, Rather PN (2008) Evidence for extracellular control of RpoS proteolysis in *Escherichia coli*. FEMS Microbiol Lett 286:50–59
6. Balandina A, Claret L, Hengge-Aronis R, et al (2001) The *Escherichia coli* histone-like protein HU regulates rpoS translation. Mol Microbiol 39:1069–1079
7. Zhou Y, Gottesman S, Hoskins JR, et al (2001) The RssB response regulator directly targets sigma (S) for degradation by ClpXP. Genes Dev 15:627–637
8. Yamashino T, Ueguchi C, Mizuno T (1995) Quantitative control of the stationary phase-specific sigma factor, sigma S, in *Escherichia coli*: involvement of the nucleoid protein H-NS. EMBO J 14:594–602
9. Jishage M, Ishihama A (1998) A stationary phase protein in *Escherichia coli* with binding activity to the major sigma subunit of RNA polymerase. Proc Natl Acad Sci U S A 95: 4953–4958
10. Jishage M, Kvint K, Shingler V, et al (2002) Regulation of sigma factor competition by the alarmone ppGpp. Genes Dev 16:1260–1270

11. Wassarman KM, Storz G (2000) 6S RNA regulates *E. coli* RNA polymerase activity. Cell 101:613–623
12. Shin M, Song M, Rhee JH, et al (2005) DNA looping-mediated repression by histone-like protein H-NS: specific requirement of Esigma70 as a cofactor for looping. Genes Dev 19:2388–2398
13. Typas A, Hengge R (2006) Role of the spacer between the -35 and -10 regions in sigmas promoter selectivity in *Escherichia coli.* Mol Microbiol 59:1037–1051
14. Typas A, Becker G, Hengge R (2007) The molecular basis of selective promoter activation by the sigmaS subunit of RNA polymerase. Mol Microbiol 63: 1296–1306
15. Wade JT, Roa DC, Grainger DC, et al (2006) Extensive functional overlap between sigma factors in *Escherichia coli.* Nat Struct Mol Biol 13:806–814
16. Waldminghaus T, Skarstad K (2010) ChIP on Chip: surprising results are often artifacts. BMC Genomics 11:414
17. Hansen UM, McClure WR (1980) Role of the sigma subunit of *Escherichia coli* RNA polymerase in initiation. II. Release of sigma from ternary complexes. J Biol Chem 255: 9564–9570
18. Travers AA, Burgess RR (1969) Cyclic re-use of the RNA polymerase sigma factor. Nature 222:537–540
19. Straney DC, Crothers DM (1985) Intermediates in transcription initiation from the *E. coli* lac UV5 promoter. Cell 43:449–459
20. Kapanidis AN, Margeat E, Laurence TA, et al (2005) Retention of transcription initiation factor sigma70 in transcription elongation: single-molecule analysis. Mol Cell 20: 347–356
21. Mukhopadhyay J, Kapanidis AN, Mekler V, et al (2001) Translocation of sigma (70) with RNA polymerase during transcription: fluorescence resonance energy transfer assay for movement relative to DNA. Cell 106:453–463
22. Reppas NB, Wade JT, Church GM, et al (2006) The transition between transcriptional initiation and elongation in *E. coli* is highly variable and often rate limiting. Mol Cell 24:747–757
23. Mooney RA, Landick R (2003) Tethering sigma70 to RNA polymerase reveals high in vivo activity of sigma factors and sigma70-dependent pausing at promoter-distal locations. Genes Dev 17:2839–2851
24. Ring BZ, Yarnell WS, Roberts JW (1996) Function of *E. coli* RNA polymerase sigma factor sigma 70 in promoter-proximal pausing. Cell 86:485–493
25. Salgado H, Martínez-Antonio A, Janga SC (2007) Conservation of transcriptional sensing systems in prokaryotes: a perspective from *Escherichia coli.* FEBS Lett 581:3499–3506
26. Finn RD, Mistry J, Tate J, et al (2010) The Pfam protein families database. Nucleic Acids Res 38:D211–D222
27. Altschul SF, Madden TL, Schäffer AA, et al (1997) Gapped BLAST and PSI-BLAST: a new generation of protein database search programs. Nucleic Acids Res 25:3389–3402
28. Pfreundt U, James DP, Tweedie S, et al (2010) FlyTF: improved annotation and enhanced functionality of the *Drosophila* transcription factor database. Nucleic Acids Res 38: D443–D447
29. Gama-Castro S, Jiménez-Jacinto V, Peralta-Gil M, et al (2008) RegulonDB (version 6.0): gene regulation model of *Escherichia coli* K-12 beyond transcription, active (experimental) annotated promoters and Textpresso navigation. Nucleic Acids Res 36:D120–D124
30. Charoensawan V, Wilson D, Teichmann SA (2010) Genomic repertoires of DNA-binding transcription factors across the tree of life. Nucleic Acids Res 38:7364–7377
31. van Nimwegen E (2003) Scaling laws in the functional content of genomes. Trends Genet 19:479–484
32. Ranea JA, Grant A, Thornton JM, et al (2005) Microeconomic principles explain an optimal genome size in bacteria. Trends Genet 21:21–25

33. Madan Babu M, Teichmann SA (2003) Evolution of transcription factors and the gene regulatory network in *Escherichia coli*. Nucleic Acids Res 31:1234–1244
34. Madan Babu M, Teichmann SA, Aravind L (2006) Evolutionary dynamics of prokaryotic transcriptional regulatory networks. J Mol Biol 358:614–633
35. Cole ST, Eiglmeier K, Parkhill J, et al (2001) Massive gene decay in the leprosy bacillus. Nature 409:1007–1011
36. Andersson SG, Zomorodipour A, Andersson JO, et al (1998) The genome sequence of *Rickettsia prowazekii* and the origin of mitochondria. Nature 396:133–140
37. Anantharaman V, Koonin EV, Aravind L (2001) Regulatory potential, phyletic distribution and evolution of ancient, intracellular small-molecule-binding domains. J Mol Biol 307: 1271–1292
38. Keseler IM, Bonavides-Martínez C, Collado-Vides J, et al (2009) EcoCyc: a comprehensive view of *Escherichia coli* biology. Nucleic Acids Res 37:D464–D470
39. Sellick CA, Reece RJ (2005) Eukaryotic transcription factors as direct nutrient sensors. Trends Biochem Sci 30:405–412
40. Martínez-Antonio A, Collado-Vides J (2003) Identifying global regulators in transcriptional regulatory networks in bacteria. Curr Opin Microbiol 6:482–489
41. Kahramanoglou C, Seshasayee ASN, Prieto AI, et al (2011) Direct and indirect effects of H-NS and Fis on global gene expression control in *Escherichia coli*. Nucleic Acids Res 39: 2073–2091
42. Seshasayee AS, Fraser GM, Babu MM, et al (2009) Principles of transcriptional regulation and evolution of the metabolic system in *E. coli*. Genome Res 19:79–91
43. Anderson JJ, Quay SC, Oxender DL (1976) Mapping of two loci affecting the regulation of branched-chain amino acid transport in *Escherichia coli* K-12. J Bacteriol 126:80–90
44. Lin R, D'Ari R, Newman EB (1992) Lambda placMu insertions in genes of the leucine regulon: extension of the regulon to genes not regulated by leucine. J Bacteriol 174: 1948–1955
45. Chen S, Hao Z, Bieniek E, et al (2001) Modulation of Lrp action in *Escherichia coli* by leucine: effects on non-specific binding of Lrp to DNA. J Mol Biol 314:1067–1075
46. Cho BK, Barrett CL, Knight EM, et al (2008) Genome-scale reconstruction of the Lrp regulatory network in *Escherichia coli*. Proc Natl Acad Sci U S A 105:19462–19467
47. Calvo JM, Matthews RG (1994) The leucine-responsive regulatory protein, a global regulator of metabolism in *Escherichia coli*. Microbiol Rev 58:466–490
48. Chen S, Rosner MH, Calvo JM (2001) Leucine-regulated self-association of leucine-responsive regulatory protein (Lrp) from *Escherichia coli*. J Mol Biol 312:625–635
49. McFarland KA, Lucchini S, Hinton JC, et al (2008) The leucine-responsive regulatory protein, Lrp, activates transcription of the *fim* operon in *Salmonella enterica* serovar typhimurium via the *fimZ* regulatory gene. J Bacteriol 190:602–612
50. Tapias A, López G, Ayora S (2000) *Bacillus subtilis* LrpC is a sequence-independent DNA-binding and DNA-bending protein which bridges DNA. Nucleic Acids Res 28:552–559
51. Beloin C, Jeusset J, Revet B, et al (2003) Contribution of DNA conformation and topology in right-handed DNA wrapping by the *Bacillus subtilis* LrpC protein. J Biol Chem 278:5333–5342
52. Zheng D, Constantinidou C, Hobman JL, et al (2004) Identification of the CRP regulon using in vitro and in vivo transcriptional profiling. Nucleic Acids Res 32:5874–5893
53. Soutourina O, Kolb A, Krin E, et al (1999) Multiple control of flagellum biosynthesis in *Escherichia coli*: role of H-NS protein and the cyclic AMP-catabolite activator protein complex in transcription of the flhDC master operon. J Bacteriol 181:7500–7508
54. Liu M, Durfee T, Cabrera JE, et al (2005) Global transcriptional programs reveal a carbon source foraging strategy by *Escherichia coli*. J Biol Chem 280:15921–15927
55. Grainger DC, Hurd D, Harrison M, et al (2005) Studies of the distribution of *Escherichia coli* cAMP-receptor protein and RNA polymerase along the *E. coli* chromosome. Proc Natl Acad Sci U S A 102:17693–17698

56. Lin SH, Lee JC (2003) Determinants of DNA bending in the DNA-cyclic AMP receptor protein complexes in *Escherichia coli*. Biochemistry 42:4809–4818
57. Napoli AA, Lawson CL, Ebright RH, et al (2006) Indirect readout of DNA sequence at the primary-kink site in the CAP-DNA complex: recognition of pyrimidine-purine and purine-purine steps. J Mol Biol 357:173–183
58. Schneider R, Travers A, Muskhelishvili G (1997) FIS modulates growth phase-dependent topological transitions of DNA in *Escherichia coli*. Mol Microbiol 26:519–530
59. Ali Azam T, Iwata A, Nishimura A, et al (1999) Growth phase-dependent variation in protein composition of the *Escherichia coli* nucleoid. J Bacteriol 181:6361–6370
60. Browning DF, Cole JA, Busby SJ (2008) Regulation by nucleoid-associated proteins at the *Escherichia coli* nir operon promoter. J Bacteriol 190:7258–7267
61. Grainger DC, Goldberg MD, Lee DJ, et al (2008) Selective repression by Fis and H-NS at the *Escherichia coli* dps promoter. Mol Microbiol 68:1366–1377
62. Squire DJ, Xu M, Cole JA, et al (2009) Competition between NarL-dependent activation and Fis-dependent repression controls expression from the *Escherichia coli* yeaR and ogt promoters. Biochem J 420:249–257
63. Bradley MD, Beach MB, de Koning AP, et al (2007) Effects of Fis on *Escherichia coli* gene expression during different growth stages. Microbiology 153:2922–2940
64. Cho BK, Knight EM, Barrett CL, et al (2008) Genome-wide analysis of Fis binding in *Escherichia coli* indicates a causative role for A-/AT-tracts. Genome Res 18:900–910
65. Maurer S, Fritz J, Muskhelishvili G (2009) A systematic in vitro study of nucleoprotein complexes formed by bacterial nucleoid-associated proteins revealing novel types of DNA organization. J Mol Biol 387:1261–1276
66. Schneider R, Lurz R, Lüder G, et al (2001) An architectural role of the *Escherichia coli* chromatin protein FIS in organising DNA. Nucleic Acids Res 29:5107–5114
67. Dorman CJ (2004) H-NS: a universal regulator for a dynamic genome. Nat Rev Microbiol 2:391–400
68. Dame RT, Luijsterburg MS, Krin E, et al (2005) DNA bridging: a property shared among H-NS-like proteins. J Bacteriol 187:1845–1848
69. Dame RT, Noom MC, Wuite GJ (2006) Bacterial chromatin organization by H-NS protein unravelled using dual DNA manipulation. Nature 444:387–390
70. Dorman CJ (2007) Probing bacterial nucleoid structure with optical tweezers. Bioessays 29:212–216
71. Noom MC, Navarre WW, Oshima T, Wuite GJ, Dame RT (2007) H-NS promotes looped domain formation in the bacterial chromosome. Curr Biol 17:R913–R914
72. Oshima T, Ishikawa S, Kurokawa K, et al (2006) *Escherichia coli* histone-like protein H-NS preferentially binds to horizontally acquired DNA in association with RNA polymerase. DNA Res 13:141–153
73. Doyle M, Fookes M, Ivens A, et al (2007) An H-NS-like stealth protein aids horizontal DNA transmission in bacteria. Science 315:251–252
74. Lucchini S, Rowley G, Goldberg MD, et al (2006) H-NS mediates the silencing of laterally acquired genes in bacteria. PLoS Pathog 2:e81
75. Schechter LM, Jain S, Akbar S, et al (2003) The small nucleoid-binding proteins H-NS, HU, and Fis affect hilA expression in *Salmonella enterica* serovar Typhimurium. Infect Immun 71:5432–5435
76. Hinton JC, Santos DS, Seirafi A, et al (1992) Expression and mutational analysis of the nucleoid-associated protein H-NS of *Salmonella typhimurium*. Mol Microbiol 6: 2327–2337
77. Baños RC, Vivero A, Aznar S, et al (2009) Differential regulation of horizontally acquired and core genome genes by the bacterial modulator H-NS. PLoS Genet 5:e1000513
78. Barth M, Marschall C, Muffler A, et al (1995) Role for the histone-like protein H-NS in growth phase-dependent and osmotic regulation of sigma S and many sigma S-dependent genes in *Escherichia coli*. J Bacteriol 177:3455–3464

79. Stoebel DM, Free A, Dorman CJ (2008) Anti-silencing: overcoming H-NS-mediated repression of transcription in Gram-negative enteric bacteria. Microbiology 154:2533–2545
80. Grillo AO, Brown MP, Royer CA (1999) Probing the physical basis for trp repressor-operator recognition. J Mol Biol 287:539–554
81. Hartwell LH, Hopfield JJ, Leibler S, et al (1999) From molecular to modular cell biology. Nature 402:C47–C52
82. Ma HW, Kumar B, Ditges U, et al (2004) An extended transcriptional regulatory network of *Escherichia coli* and analysis of its hierarchical structure and network motifs. Nucleic Acids Res 32:6643–6649
83. Balázsi G, Barabási AL, Oltvai ZN (2005) Topological units of environmental signal processing in the transcriptional regulatory network of *Escherichia coli*. Proc Natl Acad Sci U S A 102:7841–7846
84. Freyre-González JA, Alonso-Pavón JA, Treviño-Quintanilla LG, et al (2008) Functional architecture of *Escherichia coli*: new insights provided by a natural decomposition approach. Genome Biol 9:R154
85. Resendis-Antonio O, Freyre-González JA, Menchaca-Méndez R, et al (2005) Modular analysis of the transcriptional regulatory network of *E. coli*. Trends Genet 21:16–20
86. Shen-Orr SS, Milo R, Mangan S, et al (2002) Network motifs in the transcriptional regulation network of *Escherichia coli*. Nat Genet 31:64–68
87. Martínez-Antonio A, Janga SC, Thieffry D (2008) Functional organisation of *Escherichia coli* transcriptional regulatory network. J Mol Biol 381:238–247
88. Cerca N, Jefferson KK (2008) Effect of growth conditions on poly-N-acetylglucosamine expression and biofilm formation in *Escherichia coli*. FEMS Microbiol Lett 283:36–41
89. Romeo T (1998) Global regulation by the small RNA-binding protein CsrA and the non-coding RNA molecule CsrB. Mol Microbiol 29:1321–1330
90. Kaplan S, Bren A, Zaslaver A, et al (2008) Diverse two-dimensional input functions control bacterial sugar genes. Mol Cell 29:786–792
91. Lozada-Chávez I, Janga SC, Collado-Vides J (2006) Bacterial regulatory networks are extremely flexible in evolution. Nucleic Acids Res 34:3434–3445

Chapter 3
A Catalogue of Eukaryotic Transcription Factor Types, Their Evolutionary Origin, and Species Distribution

Matthew T. Weirauch and T.R. Hughes

Abstract Transcription factors (TFs) play key roles in the regulation of gene expression by binding in a sequence-specific manner to genomic DNA. In eukaryotes, DNA binding is achieved by a wide range of structural forms and motifs. TFs are typically classified by their DNA-binding domain (DBD) type. In this chapter, we catalogue and survey 91 different TF DBD types in metazoa, plants, fungi, and protists. We briefly discuss well-characterized TF families representing the major DBD superclasses. We also examine the species distributions and inferred evolutionary histories of the various families, and the potential roles played by TF family expansion and dimerization.

3.1 Introduction

Eukaryotic genomes display remarkable diversity in their transcription factor (TF) repertoires, in terms of both presence and prevalence of different TF families in different lineages. It is estimated that TFs constitute between 0.5 and 8% of the gene content of eukaryotic genomes, with both the absolute number and proportion of TFs in a genome roughly scaling with the complexity of the organism [1]. Most eukaryotic TFs tend to recognize short, degenerate DNA sequence motifs, in contrast to the larger motifs preferred by prokaryotic TFs [2]. Cooperation among TFs, rather than highly-specific sequence preferences, is believed to be a pervasive feature of eukaryotic transcriptional regulation [3].

The distinguishing feature of TFs, relative to other transcriptional regulatory proteins, is that they interact with DNA in a sequence-specific manner [4, 5]. In the vast majority of well-studied cases, these interactions are mediated by DNA binding domains (DBDs) [6], and TF families are typically defined on the basis of sequence similarity of their DBDs. Eukaryotic DBDs display a wide range of structural forms

---

M.T. Weirauch (✉)
Banting and Best Department of Medical Research, Donnelly Centre for Cellular and Biomolecular Research, University of Toronto, Toronto, ON, Canada M5S 3E1
e-mail: matt.weirauch@utoronto.ca

T.R. Hughes (ed.), *A Handbook of Transcription Factors*, Subcellular Biochemistry 52, DOI 10.1007/978-90-481-9069-0_3, spanning a diverse array of protein folds (Fig. 3.1), each of which represents a different solution to the problem of recognizing DNA sequences. Most strategies involve interactions with the major groove, although minor groove and/or phosphate and sugar backbone interactions also appear frequently.

TFs have traditionally been classified into superclasses based on broad structural similarities of their DBDs [6]. Such distinctions are somewhat arbitrary (e.g. MADS box and CSL are both members of the β-scaffold superfamily, despite sharing little structural similarity), and are not equivalent to phylogenetic classifications, since the different domain types presumably arose independently. However, they are useful for grouping TFs based on distinctive structural features. Here, we adopt the use of four major superclasses of TFs: *basic*, *zinc-coordinating*, *helix-turn-helix*, and *β-scaffold*. In addition to these four major superfamilies, a wide range of TFs employ less conventional strategies for recognizing DNA sequences, including the AT hook, which recognizes sequences in the minor groove utilizing fewer than a dozen amino acids [7], and the strongly twisted antiparallel β-sheet and four α-helices comprising the SAND domain [8]. These are grouped as *other*.

Several excellent reviews have previously covered the topic of eukaryotic TFs [6, 9–11]. In this chapter, we give an overview of the major classes of eukaryotic TFs, including the types, evolutionary history, and distribution of their DBDs. We restrict our attention to sequence-specific TFs with defined DBDs, so non-sequence specific families such as HMG box (excluding the Sox subfamily), and TFs with undefined DBDs (e.g. many possible candidates from [12] and [13]) are not covered here. It is likely that this overview is incomplete: most eukaryotic lineages are much less well-studied than yeasts, metazoans, and plants, and their genomes may encode novel sequence-specific DNA-binding proteins; in addition, novel DNA-binding activities presumably arise de novo at some frequency in all lineages, in the same way that novel protein-protein interactions arise. It is also possible that some of the proteins enumerated here are not bona fide TFs: outside of major experimental model organisms, most individual proteins have not been examined experimentally.

3.2 A Catalogue of Eukaryotic Transcription Factor Families

The repertoire of TFs carried by individual genomes varies drastically across the eukaryotic kingdom [14]. For example, whereas the majority of metazoan TFs are members of the C_2H_2 zinc finger, homeodomain, and bHLH families, the genomes of plants are predominantly populated by the AP2, MADS box, WRKY, and B3 families, and the largest family of fungal TFs is the zinc cluster (C_6 zinc finger). Figure 3.2 depicts the size of each family across a representative collection of eukaryotic organisms with sequenced genomes. Due to space limitations, we only briefly discuss some of the major families of TFs here, focusing on those that are most prevalent, or have been the subject of intense research. Several of the largest and most well-studied metazoan classes are discussed in separate chapters in this volume (see Chapters 4, 5, and 6, which encompass C_2H_2 zinc fingers, homeodomains, and nuclear receptors). A survey of known eukaryotic TF

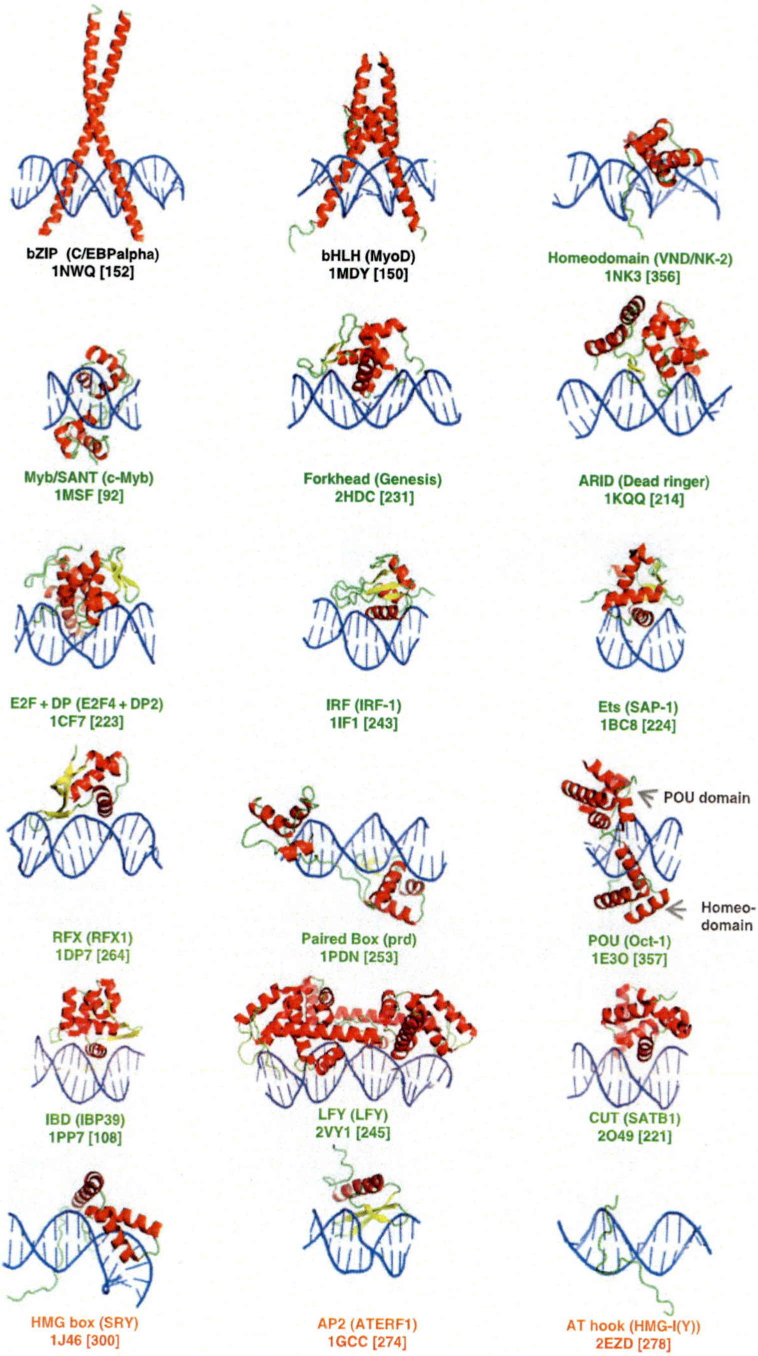

Fig. 3.1 (continued)

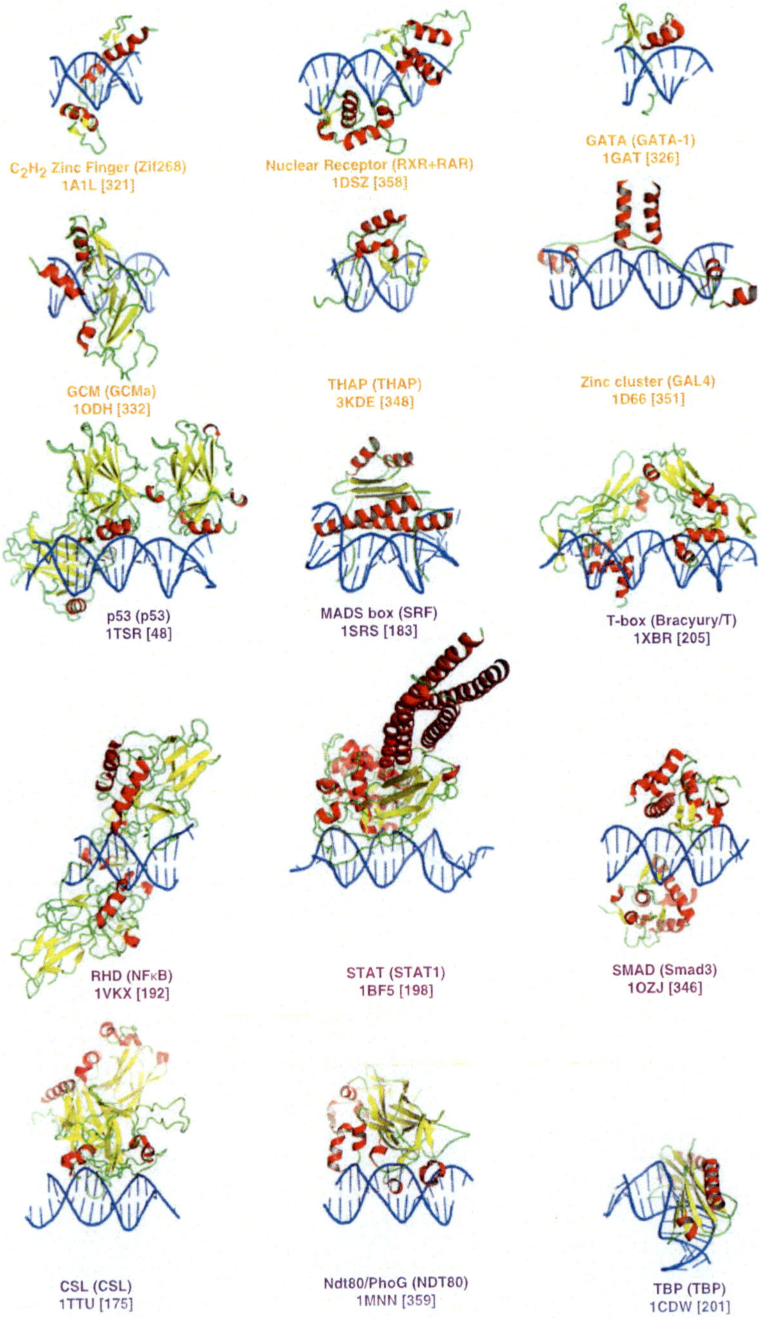

Fig. 3.1 (continued)

families is provided in Table 3.1, which provides specific examples, references, and other information for each family, including Pfam and Interpro identifiers. For inclusion in Table 3.1, a family must have experimental evidence demonstrating sequence-specific DNA binding, along with demonstration of transcriptional regulatory activity in vivo. As such, several families of putative TFs are not included, such as NiN/RWP-RK [15] and Nozzle/SPL [16].

3.2.1 Metazoan Transcription Factor Families

The diversity of forms and lifestyles displayed by members of the metazoan kingdom is thought to at least partially be a result of differences in genomic transcriptional control elements [17]. Despite a prevalence of *cis* differences [18, 19], most metazoans possess a similar basic repertoire of *trans* acting factors (i.e. TFs) (Fig. 3.2), with the largest classes including C_2H_2 zinc fingers, homeodomains, and bHLH (which together constitute over 80% of all known and predicted human TFs [20]). In this section, we briefly introduce the major metazoan families, many of which are also present in the other eukaryotic branches (Fig. 3.2).

3.2.1.1 Metazoan Basic Superfamily Transcription Factors

The basic superfamily, which includes the leucine zipper (bZIP) and helix-loop-helix (bHLH) families, is largely comprised of TFs that dimerize and bind DNA in a scissor-type grip [21]. Members of this superfamily are composed of a basic α-helical DNA-contacting region, and a dimerization interface containing a leucine zipper, HLH, or Helix-Span-Helix (HSH) (Table 3.1). Often, the expression or availability of one dimerization subunit is controlled, while the other is constitutively expressed, a strategy which allows for further fine-tuning of regulatory control [22]. Due to their dimerization upon binding DNA, most members of this superfamily recognize palindromic binding sequences (Table 3.1).

The bZIP and bHLH classes are among the largest in vertebrates, with 53 and 110 members being present in the human genome, respectively. Each class presumably arose from a single common ancestral protein, which subsequently underwent multiple periods of duplication and divergence. This series of evolutionary events resulted in the dozens of subfamilies present in extant species, with each subfamily possessing distinctive dimerization and DNA sequence preferences [23–28]. Among many other well-studied proteins in these classes, the bHLH family includes

Fig. 3.1 Gallery of DNA binding domains. Cartoon representations of transcription factor DNA binding domain interactions with DNA. Structures were obtained from the Protein Data Bank (PDB, http://www.pdb.org/; [360]). TF family name is indicated below each structure, along with the name of the protein, its PDB accession number, and the corresponding reference. DNA is colored *blue*, with α-helices in *red*, β-sheets in *yellow*, and loop regions in *green*. Binding domains are organized by superfamily, indicated by font color: *black*, basic; *green*, helix-turn-helix; *orange*, zinc-coordinating; *red*, other; *purple*, β-scaffold

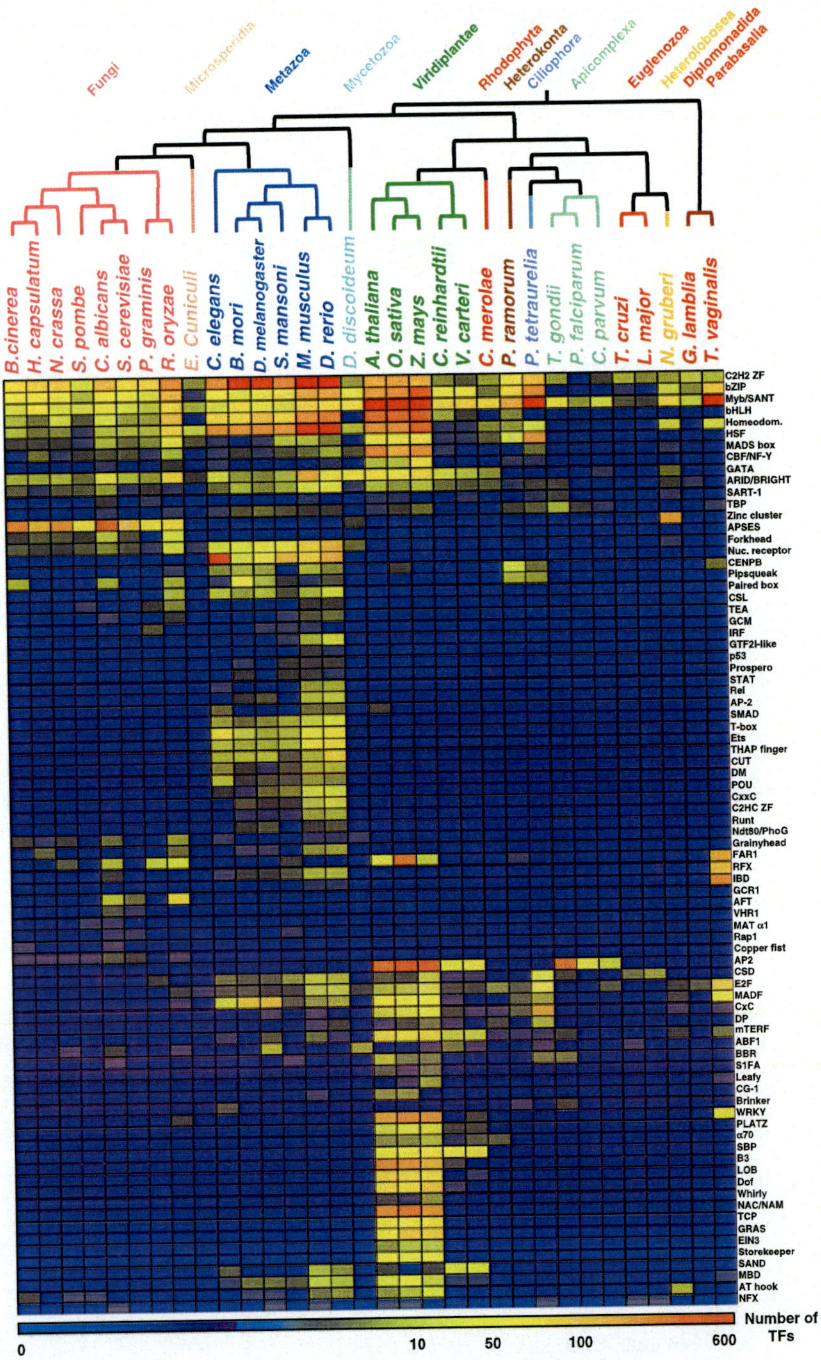

Fig. 3.2 (continued)

MyoD TFs, which control skeletal muscle development [29], and the bZIP family includes the Jun and Fos proteins, which are well known for their roles in cancer progression [30].

3.2.1.2 Metazoan Zinc-Coordinating Transcription Factors

Zinc-coordinating TFs have structures that are stabilized by the tetrahedral coordination of a zinc atom (Zn++) by histidine and cysteine residues. This superfamily includes a wide range of TF classes, including C_2H_2 zinc fingers, nuclear receptors (C_4 zinc fingers), and the mostly fungal-specific zinc cluster family. C_2H_2 zinc fingers (see Chapter 4) represent the largest family of eukaryotic TFs, with metazoan genomes housing upwards of 600 members of this family. Most C_2H_2-containing TFs possess multiple zinc fingers (and sometimes dozens), each of which recognizes three (or more) specific base pairs [31]. C_2H_2 zinc fingers have been the focus of intense study due to their potential to be engineered to recognize specific sequences [32].

Members of the metazoan-specific nuclear receptor family (see Chapter 6) control a wide range of physiological processes. In particular, nuclear receptors act as hormone and environmental sensors, responding to lipophilic molecules such as fatty acids, vitamins and steroids [33]. Nuclear receptors originally were thought to have originated in metazoans, but a recent study suggests they might have emerged early during eukaryotic evolution [34].

GATA family TFs, which contain two adjacent zinc fingers, are named after their consensus recognition sequence (WGATAR) [35, 36]. GATA binding sites have been extensively studied in the regulatory regions of the human β-globin gene cluster, using a combination of comparative genomic and experimental techniques [37]. Members of the GATA family play important regulatory roles in hematopoietic stem cells, as well as in mesoderm and endoderm-derived tissues [38]. Mutations in GATA TFs have been implicated in numerous human diseases, including cancer, congenital heart defects, and down syndrome [38, 39].

Fig. 3.2 Species distribution of transcription factor families. Heatmap depicting abundance of transcription factor families across a representative sampling of sequenced eukaryotic genomes. Phylogenetic relationships at the top are borrowed from [98] (branches not to scale). Each entry indicates the number of proteins with a match to the corresponding Pfam domain (using the Hmmer program [361] with recommended similarity thresholds of 0.01 for both the sequence e-value and the domain conditional e-value). Color key is depicted at the *bottom* (note logarithmic scale). DBD families were hierarchically clustered using average linkage clustering. Brief definition of clades: Fungi: yeasts, molds, mushrooms, etc.; Microsporidia: spore-forming unicellular parasitic fungi; Metazoa: animals; Mycetozoa: slime molds; Viridiplantae: green algae and land plants; Rhodophyta: red algae (including many seaweeds); Heterokonta: mostly diatomic algae (including plankton) and kelp; Ciliophora: protists with cilia; Apicomplexa: unicellular, spore-forming parasitic protists; Euglenozoa: flagellate unicellular protists; Heterolobosea: colorless unicellular protists, many of which can transform between multiple forms; Diplomonadida: mostly double-celled parasitic flagellate protists; Parabasalia: flagellate protists, most of which are symbiotic with animals. DBDs without a pfam model are not shown, nor are Sox domains, which share a Pfam model with the non-sequence-specific HMG class

Table 3.1 Catalog of eukaryotic transcription factor binding domains

SC[a]	Family[b]	Structural motif[c]	Dimer?[d]	Example TF(s)[e]	Sequence motif(s)[f]	Domain(s)[g]
Ba	AP-2 / Basic helix-span-helix (bHSH)	Basic α-helix + HSH [147]	Dimer [147]	AP-2 proteins	$GCCN_{3,4}GGC$, $GCCN_{3,4}GGG$ [148], CCCCAGGC [149]	TF_AP-2/IPR013854
Ba	Basic helix-loop-helix (bHLH)	Basic α-helix + HLH [150, 151]	Dimer [150, 151]	MyoD, E12, E47, Max, Sreb1, Ahr, Arnt, Hairy	CANNTG (E-box), NRCGTG (non-E-box), CACGMG (N-box) [26]	HLH/IPR001092
Ba	Basic leucine zipper (bZIP)	Basic α-helix + leucine zipper [152, 153]	Dimer [152, 153]	C/EBP, GCN4, Fos, Jun, CREB proteins, PAR, Oasis proteins, ATF proteins	ATTGCGCAAT (CCAAT), TGAGTCA (TRE), TGACGTCA (CRE), TGCTGAGTCAT (MARE), GATGACGTGKNNNWT (CRE-L), ATTACGTAAT (PAR), RRRTGCAATMCCC (CHOP), TTACTAA (YAP), CCACGTGG (G-box), CACGTGGC (ABRE) [154]	bZIP_1, bZIP_2, bZIP_Maf/IPR011616
Ba	BES1/BZR1/LAT61	HLH-like [155]	Hetero (w/bHLH BIM1) [156]	BES1, BZR1, BEH proteins	CANNTG (E-box) [156], CGTGYG (BRRE) [155]	(none)/(none)
Ba	GTF2I-like	HLH-like (longer than normal), lacking a basic region [157, 158]	No	GTF proteins	CGGATTAAC (BLM) [157], GGGATTRBR [158]	GTF2I/IPR004212
Ba	SART-1	bZIP-like [159]	No	SART proteins, HAF	CCCCCACCCCCACCCGC (EP17) [159]	SART-1/IPR005011

Table 3.1 (continued)

SC[a]	Family[b]	Structural motif[c]	Dimer?[d]	Example TF(s)[e]	Sequence motif(s)[f]	Domain(s)[g]
Ba	Storekeeper (STK) / GeBP-like	Similar to bZIP, but wider spacing between basic domain and leucine zipper [160, 161]	Dimer [162]	GEBP, GPL proteins, STK	GCTAAACAAT (B-box) [161]	DUF573/IPR007592
Ba	TCP	HLH [163, 164]	Dimer [163, 164]	CYC, TB1, PCF proteins	GGNCCCAC, GTGGNCCC [164]	TCP/IPR005333
Ba	Trihelix/GT	One or two trihelix domains (helix-loop-helix-loop-helix) [165]	Dimer [166, 167]	GT proteins, PTL, ASIL1, DF1	AT-rich sequences (GT elements) [168]	(none)/(none)
β	B3/VP1/IAA / Auxin response factor (ARF)	β-sheet + two α-helices between β-strands (similar to *Eco*RII) [72]	Dimer [169]	ARF proteins, ETT, FUS3, RAV1, ABI3	TGTCTC (AuxRE), CACCTG CATGCA (Sph/RY) [72]	B3/IPR00334
β	CG-1/CAMTA	Immunoglobulin-like [170, 171]	No	CAMTA proteins	CGCG (CG-1 element) [170, 172]	CG-1/IPR005559
β	Cold shock domain (CSD)	Five-stranded β-barrel [173]	No [174]	Y-box proteins, lin-28, DBP proteins, FRGY2	CTGATTGGCCAA (Y-box) [173]	CSD/IPR002059
β	CSL/LAG1 / suppressor of hairless	β-sandwich (similar to Rel) [175]	No [175, 176]	Su(H), lag-1, CBF1, RBP-Jκ	YGTGGGAA [175–177]	LAG1-DNAbind/IPR01535

Table 3.1 (continued)

SC[a]	Family[b]	Structural motif[c]	Dimer?[d]	Example TF(s)[e]	Sequence motif(s)[f]	Domain(s)[g]
β	Grainyhead/CP2/LSF	Immunoglobulin-like β-barrel (similar to p53) [178]	No [179], Dimer [180]	grh, gem, CP2, UBP proteins, MGR, BOM	GCNCNANCCAG [181], AAACCGGTTT [182], CTGG [182], wide range of sequences [178]	CP2/IPR007604
β	MADS box/SRF	Anti-parallel coiled coil of two α-helices, one from each subunit [183, 184]	Dimer [183, 185]	MADS proteins, MEF proteins, SRF, AGL proteins, SLM proteins, MCM1	CCW_6GG (CArG-box), AT-rich sequences [184]	SRF-TF/IPR002100
β	MBD	α/β sandwich [186]	No [187]	MBD proteins	CG dinucleotides [187]	MBD/IPR001739
β	Ndt80/PhoG	Immunoglobulin-like β-sandwich (similar to p53) [188]	No [188]	Ndt80, Gm98, Gm239, pqn-47	GNCRCAAAW (MSE) [188, 189]	NDT80_PhoG/IPR007888
β	p53	Loop-sheet-helix (β-sandwich), zinc coordinating [48]	Tetramer [190]	p53, p73	Two or more copies of RRRCWWGYYY, separated by up to 13 bases [191]	P53/IPR011615
β	Rel homology region (RHR/RHD)	Two β-sandwiches (p53-like) [192–194]	Dimer [192–194]	Rel proteins, NFκB (p50/p65), p52, Dorsal, Relish, NFAT	GGGRNYYYCC (κB element) [195]	RHD/IPR011539

Table 3.1 (continued)

SC[a]	Family[b]	Structural motif[c]	Dimer?[d]	Example TF(s)[e]	Sequence motif(s)[f]	Domain(s)[g]
β	Runt	Immunoglobulin-like β-sandwich (similar to p53) [196]	No; binding enhanced by Hetero with CBFβ [196]	Runt, AML proteins, lozenge, PEBP2α proteins, RUNX proteins, rnt-1	RACCRCA [197]	Runt/IPR013524
β	STAT	Several p53-like β-sheets [198]	Dimer [198, 199]	STAT proteins	$TTN_{5-6}AA$ (GAS-like element), $AGTTTN_3TTTCC$ (ISRE) [199]	STAT_bind/IPR013801
β	TATA binding protein (TBP)	Concave antiparallel β-sheet [200–203]	No [200–203]	TBP	AT-rich sequences (TATA box) [204]	TBP/IPR000814
β	T-box	Immunoglobulin-like β-barrel (similar to p53) [205, 206]	No [207]	TBX proteins, Brachyury, Eomes, mid, bi, ET, T	TCACACCT (T-element) [208]	T-box/IPR001699
β	Whirly/PBF2	Tetramer of crossed anti-parallel β-sheets that interact with an α-helix [74]	No [74]	PBF-2, TIF1, WHY proteins	TGACANNNNTGTCA [209], GTCAAAAW (PB element) [73]	Whirly/IPR013742
H	APSES	β-sheet interspersed with two pairs of α-helices [210–212]	Dimer [211]	SOK2, XBP1, PHD1, MBP1, SWI4	ACGCGTNA (MCB) [213]	KilA-N/IPR018004
H	ARID/BRIGHT	Modified HTH (multi-helical) [214]	No [214]	Bright, ARID proteins, osa, Dead ringer, Mrf1/2	AT-rich sequences, some non-specific [214]	ARID/IPR001606

Table 3.1 (continued)

SC[a]	Family[b]	Structural motif[c]	Dimer?[d]	Example TF(s)[e]	Sequence motif(s)[f]	Domain(s)[g]
H	Brinker	Four α-helices forming a HTH [215]	No [215]	Brinker	GGCGYY [215, 216]	BrkDBD/IPR018586
H	CENPB	HTH [217]	No [218]	CENPB, JRK, TIGD proteins, ABP1/CBP1, CBH	TTCGNNNNANNCGGG (CENP-B box) [219, 220]	CENP-B_N/IPR006695
H	CUT/ONECUT/CDP	One or more Homeodomain-like regions (similar to POU) [221]	Homo [222]	CUT, HNF-6, MCLOX proteins, ceh-39, ceh-21, SATB1	YYRAT [221]	CUT/IPR003350
H	DP	Winged helix-turn-helix (wHTH: HTH + three-stranded antiparallel β-sheet) [223]	Hetero (w/E2F) [223]	DP proteins	TTTSGCGCS [223]	DP/IPR014889
H	Ets	Unique wHTH-like structure [224, 225]	No [224, 225]	Ets proteins, ERG, FLI1, ETV proteins, ast-1	GGAW core: ACCGGAWRY (class I), CCGGAART (class II), GNGGAAGT (class III), CCGGAT (class IV) [226]	Ets/IPR000418

Table 3.1 (continued)

SC[a]	Family[b]	Structural motif[c]	Dimer?[d]	Example TF(s)[e]	Sequence motif(s)[f]	Domain(s)[g]
H	E2F	Modified wHTH [223]	Mostly Hetero (w/ DP) [223], also Homo [227–229], and Mono [230]	E2F proteins, E2L proteins	TTTSGCGCS [223]	E2F_TDP/IPR003316
H	Forkhead (FKH) / Forkhead box (Fox)	wHTH [231]	No [45, 231]	Forkhead, HNF proteins, FOX proteins, PU-1, daf-16	TGTTTA core [65, 232–234]	Fork_head/IPR001766
H	HB-PHD/ ZF-HD	Homeodomain [235]	No [236]	ZFHD proteins	AGTAATTAAANNNNNAATTA [236], AGTGTCTTGTAATTAAAA [237]	(none)/IPR006455
H	Heat shock factor (HSF)	wHTH [238]	Homo-trimers [239]	HSF proteins, MGA1	Two to six copies (usually three) of alternating and inverted NGAAN (HSEs) [239]	HSF_DNA-bind/IPR000232
H	Homeodomain / homeobox (Hox)	Homeodomain [240, 241]	No [240, 241]	Hox proteins, Ftz, Ubx, engrailed, MATα2, bicoid, Homez	TAAT, ATAAAA, TCGTAAA [242]	Homeobox/IPR00135
H	Inr binding domain (IBD)	Modified wHTH (similar to Ets) [108]	No [108]	IBP39	GTYACTTCACWT (initiator element/Inr) [108]	IBD/IPR018845
H	Interferon regulatory factor (IRF)	wHTH [243, 244]	No [243]	IRF proteins	GAAA, often occurring in repeats (IRF-E) [243, 244]	IRF/IPR001346

Table 3.1 (continued)

SC[a]	Family[b]	Structural motif[c]	Dimer?[d]	Example TF(s)[e]	Sequence motif(s)[f]	Domain(s)[g]
H	LEAFY/LFY/FLO	Seven α-helix fold (HTH-like) [245]	Homo [245]	LEAFY/LFY	CCANTG [245, 246]	FLO_LFY/IPR002910
H	MADF	Myb-like HTH [247]	Dimer [247]	Adf-1, Dip3, Stwl	GYY_N [248], CCWNNCCWNNCC [249]	MADF_DNA_bdg/IPR006578
H	MATα1	Homeodomain [250]	Hetero (with MCM1) [250]	MATα1	ACAATGACAG (Q site) [250]	MAT_Alpha1/IPR006856
H	Myb/SANT	Three HTH repeats (R1, R2, and R3)- R2 and R3 affect binding specificity [92]	No [92]	MYB proteins, MTA proteins, RERE, ZZZ3	YAACKG (canonical) [251], CNGTTR, GKTWGTTR, GKTWGGTR (MBSIIG), CTCAGCG [252]	Myb_DNA-binding/IPR014778
H	Paired box (Pax)	Paired (HTH) + Homeodomain [253]	No [253]	PAX proteins, prd, sv, toy, ey, JRKL	TCACGCWTSA [253], TAATTA [242]	PAX/IPR001523
H	Pipsqueak (Psq)	Four tandem HTH repeats [254, 255]	No	psq, pfk, rib, LCOR	Direct repeats of GAGA [254]	HTH_psq/IPR007889
H	POU	N-terminal POU-specific domain (HTH), C-terminal Homeodomain (HTH) [256]	No [256]	POU proteins, OCT proteins, BRN proteins, vv1, ceh-6, ceh-18, pdm2, pbm3, nub	Combinations of two flexibly oriented half-sites (ATGC and AAAT): monomeric [e.g. ATGCAAAT (octomer)], dimeric [e.g. ATGCATATGCAT (MORE), CTCATGAAATATGCAAAT (PORE)] [257]	Pou/IPR000327
H	Prospero/Pros/Prox	Atypical homeodomain [258]	No [259]	Prospero, Prox proteins	AAGACG [260], CWYNNCY [259]	Prox1/IPR007738

Table 3.1 (continued)

SC[a]	Family[b]	Structural motif[c]	Dimer?[d]	Example TF(s)[e]	Sequence motif(s)[f]	Domain(s)[g]
H	Rap1	Two HTH domains: homeodomain-like, and myb-like [261, 262]	No [261]	RAP1	RCAAYCCRYNCAYY [263]	Rap1-DNA-bind/IPR015280
H	Regulatory factor X (RFX)	Atypical wHTH (DNA recognition by the wing) [264]	No [265, 266]	RFX proteins	RGYAAC, often in the form of an inverted repeat: GYNRCCN$_{0-3}$RGYAAC (X-box) [266]	RFX_DNA_binding/ IPR003150
H	Sigma 70 (plants)	HTH [267–269]	No	SIG proteins, SLF proteins	TTGACAN$_{11-14}$TGTGCTATAAT [267, 268]	Sigma70_r2/IPR007627
H	TEA/ATTS/TEF	Homeodomain [270, 271]	No [270]	TEF proteins, scalloped, TEC1, AbaA	Tandem repeats of TGGAATGT/WRRWATGY/ NDGHATNT (M-CAT) [270]	TEA/IPR000818
H	VHR1	HTH (?) [272]	No	VHR1	AATCAN$_8$TGA [272]	Vhr1/IPR007147
O	AFT	(Unknown) [273]	No	AFT1, AFT2	YRCACCCR [273]	AFT/IPR014842
O	AP2/GBD/EREBP/ERF/ GCC-box binding protein	β-sheet packed parallel to an α-helix [274, 275]	No (plants) [274], Dimer (Apicomplexa) [275]	AP2, ATERF proteins, DREB proteins, ERF proteins, ApiAP2	Plants: AGCCGCC (GCC box) [276]; Apicomplexa: TGCATGCA [277]	AP2/IPR001471
O	AT hook	Crescent-shaped flap-like element with projecting basic residues; often multiple, and found with HMG domain [278]	No [278]	CACNA1A, AKNA, HMGA proteins, MLL, TCF1, LEF1	AT-rich sequences [7]	AT_hook/IPR017956

Table 3.1 (continued)

SC[a]	Family[b]	Structural motif[c]	Dimer?[d]	Example TF(s)[e]	Sequence motif(s)[f]	Domain(s)[g]
O	CBF/NF-Y	Amphipatic α-helix [279, 280]	Hetero-trimeric (CBFA/B/C) [56, 281]	NF-Y proteins	CCAAT [59, 282]	CBFB_NFYA/IPR001289
O	DAL82	Unique N-terminal DBD with unknown fold [283]	No	DAL82	GAAAATTGCGTT (UIS_{ALL}) [284]	(none)/(none)
O	DBP/DNC	(Unknown) [285]	No	DBP1	TAATATTTGCCTTT [286]	(none)/(none)
O	EIN3/EIN3-like (EIL)	Five α-helices [287]	Homo [287, 288]	EIL proteins, TEIL	AYGWAYCT [289]	EIN3/IPR006957
O	GCR1	Three small α-helices [290, 291]	No	GCR1, MSN1, HOT1, hSGT1	GGCTTCCWC (CT box) [290, 292, 293]	GCR1_C/IPR022210
O	GRAS/Scarecrow	Similar to STAT (?) [294]	Dimer [295]	SCR, SHR, LAS, GAI, NSP1, NSP2, DELLA proteins	AATTT [295]	GRAS/IPR005202
O	NAC/NAM	Twisted β-sheet and three α-helices [91]	Dimer [91, 296]	NAC1, NAM, NAP, TIP, CUP proteins	CATGTG (NACRS) [297]	NAM/IPR003441
O	SAND/KDWK	SH3-like β-barrel [8, 298]	No [298]	Sp100, NUDR, AIRE, GMEB proteins, ULTRA PETALA1	RCGY [298], TTCG repeats [299]	SAND/IPR000770

Table 3.1 (continued)

SC[a]	Family[b]	Structural motif[c]	Dimer?[d]	Example TF(s)[e]	Sequence motif(s)[f]	Domain(s)[g]
O	Sox	Multiple HMG boxes; simple trihelical fold [300]	Mostly no, some require dimers [64]	SOX proteins, SRY	WWCAAWG [301]	HMG_box/IPR000910
O	S1FA	α-helix (?) [286, 302]	No	S1FA proteins	ATGGTAACAATT (S1F) [302]	S1FA/IPR006779
Z	ABF1/ARS1 binding factor	Atypical zinc finger [303]	No	ABF1, BAF1	RTCRYNNNNNACG [304]	BAF1_ABF1/IPR006774
Z	Alfin-like	Novel C_4 + HC_3 zinc-fingers [305]	No	Alfin1	GNGGTG, GTGGNG [305]	(none)/(none)
Z	BBR/BPC	Zinc finger (?) [306]	No	GBP, BBR, BPC proteins	GA repeats [306, 307], RGARAGRRA [307]	GAGA_bind/IPR010409
Z	COE/EBF	HC_3 zinc finger [308]	Homo [308, 309]	Collier, Olf1, EBF	Inverted repeat of GGGAWT separated by two bases [309, 310]	(none)/IPR003523
Z	Copper fist	Three-stranded antiparallel β-sheet w/ two α-helical segments projecting from one end [311]	No [312]	MAC1, CUP2, HAA1, ACE1, LPZ8	GTCTTTTYYGCTGA [312, 313]	Copper-fist/IPR001083
Z	CxC/CRC/CPP-like	Trinuclear zinc cluster [314, 315]	No	TSO1, CPP1, MSL2	GA-rich elements [315]	CXC/IPR005172
Z	CxxC	Two zinc ions coordinated tetrahedrally by four conserved cysteines [316]	No	PCM1, HRX, DNMT proteins, CGBP	CG dinucleotides [317, 318]	zf-CXXC/IPR002857

Table 3.1 (continued)

SC[a]	Family[b]	Structural motif[c]	Dimer?[d]	Example TF(s)[e]	Sequence motif(s)[f]	Domain(s)[g]
Z	C_2HC/CCHC zinc finger	Multiple C_2HC fingers (single C_2HCs do not bind DNA) [319]	No	ST18, MYT1, MYT1L	G-rich sequences [320]	zf-C2HC/IPR002515
Z	C_2H_2 zinc finger / ββα zinc finger / Krüppel	Multiple (between 1 and 37) ββα zinc fingers [321]	No [321]	TFIIIA, Krüppel, Sp1, Hunchback, Snail, ADR1, Evi1, SWI5, Xfin, Egr1, Zif268, GLI	Three distinct bases for each zinc finger [31]	zf-C2H2/IPR007087
Z	DM/Doublesex	"Intertwined" CCHC and HCCC zinc-binding sites [322]	Dimer [323]	Doublesex, mab-3, DMRT proteins	RNNACWAWGTNNY (palindromic) [323]	DM/IPR001275
Z	Dof	Multiple C_2C_2-like zinc fingers [324]	No	OBP1, PBF1, NtBBF1	AAAG, AGTA [83]	zf-Dof/IPR003851
Z	FAR1/FRS	Zinc-binding (related to WRKY) [325]	No	FHY3, FAR1	CACGCGC (FBS) [325]	FAR1/IPR004330
Z	GATA	Two C_2C_2 ββα zinc fingers [326, 327]	Any [36, 327, 328]	GATA proteins, MTA proteins, AREA, egr-1, ZIM	WGATAR (and variations) [35]	GATA/IPR000679
Z	GBF zinc finger (Dictyostelium)	Two novel C_4 zinc fingers [329, 330]	No [330]	gbfA	Two half sites of KGKGKGK (GBRE) [329, 331]	(none)/(none)

Table 3.1 (continued)

SC[a]	Family[b]	Structural motif[c]	Dimer?[d]	Example TF(s)[e]	Sequence motif(s)[f]	Domain(s)[g]
Z	GCM	Three α-helices and a β-pleated sheet (related to WRKY) [332]	No [332]	GCM proteins	RCCCGNAT [333, 334]	GCM/IPR003902
Z	HRT-like zinc finger	One or more C_3H zinc fingers [335]	No	HRT proteins	GGCCGATAACAAACTCCGGCC (GARE) [335]	(none)/(none)
Z	LOB/LBD/AS2	Atypical C_2C_2 zinc finger (?) [336]	Homo [336]	LOB proteins	GCGGCG (LBD motif) [336]	DUF260/IPR004883
Z	mTERF	Novel triangular three-helix motif [337]	No [338]	mTERF proteins	TGGTARARGTYGGT [339]	mTERF/IPR003690
Z	NFX zinc finger/NF-X1	Cysteine-rich motif repeated 7 times [340]	No	NFX1, FAP1, shuttle craft (stc)	CCTAGCAACAGATG (X1 box) [340]	zf-NF-X1/IPR000967
Z	Nuclear receptor / C_4 zinc finger	Pair of multi-cysteine ββα zinc fingers [341]	Any [341, 342]	Nuclear hormone receptor proteins, AR, ESR1	Direct, inverted, and everted repeats of RGGTCA (HREs) [343]	zf-C4/IPR001628
Z	PLATZ	Double zinc finger motif [344]	No	PLATZ1	AT-rich sequences [344]	PLATZ/IPR006734
Z	SBP/SQUAMOSA	C_3H/C_4 and C_2HC zinc fingers [81]	No [81]	SBP proteins, SPL proteins	TNCGTACAA [79]	SBP/IPR004333
Z	SMAD/MH1/DWA/NF-1	Zinc coordinating β-hairpin [345, 346]	No [345, 346]	SMAD proteins, MAD proteins, daf-3, daf-8, daf-14	GTCT (Smad box/SBE) [345]; in vivo, multiple SBEs and/or cofactors usually required [347]	MH1/IPR003619
Z	THAP finger	C_2CH zinc finger [348]	No [348]	THAP proteins	TNNGGGNW [348]	THAP/IPR006612

Table 3.1 (continued)

SC[a]	Family[b]	Structural motif[c]	Dimer?[d]	Example TF(s)[e]	Sequence motif(s)[f]	Domain(s)[g]
Z	VOZ	Single atypical C_3H zinc finger + basic region [349]	Dimer [349]	VOZ proteins	$GCGTN_7ACGC$ [349]	(none)/(none)
Z	WRKY	Four-stranded β-sheet [75]	No [75]	WRKY proteins, TIZZ, WIZZ, ZAP1, mod	TTGACY (W-box), TGCGCTT (PRE4 element), TAAAGATTACTAATAGGAA (SURE element), TTTTCCAC (WK box) [350]	WRKY/IPR003657
Z	Zinc cluster / C_6 zinc finger	Pair of multi-cysteine ββα zinc fingers [351]	Any [351–353]	GAL4, LAC9, PPRI, HAP1, PDR proteins	Direct, inverted, and everted repeats of CGG (different spacings and orientations for different proteins) [99]	Zn_clus/IPR001138

[a]Superclass, based on DBD structure. Key: Ba, basic; β, β-scaffold; H, helix-turn-helix; O, other; Z, zinc-coordinating.
[b]Transcription factor family name(s).
[c]Brief description of structural motif used to bind DNA. Where available, references are provided for solved protein structures complexed with DNA. In lieu of such information, references are for (in decreasing order of priority): protein structures without DNA, or experimental evidence of sequence-specific binding.
[d]Do family members bind as obligate dimers? Key: Homo, most members of this family bind DNA as obligate homodimers; Hetero, most bind as obligate heterodimers; Dimer, members bind as homo- or heterodimers; Any, some family members bind as monomers, and others as obligate dimers; No, members of the family are not known to require dimerization.
[e]Sampling of transcription factor members of the family.
[f]Sequence motifs known to be recognized by members of the family. IUPAC codes are used for consensus sequences. Numeric subscripts indicate multiple nucleotides (e.g. N_{1-4} means a run of one to four bases of any type). Regulatory element names (where available) are indicated in parentheses.
[g]Pfam [354] and Interpro [355] domain models for the corresponding family.

3.2.1.3 Metazoan Helix-Turn-Helix Transcription Factors

The helix-turn-helix (HTH) superfamily is composed of proteins that bind the major groove of DNA using an open tri-helical bundle [40]. In addition to their widespread prevalence in nearly all eukaryotic genomes, members of the HTH superfamily constitute the majority of TFs in both the archaeal and bacterial superkingdoms, although almost no eukaryotic TF shares strong sequence similarity with a prokaryotic counterpart [40]. In eukaryotes, homeodomains represent the largest family of TFs that bind DNA utilizing a HTH structure (see Chapter 5). The homeodomain family consists of dozens of subfamilies, including (among many others) Hox proteins, which are well-known for their control of developmental processes in metazoan genomes [41], and POU proteins (such as Oct4, the classical marker of Embryonic Stem cells and one of the four "Yamanaka factors" [42]), which contain a separate, structurally homologous POU HTH domain along with a homeodomain [43].

In addition to the basic tri-helical core, a wide variety of modifications and additions to the HTH structural motif have arisen, most notably the winged helix-turn-helix (wHTH) motif, which adds a β-strand hairpin unit [44]. TF families that utilize the wHTH motif for DNA binding include Forkhead box (Fox), Ets, IRF, RFX, Heat shock factor (HSF), and E2F (Table 3.1). Fox TFs, which originated in unicellular eukaryotes and subsequently expanded in the mammalian lineage, play important roles in a wide range of developmental processes, including organogenesis and speech acquisition [45]. Mutations in Fox genes have also been implicated in a wide range of human diseases, including cancer, glaucoma, and various language disorders [45].

3.2.1.4 Metazoan β-Scaffold Transcription Factors

TFs that bind DNA utilizing a β-scaffold-like structure include the p53 tumor suppressor, which has been referred to as "the guardian of the genome" due to its important mutation prevention role in eukaryotic genomes [46]. p53 is activated by DNA damage or hypoxia, and controls genes involved in cell cycle arrest, DNA repair and programmed cell death [47]. Mutation or inactivation of p53 results in a range of cancer-favoring scenarios, including errors in cell-cycle checkpoints and apoptosis [47]. p53 binds DNA with a 200 amino acid DBD consisting of a zinc-coordinating β-sandwich [48]. p53 is structurally related to several major eukaryotic TF families, including the Rel Homology Region (RHR) family, which includes the immune-responsive NFκB complex [49], and the STAT family, whose members control a variety of cellular events upon their activation by a series of extracellular signaling proteins [50].

The cold shock domain (CSD) is found in all three superkingdoms, and is thus thought to represent one of the most ancient TF families [51]. In addition to the transcriptional regulation roles played by CSD Y-box TFs such as YB1 [52] and FRGY2 [53], CSD-containing proteins are involved in a wide range of other processes, including translation initiation, RNA degradation, and pre-mRNA splicing

[54]. Members of the Y-box family are capable of a variety of functions beyond DNA binding, including DNA and RNA melting, annealing, and strand exchange activities [55].

3.2.1.5 Other Metazoan Transcription Factor Families

CBF/NF-Y binds DNA as a complex that is minimally composed of the CBF-A (NFY-B/HAP3), CBF-B (NFY-A/HAP2), and CBF-C (NF-YC/HAP5) subunits [56, 57]. The resulting complex recognizes the CCAAT box, which is located within 100 bases of the transcription start site in an estimated 30% of eukaryotic promoters [58, 59] (see Chapter 10, which provides an overview of sequences commonly found in human and *Drosophila* promoters). Subunits of CBF/NF-Y are subject to a variety of regulatory control methods, including transcriptional, post-transcriptional, and post-translational means [60].

Sox family TFs may have played a role in the development of eukaryotic multicellularity, due to their importance in the control of genes involved in the extracellular matrix, cell adhesion, and signaling [61]. A typical mammalian genome contains ~20 Sox TFs, each of which participates in multiple developmental processes [62]. A well-known member of the Sox family, SRY, initiates male sex determination in mammals [63]. Unlike other HMG-containing proteins, most Sox TFs bind to specific DNA sequences (WWCAAWG), with the various Sox proteins recognizing different flanking sequences [64]. Recent data indicate that some Sox proteins prefer alternative sequences, including TGAATG (Hbp and Bbx subfamilies), and TCAAAG (Tcf and Lef subfamilies) [65].

3.2.2 Plant Transcription Factor Families

The plant kingdom is comprised of the Viridiplantae (green plants), Rhodophyta (red algae) and Glaucophyta (simple glaucophyte algae). Viridiplantae include both non-vascular and vascular plants, with vascular plants including the seed plants, of which flowering plants are a subgroup. In general, between 3 and 6% of green plant genes encode TFs, while the genes of algae are comprised of only between 0.5 and 2% [66], although the smaller proportion of algae TFs may be exaggerated due to unknown algae-specific TF families. Interestingly, many plant-specific DBDs bind DNA utilizing β-sheets, in contrast to other eukaryotes and prokaryotes, which largely bind DNA utilizing α-helices [67]. Two major bursts of gains and expansions in the repertoire of plant TFs have been identified, coincident with the water-to-land transition and the radiation of flowering plants [66]. Major eukaryotic TF families present in plants include Myb/SANT, bHLH, bZIP, Homeodomain, and C_2H_2 zinc fingers (Fig. 3.2), in addition to several largely plant-specific TF families, which we discuss here.

3.2.2.1 Plant β-Scaffold Transcription Factors

MADS box proteins are found in a variety of organisms (Fig. 3.2), but are most prominent in the genomes of flowering plants, where they have undergone multiple expansions stemming from whole genome duplication events [68]. MADS box TFs control all major aspects of the life of green plants, and the timing of the expansion of this family suggests that it might have played a key role in the evolution of flowering plants [69]. MADS box TFs are categorized as either Type I (animal SRF-like) or Type II (fungal MEF2-like), with members of each class recognizing different DNA sequences and inducing different degrees of DNA bending [70]. It has recently been suggested, based on sequence homology across the entire DBD, that the MADS box domain originated from subunit A of topoisomerase IIA enzymes [71].

B3 domains are present in three families of plant TFs: Auxin response factors (ARFs), VP1, and RAV-like AP2 TFs (which also have AP2 domains). The B3 domain consists of a β-sheet and two α-helices situated between β-strands, with loop residues predicted to make deep contacts with the major groove of DNA [72] (there is, as yet, no structure of a B3 domain in complex with DNA). Although believed to be plant-specific, B3 domains share structural similarities with the *Escherichia coli* EcoRII restriction enzyme, suggesting a possible horizontal transfer event between a eubacterial ancestor to an ancestral plant (or vice versa) [67].

Members of the Whirly TF family are found throughout the plant kingdom, and play roles in the regulation of genes involved in defense response [73]. Whirly TFs bind DNA as tetramers, with each unit consisting of two anti-parallel β-sheets packed perpendicularly against each other [74]. The four resulting blade-like extensions adopt a striking "whirligig-like" appearance, providing the namesake for this family of TFs. Whirly protein complexes bind single stranded DNA, in contrast to the majority of eukaryotic TFs [74].

3.2.2.2 Plant Zinc-Coordinating Transcription Factors

WRKY TFs, named after a conserved WRKYGQK sequence in their DBDs, are zinc-coordinating proteins that adopt a β-sheet fold [75]. Most WRKY TFs are classified as group I, whose members contain two WRKY domains (with the C-terminal domain binding DNA), while group II and III TFs posses a single WRKY domain [76]. Although initially believed to be plant-specific, WRKY proteins have a similar fold to metazoan GCM family TFs and fungal Rcs1p and Rbf1p proteins [77], and group I WRKY proteins have been identified in the protist *Giardia lamblia* and the slime mold *Dictyostelium discoideum* [78], suggesting a more ancient evolutionary origin.

SBP family TFs are involved in a variety of developmental processes, including flower development in particular [79]. Several members of this family are post-transcriptionally controlled by the microRNA miR156, whose overexpression results in a delay in flowering [80]. SBP DBDs include a pair of zinc binding sites consisting of eight residues in a novel C_3HC_2HC or C_6HC configuration, with the first four residues coordinating one zinc atom, and the last four coordinating the other [81] (there is, as yet, no structure of an SBP domain in complex with DNA).

The Dof family is composed of a diverse range of proteins that bind DNA utilizing a C_2C_2 zinc finger [82]. Like many other zinc fingers, the Dof domain can also function in the mediation of protein-protein interactions, often with members of the bZIP family [83]. Dof TFs can be classified into six distinct subfamilies, each with a unique domain architecture [84]. A wide range of plant-specific processes are controlled by members of the Dof family, including light-regulated gene expression, germination, dormancy, and flowering time [85].

3.2.2.3 Additional Plant Transcription Factor Families

AP2 proteins comprise the largest family of TFs that are mostly specific to plants, although members have recently been identified in a wide range of organisms, including TFs in apicomplexans (unicellular animal parasites) [86], and endonucleases in prokaryotes [87], bacteriophage [88], and yeast [89]. The AP2 family is further divided into several subfamilies, with type A TFs being largely involved in the regulation of abiotic stress responses and type B subfamily members participating in disease resistance responses [90]. The unusual phylogenetic distribution of the AP2 family (Fig. 3.2) has been proposed to be a result of their origins in mobile elements [86].

NAC proteins comprise the second-largest family of plant TFs, with the *Arabidopsis thaliana* genome containing over one hundred putative NAC TFs. NAC TFs control a wide range of plant processes, including root formation, floral development, and stress response, and bind DNA as symmetric homodimers [91]. Structurally, the central four strands of the NAC monomer are highly similar to the four-stranded β-sheet of the WRKY domain, suggesting an ancient evolutionary relationship [67, 77].

The Myb/SANT family, though structurally distinct from homeodomains, also binds DNA utilizing an HTH-based DBD. Myb/SANT proteins can contain up to three imperfect repeating DNA-binding α-helical sections (designated R1, R2, and R3), with most TFs containing the R2 and R3 domains [92]. In vascular plants, Myb/SANT TFs play key roles in development, with several members displaying striking tissue-specific expression patterns (reviewed in [93]). In plants, Myb/SANT TFs often interact cooperatively with members of the bHLH family, as exemplified by the regulatory control of the phenylpropanoid biosynthetic pathways [94, 95]. Myb TFs are most abundant in plants, but are prevalent across the eukaryotic kingdom (Fig. 3.2). Members of the Myb family are known to be factors in several human cancer subtypes [96].

3.2.3 Fungal Transcription Factors

The fungal kingdom encompasses a wide evolutionary distance [97, 98], and the majority of its transcriptional regulators remain largely unexplored outside of the yeast clade. The largest family of fungal TFs is the zinc cluster (C_6 zinc finger) family, a multifunctional class that controls several crucial fungal pathways [99], with the galactose-uptake regulator GAL4 being among the more well-characterized examples [100]. Originally thought to be fungal specific, zinc clusters have recently been discovered in the slime mold *D. discoideum*, the marine phytoplankton *Thalassiosira pseudonana,* and the amoeba *Naegleria gruberi*, with the latter possessing a major lineage-specific expansion [101]. Other major fungal TF families include C_2H_2 zinc fingers, bZIP, bHLH, Myb/SANT, GATA, and homeodomain. The copper fist domain, which is stabilized by multiple copper ions [102], is thought to be fungal-specific. APSES, another fungal-specific TF domain, is believed to have evolved through the capture of a viral KilA-N-like precursor early in fungal evolution [103], and is also present in transposable elements of the parasitic protist *Trichomonas vaginalis* [104]. Recent findings suggest that fungal genomes might contain nuclear receptor-like TFs, a family classically thought to be specific to the metazoan lineage [34].

3.2.4 Transcription Factors in Protists

Protist is, in practice, a catch-all term for eukaryotic microorganisms, which represent the majority of diversity found in the eukaryotic kingdom [105]. Members of this taxonomic group are largely understudied at the molecular level, relative to the more prevalent model systems, raising the possibility that additional TF families have yet to be discovered and characterized. Despite a recent increase in the availability of sequenced protist genomes, identifying novel protist TF families using sequence-based techniques remains challenging [106, 107]. One example of a protist-specific TF family is the *T. vaginalis* IBD family [108, 109], which has expanded to include over 100 members in the genome of this parasite [101].

Many parasitic protists, including members of the apicomplexans, diplomonadida, and parabasala, appear to have a reduced set of TF families, for which they seem to compensate by the lineage-specific expansion of individual TF families, along with the presence of well-developed chromatin modification, basal transcription, and post-transcriptional systems [110–113]. For example, the genome of *Cryptosporidium parvum* encodes several E2F TFs, while other apicomplexans have completely lost this family, and *Toxoplasma* genomes have many more AP2 TFs than other apicomplexans [101]. Major apicomplexan protist TF families include Myb/SANT, AP2 [86], C_2H_2 zinc fingers, and Whirly [114]. Families undergoing major expansion events in protists include bZIPs (in both phytophthora (water molds) and paramecium), and heat shock factors (in heterokonts/stramenopiles and paramecium [101]).

3.3 A Brief History of Eukaryotic Transcription Factor Families

The distribution of TF families across the extant eukaryotic lineages presumably arose through a series of varied evolutionary events, including the de novo appearance of functional peptides and protein domains, adaptation of ancestral proteins into novel TFs, family expansions in the form of duplication and divergence, lineage-specific losses (e.g. E2F, which is missing in most fungal genomes [101]), and the acquisition of novel TFs through horizontal transfer (e.g. the transfer of WRKY from plants to phytophthora (water molds) and *Giardia lamblia* [77] and of zinc clusters from fungi to *Dictyostelium*, *Thalassiosire*, and *Naegliria* [101]). The evolutionary history of protein families can be inferred by considering their phylogenetic distribution, providing an estimate of when (and from whence) the various families arose. In this section, we first give brief evolutionary histories of the major TF families in ancestral eukaryotes, as well as those present in extant animal and plant lineages. We then describe how the evolution and diversification of the eukaryotic kingdom was apparently aided by expansions of TF families, as well as increases in cooperative interactions (facilitated by dimerization). A summary of the origins and expansions of the major eukaryotic TF families is depicted in Fig. 3.3. Figure 3.3 is anthropocentric, or at least metazoan-centric, mainly because molecular biology tends to be anthropocentric, due to the fact that it is conducted by humans, and is often funded by (and conducted at) medically-oriented organizations. As the number and variety of sequenced genomes continues to increase (particularly for under-sampled clades such as protists, demosponges, ctenophores (comb jellies), cnidarians (corals/sea anemones/jellyfish), and non-metazoan/fungal opisthokont genomes), our ability to more broadly infer the natural history of TF

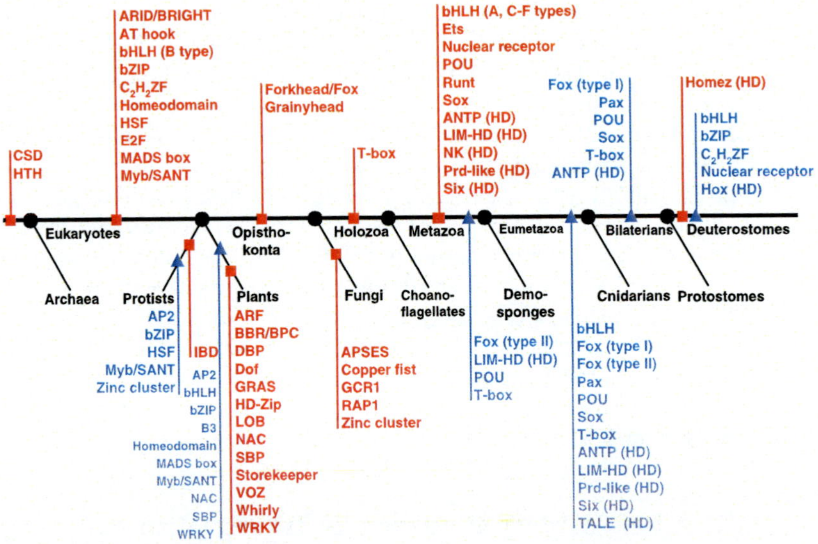

Fig. 3.3 Evolutionary timeline of transcription factor families. Estimated timing of the appearance (*red*) of major TF families and subfamilies, as well as large-scale family expansions (*blue*). Homeodomain subfamilies are indicated as "HD". Brief definition of select clades: Opisthokonta: fungi and metazoa; Holozoa: animals and choanoflagellates; Choanoflagellates: free-living unicellular and colonial flagellate organisms; Eumetazoa: animals, excluding sponges and some other simple animals; Demosponges: group containing the majority of sponges; Bilaterians: all animals with bilaterial symmetry; Cnidarians: aquatic jelly-like organisms, including jelly fish, corals, and sea anemones; Deuterostomes: vertebrates, echinoderms (e.g. sea stars and sea urchins), tunicates (e.g. sea squirts) and others; Protostomes: insects, crustaceans, nematodes, flatworms, mollusks, brachiopods, and others

families will likewise continue to improve. We refer readers who wish to learn more about the eukaryotic evolutionary tree and taxonomic groupings to the following sources [98, 115, 116].

3.3.1 Events Pre-dating the Emergence of Metazoans

The genomes of the earliest eukaryotes are thought to have contained members of the bZIP, C_2H_2 zinc finger, Myb/SANT, ARID/BRIGHT, E2F, and homeodomain families [101, 117, 118] (Fig. 3.3). The HTH fold is present in the DBDs of the majority of prokaryotic TFs (see Chapter 2), suggesting that ancient HTH-containing eukaryotic-specific TF families, such as Myb/SANT and ARID/BRIGHT, may have emerged through the divergence of inherited TFs [40], although their lack of strong sequence similarity allows for the possibility that they might have arisen independently.

Early in the evolution of eukaryotes, it would appear that a wide range of diverse structures were recruited for DNA binding. Most innovations contained structures whose de novo innovation is simpler, such as α-helical or metal-chelation supported DBDs [119]. In addition to direct inheritance, transposons have been proposed to be an important source of novel TFs [77]. Bacterial transposases largely bind DNA utilizing the HTH structural motif, and several eukaryotic HTH domains are thought to have originated from transposases, including WRKY [77], B3 [72], Pax [120], CENPB [121], and AP2 [86, 122]. The presence of these DBDs in transposases suggests a potential means for their lateral transfer to eukaryotic lineages. Furthermore, the ability of transposons to regulate their own expression by binding to specific internal sequences suggests that they might also be capable of transferring the raw material for *cis* regulatory elements upon their integration into a host genome [123, 124].

3.3.2 The Evolution of Metazoan Transcription Factor Families

Many bilaterian TF families originated at the dawn of the animal kingdom, before the divergence of contemporary lineages (Fig. 3.3, branch point of the demosponges and eumetazoa). The regulatory complexity afforded by the explosion in the size and number of eukaryotic TFs presumably provided a foundation for the development of multicellularity and embryogenesis. Major metazoan TF family innovations include bHLHs of types A and C-F (e.g. MyoD, Twist, Arnt, Hairy, and Clock [26]), Ets, Runt, and Sox, along with a variety of homeodomain subfamilies [125] (Fig. 3.3). The development of these novel DBDs occurred through a variety of means, including the combination of an animal-specific domain with a more ancient one (e.g. combining POU, Pax, or Six with a homeodomain) and the apparent de novo creation of a novel domain (e.g. Ets, which possesses no clear relatives outside the metazoa [125]).

Based on analyses of the genome of the sponge *Amphimedon queenslandica*, the common ancestor to the metazoa likely possessed only a limited number of

TFs within each family, suggesting that multiple independent expansion events occurred after the branching off of demosponges [26, 126]. Because of differences in timing (and differences in fold increases), the majority of metazoan TF family expansions (e.g. bZIP, bHLH, and nuclear receptors [26, 125]) were likely to have been the result of single gene duplication events, as opposed to the whole genome duplication-based expansions of the MADS box family in plants [68] and many TFs in *Saccharomyces cerevisiae* [127].

Several major phases of TF family expansion have been identified along the eukaryotic lineage. The first occurred before the divergence of the demosponge and eumetazoan lineages, and included a threefold increase in the sizes of the POU, T-box, and Fox (type II) families (Fig. 3.3). This period also saw the expansion and divergence of ancestral B-type bHLH TFs into types A and C-F, which was followed by a subsequent bHLH expansion phase after the split between cnidarians and bilaterians [125]. A second period of expansion occurred early in eumetazoan evolution, resulting in a two- to four-fold increase in the sizes of several families, including a handful of homeodomain subfamilies [125] (Fig. 3.3), enabling the co-option of these TF classes into new roles, and presumably facilitating the evolution of complex body plans and lifestyles. Subsequently, a third and less extensive series of expansions occurred along the bilaterian stem [26, 125] (Fig. 3.3). Members of the homeodomain family underwent rapid expansion in the vertebrate lineage, expanding from ~80 proteins before the protostome-deutrostome split to ~170 in mammals [128], with the Hox gene clusters, which control segmental patterning during development, originating sometime before the split of bilaterians and cnidarians [128–130]. Other major expansion events consist of C_2H_2 zinc fingers in multiple lineages, including during the appearance of vertebrates and during the emergence of mammals and primates [20, 131], and the nuclear receptor family in *Caenorhabditis elegans* [132, 133].

3.3.3 The Evolution of Plant Transcription Factor Families

Due to their immobility, members of the plant kingdom have developed unique systems of adaptation, including the presence of many TF families thought to be unique to the plant lineage. For example, of the ~1,500 predicted TFs in the *Arabidopsis thaliana* genome, almost half were initially thought to be plant specific [134], although recent work suggests that the majority of plant-specific DBD families likely originated in non-plant species [67]. Notwithstanding, a number of TF families likely arose within the earliest land plants over 500 million years ago, including the ARF, LOB, GRAS, HD-Zip, NAC, and VOZ domains [66] (Fig. 3.3). Subsequently, the BBR/BPC and DBP families appeared in the last common ancestor (LCA) of vascular plants (~470 million years ago) and storekeeper arose 210 million years ago in the LCA of extant flowering plants [66]. Although most land plant TF families have been retained, CG-1/CAMTA and Trihelix/GT were lost from the genomes of all algae around 600 million years ago [135].

Similar to the explosion of TF family sizes seen at the dawn of the metazoa, dozens of TF families expanded after the divergence of land plants and algae, including the AP2, Myb/SANT, bHLH, bZIP, homeodomain, MADS box, and WRKY families [66]. The rate of expansion in plant TFs is higher than that seen in animals, possibly due to their relatively high frequency of polyploidization [136]. (Among major plant expansions, the AP2 family has also independently expanded in apicomplexa [86], and Myb/SANT has been a part of major expansion events in paramecium and a wide range of parasites [101].) Three TF subfamilies (EIN3/EIL, WRC/C3H/GRF, and SRS/SHI) increased with the onset of plant vascularity, with over a dozen families expanding in flowering plants, including MADS box, Myb/SANT, NAC, Homeodomain, WRKY, and B3 [66]. Although not as much is known about members of the green algae lineage, recent work has identified expanded repertoires of the NiN/RWP-RK (a putative family of TFs) and SBP families in most organisms in this group [66].

3.3.4 Evolutionary Contributions of Transcription Factor Family Expansions

Duplication of an already-existing gene is a useful evolutionary mechanism. Novel protein domains appear to arise infrequently, and altering an existing protein can have drastic consequences. In contrast, duplication introduces redundancy (at least transiently) and thus freedom. In the case of a TF, both copies will presumably regulate the same set of target genes. Due to this redundancy, one or both copies of the TF are then less constrained, facilitating the modification of their regulatory control systems. Alterations to a DBD can modify binding specificities, thus affecting interactions with existing regulatory targets as well as creating new ones. Alternatively, alterations to other domains can enable new biochemical functions, and novel interactions with other TFs (see next section). Changes to the *cis* region controlling the new TF can result in its expression in a new tissue or developmental stage, as is thought to have been the case for lineage-specific expansions of WRKY proteins in plants [77].

Mounting evidence suggests that many, if not most, eukaryotic genomes include in their evolutionary histories multiple large-scale duplication events [137–139]. The dawn of the eukaryotic kingdom saw the expansion and divergence of a range of gene families, including the spliceosome, the nuclear pore complex, and genes involved in protein stability [119], with the number of genes present in the common ancestor to all eukaryotes thought to be almost twice that of the first eukaryotes [140]. A major factor in the drastic increase in the complexity of organisms is thought to have been the expansion of ancestral TFs into the larger families and subfamilies present in modern extant species [1, 17, 141]. The resulting increase in regulatory complexity provides a plausible explanation for increases in developmental potential and the ability to create new cell types.

3.3.5 The Contribution of TF Dimerization

In addition to the expansion and diversification of TF families, changes in dimerization interactions between TFs have likely contributed substantially to the evolution of transcriptional regulation. Many families of TFs bind DNA as obligate dimers, with the largest of these classes (e.g. bHLH, bZIP, MADS box, and STAT) forming both homo- and hetero-dimers (Table 3.1) [22]. This mechanism appears to be distinct from interactions among structurally different TFs, which in general do not appear to be obligate. Eukaryotic TF dimerization is thought to have arisen several times in a structurally diverge range of protein families, with some TF dimers possibly originating as binding proteins possessing the ability to bind DNA as monomers [142–144]. In such cases, the spatial clustering of TF binding sites within promoters and enhancers might have brought TFs within close proximity of one another upon binding DNA, providing them with an opportunity to participate in physical interactions [22]. Upon obtaining the ability to function as homodimers, the ability to heterodimerize often emerges through the duplication and divergence of homodimerizing ancestors [22, 68]. Expansions of TF families capable of dimerization result in combinatorial increases in regulatory complexity, with the potential to equal or even exceed the actual number of TF genes in a genome [22]. Dimerization may have played a key role in the early evolution of bilaterians, with a core highly conserved dimerization network already being present in the urbilaterian ancestor [22]. Two subsequent rounds of whole genome duplication events around the origin of vertebrates provided further increases in the complexity of dimerization networks, particularly for members of the bZIP, bHLH and nuclear receptor families [23, 145, 146].

3.4 Summary and Future Directions

The TF families of the eukaryotic kingdom represent a diverse group, with different lineages containing different complements and numbers of TFs. The protists in particular may contain as-yet undiscovered TF classes, which could dramatically expand our catalogue. Duplication and divergence of TFs is clearly a theme in eukaryotic evolution, and is widely believed to provide a mechanistic basis for many developmental and physiological innovations. However, despite extensive progress in the identification and cataloguing of the various TF classes, the vast majority of eukaryotic TFs are uncharacterized (e.g. unknown binding preference, unknown biological role, and/or unknown regulatory targets). This is especially true in the protists, for which the TF repertoires of the majority of organisms remain largely unexplored. The continued development of high-throughput methods for the sequencing of genomes, the discovery of tissue and developmental stage expression patterns, the determination of sequence binding preferences (Chapter 7), and the identification of genomic binding locations (Chapter 8), will provide valuable information for deciphering gene regulatory interaction networks. Better knowledge of how these networks operate should, in turn, facilitate better understanding of the function and evolution of the individual TFs.

References

1. Levine M, Tjian R (2003) Transcription regulation and animal diversity. Nature 424(6945):147–151
2. Wunderlich Z, Mirny LA (2009) Different gene regulation strategies revealed by analysis of binding motifs. Trends Genet 25(10):434–440
3. Arnosti DN, Kulkarni MM (2005) Transcriptional enhancers: Intelligent enhanceosomes or flexible billboards? J Cell Biochem 94(5):890–898
4. Karin M (1990) Too many transcription factors: Positive and negative interactions. New Biol 2(2):126–131
5. Latchman DS (1997) Transcription factors: An overview. Int J Biochem Cell Biol 29(12):1305–1312
6. Luscombe NM, Austin SE, Berman HM, Thornton JM (2000) An overview of the structures of protein-DNA complexes. Genome Biol 1(1):REVIEWS001
7. Reeves R, Nissen MS (1990) The A.T-DNA-binding domain of mammalian high mobility group I chromosomal proteins. A novel peptide motif for recognizing DNA structure. J Biol Chem 265(15):8573–8582
8. Bottomley MJ, Collard MW, Huggenvik JI, Liu Z, Gibson TJ, Sattler M (2001) The SAND domain structure defines a novel DNA-binding fold in transcriptional regulation. Nat Struct Biol 8(7):626–633
9. Latchman DS (2008) Eukaryotic transcription factors. Elsevier/Academic Press
10. Papavassiliou A (1997) Transcription factors in eukaryotes. Landes Bioscience, Austin, TX
11. Charoensawan V, Wilson D, Teichmann SA (2010) Genomic repertoires of DNA-binding transcription factors across the tree of life. Nucleic Acids Res 38:7364–7377
12. Fulton DL, Sundararajan S, Badis G, Hughes TR, Wasserman WW, Roach JC, Sladek R (2009) TFCat: The curated catalog of mouse and human transcription factors. Genome Biol 10(3):R29
13. Hu S, Xie Z, Onishi A, Yu X, Jiang L, Lin J, Rho HS, Woodard C, Wang H, Jeong JS, Long S, He X, Wade H, Blackshaw S, Qian J, Zhu H (2009) Profiling the human protein-DNA interactome reveals ERK2 as a transcriptional repressor of interferon signaling. Cell 139(3):610–622
14. Charoensawan V, Wilson D, Teichmann SA (2010) Lineage-specific expansion of DNA-binding transcription factor families. Trends Genet 26(9):388–393
15. Schauser L, Wieloch W, Stougaard J (2005) Evolution of NIN-like proteins in *Arabidopsis*, rice, and *Lotus japonicus*. J Mol Evol 60(2):229–237
16. Liu X, Huang J, Parameswaran S, Ito T, Seubert B, Auer M, Rymaszewski A, Jia G, Owen HA, Zhao D (2009) The SPOROCYTELESS/NOZZLE gene is involved in controlling stamen identity in *Arabidopsis*. Plant Physiol 151(3):1401–1411
17. Carroll SB (2005) Evolution at two levels: On genes and form. PLoS Biol 3(7):e245
18. Sanges R, Kalmar E, Claudiani P, D'Amato M, Muller F, Stupka E (2006) Shuffling of cis-regulatory elements is a pervasive feature of the vertebrate lineage. Genome Biol 7(7):R56
19. Dermitzakis ET, Clark AG (2002) Evolution of transcription factor binding sites in mammalian gene regulatory regions: Conservation and turnover. Mol Biol Evol 19(7): 1114–1121
20. Vaquerizas JM, Kummerfeld SK, Teichmann SA, Luscombe NM (2009) A census of human transcription factors: Function, expression and evolution. Nat Rev Genet 10(4): 252–263
21. Vinson CR, Sigler PB, McKnight SL (1989) Scissors-grip model for DNA recognition by a family of leucine zipper proteins. Science 246(4932):911–916
22. Amoutzias GD, Robertson DL, Van de Peer Y, Oliver SG (2008) Choose your partners: Dimerization in eukaryotic transcription factors. Trends Biochem Sci 33(5): 220–229
23. Amoutzias GD, Veron AS, Weiner J 3rd, Robinson-Rechavi M, Bornberg-Bauer E, Oliver SG, Robertson DL (2007) One billion years of bZIP transcription factor evolution:

Conservation and change in dimerization and DNA-binding site specificity. Mol Biol Evol 24(3):827–835
24. Vinson C, Myakishev M, Acharya A, Mir AA, Moll JR, Bonovich M (2002) Classification of human B-ZIP proteins based on dimerization properties. Mol Cell Biol 22(18):6321–6335
25. Newman JR, Keating AE (2003) Comprehensive identification of human bZIP interactions with coiled-coil arrays. Science 300(5628):2097–2101
26. Simionato E, Ledent V, Richards G, Thomas-Chollier M, Kerner P, Coornaert D, Degnan BM, Vervoort M (2007) Origin and diversification of the basic helix-loop-helix gene family in metazoans: Insights from comparative genomics. BMC Evol Biol 7:33
27. Murre C, Bain G, van Dijk MA, Engel I, Furnari BA, Massari ME, Matthews JR, Quong MW, Rivera RR, Stuiver MH (1994) Structure and function of helix-loop-helix proteins. Biochim Biophys Acta 1218(2):129–135
28. Atchley WR, Fitch WM (1997) A natural classification of the basic helix-loop-helix class of transcription factors. Proc Natl Acad Sci U S A 94(10):5172–5176
29. Rudnicki MA, Jaenisch R (1995) The MyoD family of transcription factors and skeletal myogenesis. Bioessays 17(3):203–209
30. van Dam H, Castellazzi M (2001) Distinct roles of Jun: Fos and Jun: ATF dimers in oncogenesis. Oncogene 20(19):2453–2464
31. Wolfe SA, Nekludova L, Pabo CO (2000) DNA recognition by Cys2His2 zinc finger proteins. Annu Rev Biophys Biomol Struct 29:183–212
32. Papworth M, Kolasinska P, Minczuk M (2006) Designer zinc-finger proteins and their applications. Gene 366(1):27–38
33. Biddie SC, John S, Hager GL (2010) Genome-wide mechanisms of nuclear receptor action. Trends Endocrinol Metab 21(1):3–9
34. Naar AM, Thakur JK (2009) Nuclear receptor-like transcription factors in fungi. Genes Dev 23(4):419–432
35. Merika M, Orkin SH (1993) DNA-binding specificity of GATA family transcription factors. Mol Cell Biol 13(7):3999–4010
36. Evans T, Felsenfeld G (1989) The erythroid-specific transcription factor Eryf1: A new finger protein. Cell 58(5):877–885
37. King DC, Taylor J, Elnitski L, Chiaromonte F, Miller W, Hardison RC (2005) Evaluation of regulatory potential and conservation scores for detecting cis-regulatory modules in aligned mammalian genome sequences. Genome Res 15(8):1051–1060
38. Viger RS, Guittot SM, Anttonen M, Wilson DB, Heikinheimo M (2008) Role of the GATA family of transcription factors in endocrine development, function, and disease. Mol Endocrinol 22(4):781–798
39. Wechsler J, Greene M, McDevitt MA, Anastasi J, Karp JE, Le Beau MM, Crispino JD (2002) Acquired mutations in GATA1 in the megakaryoblastic leukemia of down syndrome. Nat Genet 32(1):148–152
40. Aravind L, Anantharaman V, Balaji S, Babu MM, Iyer LM (2005) The many faces of the helix-turn-helix domain: Transcription regulation and beyond. FEMS Microbiol Rev 29(2):231–262
41. Maconochie M, Nonchev S, Morrison A, Krumlauf R (1996) Paralogous Hox genes: Function and regulation. Annu Rev Genet 30:529–556
42. Takahashi K, Yamanaka S (2006) Induction of pluripotent stem cells from mouse embryonic and adult fibroblast cultures by defined factors. Cell 126(4): 663–676
43. Phillips K, Luisi B (2000) The virtuoso of versatility: POU proteins that flex to fit. J Mol Biol 302(5):1023–1039
44. Gajiwala KS, Burley SK (2000) Winged helix proteins. Curr Opin Struct Biol 10(1): 110–116
45. Hannenhalli S, Kaestner KH (2009) The evolution of Fox genes and their role in development and disease. Nat Rev Genet 10(4):233–240

46. Lane DP (1992) Cancer. p53, guardian of the genome. Nature 358(6381):15–16
47. Smith ND, Rubenstein JN, Eggener SE, Kozlowski JM (2003) The p53 tumor suppressor gene and nuclear protein: Basic science review and relevance in the management of bladder cancer. J Urol 169(4):1219–1228
48. Cho Y, Gorina S, Jeffrey PD, Pavletich NP (1994) Crystal structure of a p53 tumor suppressor-DNA complex: Understanding tumorigenic mutations. Science 265(5170): 346–355
49. Gilmore TD (2006) Introduction to NF-kappaB: Players, pathways, perspectives. Oncogene 25(51):6680–6684
50. Calo V, Migliavacca M, Bazan V, Macaluso M, Buscemi M, Gebbia N, Russo A (2003) STAT proteins: From normal control of cellular events to tumorigenesis. J Cell Physiol 197(2): 157–168
51. Bycroft M, Hubbard TJ, Proctor M, Freund SM, Murzin AG (1997) The solution structure of the S1 RNA binding domain: A member of an ancient nucleic acid-binding fold. Cell 88(2):235–242
52. MacDonald GH, Itoh-Lindstrom Y, Ting JP (1995) The transcriptional regulatory protein, YB-1, promotes single-stranded regions in the DRA promoter. J Biol Chem 270(8): 3527–3533
53. Tafuri SR, Wolffe AP (1992) DNA binding, multimerization, and transcription stimulation by the *Xenopus* Y box proteins in vitro. New Biol 4(4):349–359
54. Mihailovich M, Militti C, Gabaldon T, Gebauer F (2010) Eukaryotic cold shock domain proteins: Highly versatile regulators of gene expression. Bioessays 32(2):109–118
55. Skabkin MA, Evdokimova V, Thomas AA, Ovchinnikov LP (2001) The major messenger ribonucleoprotein particle protein p50 (YB-1) promotes nucleic acid strand annealing. J Biol Chem 276(48):44841–44847
56. Sinha S, Maity SN, Lu J, de Crombrugghe B (1995) Recombinant rat CBF-C, the third subunit of CBF/NFY, allows formation of a protein-DNA complex with CBF-A and CBF-B and with yeast HAP2 and HAP3. Proc Natl Acad Sci U S A 92(5):1624–1628
57. McNabb DS, Xing Y, Guarente L (1995) Cloning of yeast HAP5: A novel subunit of a heterotrimeric complex required for CCAAT binding. Genes Dev 9(1):47–58
58. Bucher P, Trifonov EN (1988) CCAAT box revisited: Bidirectionality, location and context. J Biomol Struct Dyn 5(6):1231–1236
59. Mantovani R (1998) A survey of 178 NF-Y binding CCAAT boxes. Nucleic Acids Res 26(5):1135–1143
60. Mantovani R (1999) The molecular biology of the CCAAT-binding factor NF-Y. Gene 239(1):15–27
61. Guth SI, Wegner M (2008) Having it both ways: Sox protein function between conservation and innovation. Cell Mol Life Sci 65(19):3000–3018
62. Schepers GE, Teasdale RD, Koopman P (2002) Twenty pairs of sox: Extent, homology, and nomenclature of the mouse and human sox transcription factor gene families. Dev Cell 3(2):167–170
63. Wallis MC, Waters PD, Graves JA (2008) Sex determination in mammals – before and after the evolution of SRY. Cell Mol Life Sci 65(20):3182–3195
64. Wegner M (2010) All purpose Sox: The many roles of Sox proteins in gene expression. Int J Biochem Cell Biol 42(3):381–390
65. Badis G, Berger MF, Philippakis AA, Talukder S, Gehrke AR, Jaeger SA, Chan ET, Metzler G, Vedenko A, Chen X, Kuznetsov H, Wang CF, Coburn D, Newburger DE, Morris Q, Hughes TR, Bulyk ML (2009) Diversity and complexity in DNA recognition by transcription factors. Science 324(5935):1720–1723
66. Lang D, Weiche B, Timmerhaus G, Richardt S, Riano-Pachon DM, Correa LG, Reski R, Mueller-Roeber B, Rensing SA (2010) Genome-wide phylogenetic comparative analysis of plant transcriptional regulation: A timeline of loss, gain, expansion, and correlation with complexity. Genome Biol Evol 2:488–503

67. Yamasaki K, Kigawa T, Inoue M, Watanabe S, Tateno M, Seki M, Shinozaki K, Yokoyama S (2008) Structures and evolutionary origins of plant-specific transcription factor DNA-binding domains. Plant Physiol Biochem 46(3):394–401
68. Veron AS, Kaufmann K, Bornberg-Bauer E (2007) Evidence of interaction network evolution by whole-genome duplications: A case study in MADS-box proteins. Mol Biol Evol 24(3):670–678
69. Gramzow L, Theissen G (2010) A hitchhiker's guide to the MADS world of plants. Genome Biol 11(6):214
70. Messenguy F, Dubois E (2003) Role of MADS box proteins and their cofactors in combinatorial control of gene expression and cell development. Gene 316:1–21
71. Gramzow L, Ritz MS, Theissen G (2010) On the origin of MADS-domain transcription factors. Trends Genet 26(4):149–153
72. Yamasaki K, Kigawa T, Inoue M, Tateno M, Yamasaki T, Yabuki T, Aoki M, Seki E, Matsuda T, Tomo Y, Hayami N, Terada T, Shirouzu M, Osanai T, Tanaka A, Seki M, Shinozaki K, Yokoyama S (2004) Solution structure of the B3 DNA binding domain of the *Arabidopsis* cold-responsive transcription factor RAV1. Plant Cell 16(12):3448–3459
73. Desveaux D, Marechal A, Brisson N (2005) Whirly transcription factors: Defense gene regulation and beyond. Trends Plant Sci 10(2):95–102
74. Desveaux D, Allard J, Brisson N, Sygusch J (2002) A new family of plant transcription factors displays a novel ssDNA-binding surface. Nat Struct Biol 9(7):512–517
75. Yamasaki K, Kigawa T, Inoue M, Tateno M, Yamasaki T, Yabuki T, Aoki M, Seki E, Matsuda T, Tomo Y, Hayami N, Terada T, Shirouzu M, Tanaka A, Seki M, Shinozaki K, Yokoyama S (2005) Solution structure of an *Arabidopsis* WRKY DNA binding domain. Plant Cell 17(3):944–956
76. Eulgem T, Rushton PJ, Robatzek S, Somssich IE (2000) The WRKY superfamily of plant transcription factors. Trends Plant Sci 5(5):199–206
77. Babu MM, Iyer LM, Balaji S, Aravind L (2006) The natural history of the WRKY-GCM1 zinc fingers and the relationship between transcription factors and transposons. Nucleic Acids Res 34(22):6505–6520
78. Ulker B, Somssich IE (2004) WRKY transcription factors: From DNA binding towards biological function. Curr Opin Plant Biol 7(5):491–498
79. Cardon G, Hohmann S, Klein J, Nettesheim K, Saedler H, Huijser P (1999) Molecular characterisation of the *Arabidopsis* SBP-box genes. Gene 237(1):91–104
80. Schwarz S, Grande AV, Bujdoso N, Saedler H, Huijser P (2008) The microRNA regulated SBP-box genes SPL9 and SPL15 control shoot maturation in *Arabidopsis*. Plant Mol Biol 67(1–2):183–195
81. Yamasaki K, Kigawa T, Inoue M, Tateno M, Yamasaki T, Yabuki T, Aoki M, Seki E, Matsuda T, Nunokawa E, Ishizuka Y, Terada T, Shirouzu M, Osanai T, Tanaka A, Seki M, Shinozaki K, Yokoyama S (2004) A novel zinc-binding motif revealed by solution structures of DNA-binding domains of *Arabidopsis* SBP-family transcription factors. J Mol Biol 337(1): 49–63
82. Umemura Y, Ishiduka T, Yamamoto R, Esaka M (2004) The Dof domain, a zinc finger DNA-binding domain conserved only in higher plants, truly functions as a Cys2/Cys2 Zn finger domain. Plant J 37(5):741–749
83. Yanagisawa S (2002) The Dof family of plant transcription factors. Trends Plant Sci 7(12):555–560
84. Moreno-Risueno MA, Martinez M, Vicente-Carbajosa J, Carbonero P (2007) The family of DOF transcription factors: From green unicellular algae to vascular plants. Mol Genet Genomics 277(4):379–390
85. Lijavetzky D, Carbonero P, Vicente-Carbajosa J (2003) Genome-wide comparative phylogenetic analysis of the rice and *Arabidopsis* Dof gene families. BMC Evol Biol 3:17
86. Balaji S, Babu MM, Iyer LM, Aravind L (2005) Discovery of the principal specific transcription factors of Apicomplexa and their implication for the evolution of the AP2-integrase DNA binding domains. Nucleic Acids Res 33(13):3994–4006

87. Wojciak JM, Connolly KM, Clubb RT (1999) NMR structure of the Tn916 integrase-DNA complex. Nat Struct Biol 6(4):366–373
88. Wojciak JM, Sarkar D, Landy A, Clubb RT (2002) Arm-site binding by lambda -integrase: Solution structure and functional characterization of its amino-terminal domain. Proc Natl Acad Sci U S A 99(6):3434–3439
89. Moure CM, Gimble FS, Quiocho FA (2002) Crystal structure of the intein homing endonuclease PI-SCEI bound to its recognition sequence. Nat Struct Biol 9(10): 764–770
90. Gutterson N, Reuber TL (2004) Regulation of disease resistance pathways by AP2/ERF transcription factors. Curr Opin Plant Biol 7(4):465–471
91. Ernst HA, Olsen AN, Larsen S, Lo Leggio L (2004) Structure of the conserved domain of ANAC, a member of the NAC family of transcription factors. EMBO Rep 5(3):297–303
92. Ogata K, Morikawa S, Nakamura H, Sekikawa A, Inoue T, Kanai H, Sarai A, Ishii S, Nishimura Y (1994) Solution structure of a specific DNA complex of the Myb DNA-binding domain with cooperative recognition helices. Cell 79(4):639–648
93. Du H, Zhang L, Liu L, Tang XF, Yang WJ, Wu YM, Huang YB, Tang YX (2009) Biochemical and molecular characterization of plant MYB transcription factor family. Biochemistry (Mosc) 74(1):1–11
94. Aharoni A, De Vos CH, Wein M, Sun Z, Greco R, Kroon A, Mol JN, O'Connell AP (2001) The strawberry FaMYB1 transcription factor suppresses anthocyanin and flavonol accumulation in transgenic tobacco. Plant J 28(3):319–332
95. Quattrocchio F, Wing J, van der Woude K, Souer E, de Vetten N, Mol J, Koes R (1999) Molecular analysis of the anthocyanin2 gene of petunia and its role in the evolution of flower color. Plant Cell 11(8):1433–1444
96. Ramsay RG, Gonda TJ (2008) MYB function in normal and cancer cells. Nat Rev Cancer 8(7):523–534
97. Dujon B, Sherman D, Fischer G, Durrens P, Casaregola S, Lafontaine I, De Montigny J, Marck C, Neuveglise C, Talla E, Goffard N, Frangeul L, Aigle M, Anthouard V, Babour A, Barbe V, Barnay S, Blanchin S, Beckerich JM, Beyne E, Bleykasten C, Boisrame A, Boyer J, Cattolico L, Confanioleri F, De Daruvar A, Despons L, Fabre E, Fairhead C, Ferry-Dumazet H, Groppi A, Hantraye F, Hennequin C, Jauniaux N, Joyet P, Kachouri R, Kerrest A, Koszul R, Lemaire M, Lesur I, Ma L, Muller H, Nicaud JM, Nikolski M, Oztas S, Ozier-Kalogeropoulos O, Pellenz S, Potier S, Richard GF, Straub ML, Suleau A, Swennen D, Tekaia F, Wesolowski-Louvel M, Westhof E, Wirth B, Zeniou-Meyer M, Zivanovic I, Bolotin-Fukuhara M, Thierry A, Bouchier C, Caudron B, Scarpelli C, Gaillardin C, Weissenbach J, Wincker P, Souciet JL (2004) Genome evolution in yeasts. Nature 430(6995):35–44
98. Baldauf SL, Roger AJ, Wenk-Siefert I, Doolittle WF (2000) A kingdom-level phylogeny of eukaryotes based on combined protein data. Science 290(5493):972–977
99. MacPherson S, Larochelle M, Turcotte B (2006) A fungal family of transcriptional regulators: The zinc cluster proteins. Microbiol Mol Biol Rev 70(3):583–604
100. Traven A, Jelicic B, Sopta M (2006) Yeast Gal4: A transcriptional paradigm revisited. EMBO Rep 7(5):496–499
101. Iyer LM, Anantharaman V, Wolf MY, Aravind L (2008) Comparative genomics of transcription factors and chromatin proteins in parasitic protists and other eukaryotes. Int J Parasitol 38(1):1–31
102. Dameron CT, Winge DR, George GN, Sansone M, Hu S, Hamer D (1991) A copper-thiolate polynuclear cluster in the ACE1 transcription factor. Proc Natl Acad Sci U S A 88(14): 6127–6131
103. Iyer LM, Koonin EV, Aravind L (2002) Extensive domain shuffling in transcription regulators of DNA viruses and implications for the origin of fungal APSES transcription factors. Genome Biol 3(3):RESEARCH0012
104. Pritham EJ, Putliwala T, Feschotte C (2007) Mavericks, a novel class of giant transposable elements widespread in eukaryotes and related to DNA viruses. Gene 390(1–2):3–17

105. Simpson AG, Inagaki Y, Roger AJ (2006) Comprehensive multigene phylogenies of excavate protists reveal the evolutionary positions of "Primitive" Eukaryotes. Mol Biol Evol 23(3):615–625
106. Gardner MJ, Hall N, Fung E, White O, Berriman M, Hyman RW, Carlton JM, Pain A, Nelson KE, Bowman S, Paulsen IT, James K, Eisen JA, Rutherford K, Salzberg SL, Craig A, Kyes S, Chan MS, Nene V, Shallom SJ, Suh B, Peterson J, Angiuoli S, Pertea M, Allen J, Selengut J, Haft D, Mather MW, Vaidya AB, Martin DM, Fairlamb AH, Fraunholz MJ, Roos DS, Ralph SA, McFadden GI, Cummings LM, Subramanian GM, Mungall C, Venter JC, Carucci DJ, Hoffman SL, Newbold C, Davis RW, Fraser CM, Barrell B (2002) Genome sequence of the human malaria parasite Plasmodium falciparum. Nature 419(6906):498–511
107. Templeton TJ, Iyer LM, Anantharaman V, Enomoto S, Abrahante JE, Subramanian GM, Hoffman SL, Abrahamsen MS, Aravind L (2004) Comparative analysis of apicomplexa and genomic diversity in eukaryotes. Genome Res 14(9):1686–1695
108. Schumacher MA, Lau AO, Johnson PJ (2003) Structural basis of core promoter recognition in a primitive eukaryote. Cell 115(4):413–424
109. Lau AO, Smith AJ, Brown MT, Johnson PJ (2006) *Trichomonas vaginalis* initiator binding protein (IBP39) and RNA polymerase II large subunit carboxy terminal domain interaction. Mol Biochem Parasitol 150(1):56–62
110. Clayton CE (2002) Life without transcriptional control? From fly to man and back again. Embo J 21(8):1881–1888
111. Bozdech Z, Llinas M, Pulliam BL, Wong ED, Zhu J, DeRisi JL (2003) The transcriptome of the intraerythrocytic developmental cycle of Plasmodium falciparum. PLoS Biol 1(1):E5
112. Abrahamsen MS, Templeton TJ, Enomoto S, Abrahante JE, Zhu G, Lancto CA, Deng M, Liu C, Widmer G, Tzipori S, Buck GA, Xu P, Bankier AT, Dear PH, Konfortov BA, Spriggs HF, Iyer L, Anantharaman V, Aravind L, Kapur V (2004) Complete genome sequence of the apicomplexan, *Cryptosporidium parvum*. Science 304(5669):441–445
113. Hall N, Karras M, Raine JD, Carlton JM, Kooij TW, Berriman M, Florens L, Janssen CS, Pain A, Christophides GK, James K, Rutherford K, Harris B, Harris D, Churcher C, Quail MA, Ormond D, Doggett J, Trueman HE, Mendoza J, Bidwell SL, Rajandream MA, Carucci DJ, Yates JR 3rd, Kafatos FC, Janse CJ, Barrell B, Turner CM, Waters AP, Sinden RE (2005) A comprehensive survey of the Plasmodium life cycle by genomic, transcriptomic, and proteomic analyses. Science 307(5706):82–86
114. Oakley MS, Kumar S, Anantharaman V, Zheng H, Mahajan B, Haynes JD, Moch JK, Fairhurst R, McCutchan TF, Aravind L (2007) Molecular factors and biochemical pathways induced by febrile temperature in intraerythrocytic Plasmodium falciparum parasites. Infect Immun 75(4):2012–2025
115. Roger AJ, Hug LA (2006) The origin and diversification of eukaryotes: Problems with molecular phylogenetics and molecular clock estimation. Philos Trans R Soc Lond B Biol Sci 361(1470):1039–1054
116. Adl SM, Simpson AG, Farmer MA, Andersen RA, Anderson OR, Barta JR, Bowser SS, Brugerolle G, Fensome RA, Fredericq S, James TY, Karpov S, Kugrens P, Krug J, Lane CE, Lewis LA, Lodge J, Lynn DH, Mann DG, McCourt RM, Mendoza L, Moestrup O, Mozley-Standridge SE, Nerad TA, Shearer CA, Smirnov AV, Spiegel FW, Taylor MF (2005) The new higher level classification of eukaryotes with emphasis on the taxonomy of protists. J Eukaryot Microbiol 52(5):399–451
117. Derelle R, Lopez P, Le Guyader H, Manuel M (2007) Homeodomain proteins belong to the ancestral molecular toolkit of eukaryotes. Evol Dev 9(3):212–219
118. Babu MM, Luscombe NM, Aravind L, Gerstein M, Teichmann SA (2004) Structure and evolution of transcriptional regulatory networks. Curr Opin Struct Biol 14(3): 283–291
119. Aravind L, Iyer LM, Koonin EV (2006) Comparative genomics and structural biology of the molecular innovations of eukaryotes. Curr Opin Struct Biol 16(3):409–419

120. Izsvak Z, Khare D, Behlke J, Heinemann U, Plasterk RH, Ivics Z (2002) Involvement of a bifunctional, paired-like DNA-binding domain and a transpositional enhancer in Sleeping Beauty transposition. J Biol Chem 277(37):34581–34588
121. Tanaka Y, Nureki O, Kurumizaka H, Fukai S, Kawaguchi S, Ikuta M, Iwahara J, Okazaki T, Yokoyama S (2001) Crystal structure of the CENP-B protein-DNA complex: The DNA-binding domains of CENP-B induce kinks in the CENP-B box DNA. Embo J 20(23): 6612–6618
122. Magnani E, Sjolander K, Hake S (2004) From endonucleases to transcription factors: Evolution of the AP2 DNA binding domain in plants. Plant Cell 16(9):2265–2277
123. Marino-Ramirez L, Lewis KC, Landsman D, Jordan IK (2005) Transposable elements donate lineage-specific regulatory sequences to host genomes. Cytogenet Genome Res 110(1–4):333–341
124. Jordan IK, Rogozin IB, Glazko GV, Koonin EV (2003) Origin of a substantial fraction of human regulatory sequences from transposable elements. Trends Genet 19(2):68–72
125. Degnan BM, Vervoort M, Larroux C, Richards GS (2009) Early evolution of metazoan transcription factors. Curr Opin Genet Dev 19(6):591–599
126. Larroux C, Fahey B, Degnan SM, Adamski M, Rokhsar DS, Degnan BM (2007) The NK homeobox gene cluster predates the origin of Hox genes. Curr Biol 17(8):706–710
127. Wolfe KH, Shields DC (1997) Molecular evidence for an ancient duplication of the entire yeast genome. Nature 387(6634):708–713
128. Garcia-Fernàndez J (2005) The genesis and evolution of homeobox gene clusters. Nat Rev Genet 6(12):881–892
129. Ferrier DE, Holland PW (2001) Ancient origin of the Hox gene cluster. Nat Rev Genet 2(1):33–38
130. Ogishima S, Tanaka H (2007) Missing link in the evolution of Hox clusters. Gene 387(1–2): 21–30
131. Lespinet O, Wolf YI, Koonin EV, Aravind L (2002) The role of lineage-specific gene family expansion in the evolution of eukaryotes. Genome Res 12(7):1048–1059
132. Robinson-Rechavi M, Maina CV, Gissendanner CR, Laudet V, Sluder A (2005) Explosive lineage-specific expansion of the orphan nuclear receptor HNF4 in nematodes. J Mol Evol 60(5):577–586
133. Reece-Hoyes JS, Deplancke B, Shingles J, Grove CA, Hope IA, Walhout AJ (2005) A compendium of *Caenorhabditis elegans* regulatory transcription factors: A resource for mapping transcription regulatory networks. Genome Biol 6(13):R110
134. Riechmann JL, Heard J, Martin G, Reuber L, Jiang C, Keddie J, Adam L, Pineda O, Ratcliffe OJ, Samaha RR, Creelman R, Pilgrim M, Broun P, Zhang JZ, Ghandehari D, Sherman BK, Yu G (2000) *Arabidopsis* transcription factors: Genome-wide comparative analysis among eukaryotes. Science 290(5499):2105–2110
135. Riano-Pachon DM, Correa LG, Trejos-Espinosa R, Mueller-Roeber B (2008) Green transcription factors: A chlamydomonas overview. Genetics 179(1):31–39
136. Shiu SH, Shih MC, Li WH (2005) Transcription factor families have much higher expansion rates in plants than in animals. Plant Physiol 139(1):18–26
137. Paterson AH, Chapman BA, Kissinger JC, Bowers JE, Feltus FA, Estill JC (2006) Many gene and domain families have convergent fates following independent whole-genome duplication events in *Arabidopsis, Oryza, Saccharomyces* and Tetraodon. Trends Genet 22(11):597–602
138. Edger PP, Pires JC (2009) Gene and genome duplications: The impact of dosage-sensitivity on the fate of nuclear genes. Chromosome Res 17(5):699–717
139. Van de Peer Y, Maere S, Meyer A (2009) The evolutionary significance of ancient genome duplications. Nat Rev Genet 10(10):725–732
140. Makarova KS, Wolf YI, Mekhedov SL, Mirkin BG, Koonin EV (2005) Ancestral paralogs and pseudoparalogs and their role in the emergence of the eukaryotic cell. Nucleic Acids Res 33(14):4626–4638

141. Hsia CC, McGinnis W (2003) Evolution of transcription factor function. Curr Opin Genet Dev 13(2):199–206
142. Moraitis AN, Giguere V (1999) Transition from monomeric to homodimeric DNA binding by nuclear receptors: Identification of RevErbAalpha determinants required for RORalpha homodimer complex formation. Mol Endocrinol 13(3):431–439
143. Tron AE, Welchen E, Gonzalez DH (2004) Engineering the loop region of a homeodomain-leucine zipper protein promotes efficient binding to a monomeric DNA binding site. Biochemistry 43(50):15845–15851
144. de Lumley M, Hart DJ, Cooper MA, Symeonides S, Blackburn JM (2004) A biophysical characterisation of factors controlling dimerisation and selectivity in the NF-kappaB and NFAT families. J Mol Biol 339(5):1059–1075
145. Amoutzias GD, Robertson DL, Bornberg-Bauer E (2004) The evolution of protein interaction networks in regulatory proteins. Comp Funct Genomics 5(1):79–84
146. Amoutzias GD, Pichler EE, Mian N, De Graaf D, Imsiridou A, Robinson-Rechavi M, Bornberg-Bauer E, Robertson DL, Oliver SG (2007) A protein interaction atlas for the nuclear receptors: Properties and quality of a hub-based dimerisation network. BMC Syst Biol 1:34
147. Williams T, Tjian R (1991) Characterization of a dimerization motif in AP-2 and its function in heterologous DNA-binding proteins. Science 251(4997):1067–1071
148. Mohibullah N, Donner A, Ippolito JA, Williams T (1999) SELEX and missing phosphate contact analyses reveal flexibility within the AP-2[alpha] protein: DNA binding complex. Nucleic Acids Res 27(13):2760–2769
149. Mitchell PJ, Wang C, Tjian R (1987) Positive and negative regulation of transcription in vitro: Enhancer-binding protein AP-2 is inhibited by SV40 T antigen. Cell 50(6): 847–861
150. Ma PC, Rould MA, Weintraub H, Pabo CO (1994) Crystal structure of MyoD bHLH domain-DNA complex: Perspectives on DNA recognition and implications for transcriptional activation. Cell 77(3):451–459
151. Longo A, Guanga GP, Rose RB (2008) Crystal structure of E47-NeuroD1/beta2 bHLH domain-DNA complex: Heterodimer selectivity and DNA recognition. Biochemistry 47(1): 218–229
152. Miller M, Shuman JD, Sebastian T, Dauter Z, Johnson PF (2003) Structural basis for DNA recognition by the basic region leucine zipper transcription factor CCAAT/enhancer-binding protein alpha. J Biol Chem 278(17):15178–15184
153. Glover JN, Harrison SC (1995) Crystal structure of the heterodimeric bZIP transcription factor c-Fos-c-Jun bound to DNA. Nature 373(6511):257–261
154. Deppmann CD, Alvania RS, Taparowsky EJ (2006) Cross-species annotation of basic leucine zipper factor interactions: Insight into the evolution of closed interaction networks. Mol Biol Evol 23(8):1480–1492
155. He JX, Gendron JM, Sun Y, Gampala SS, Gendron N, Sun CQ, Wang ZY (2005) BZR1 is a transcriptional repressor with dual roles in brassinosteroid homeostasis and growth responses. Science 307(5715):1634–1638
156. Yin Y, Vafeados D, Tao Y, Yoshida S, Asami T, Chory J (2005) A new class of transcription factors mediates brassinosteroid-regulated gene expression in *Arabidopsis*. Cell 120(2): 249–259
157. Vullhorst D, Buonanno A (2003) Characterization of general transcription factor 3, a transcription factor involved in slow muscle-specific gene expression. J Biol Chem 278(10):8370–8379
158. Vullhorst D, Buonanno A (2005) Multiple GTF2I-like repeats of general transcription factor 3 exhibit DNA binding properties. Evidence for a common origin as a sequence-specific DNA interaction module. J Biol Chem 280(36):31722–31731
159. Gupta M, Mungai PT, Goldwasser E (2000) A new transacting factor that modulates hypoxia-induced expression of the erythropoietin gene. Blood 96(2):491–497

160. Curaba J, Herzog M, Vachon G (2003) GeBP, the first member of a new gene family in *Arabidopsis*, encodes a nuclear protein with DNA-binding activity and is regulated by KNAT1. Plant J 33(2):305–317
161. Zourelidou M, de Torres-Zabala M, Smith C, Bevan MW (2002) Storekeeper defines a new class of plant-specific DNA-binding proteins and is a putative regulator of patatin expression. Plant J 30(4):489–497
162. Chevalier F, Perazza D, Laporte F, Le Henanff G, Hornitschek P, Bonneville JM, Herzog M, Vachon G (2008) GeBP and geBP-like proteins are noncanonical leucine-zipper transcription factors that regulate cytokinin response in *Arabidopsis*. Plant Physiol 146(3):1142–1154
163. Kosugi S, Ohashi Y (1997) PCF1 and PCF2 specifically bind to cis elements in the rice proliferating cell nuclear antigen gene. Plant Cell 9(9):1607–1619
164. Kosugi S, Ohashi Y (2002) DNA binding and dimerization specificity and potential targets for the TCP protein family. Plant J 30(3):337–348
165. Dehesh K, Hung H, Tepperman JM, Quail PH (1992) GT-2: A transcription factor with twin autonomous DNA-binding domains of closely related but different target sequence specificity. Embo J 11(11):4131–4144
166. Ayadi M, Delaporte V, Li YF, Zhou DX (2004) Analysis of GT-3a identifies a distinct subgroup of trihelix DNA-binding transcription factors in *Arabidopsis*. FEBS Lett 562(1–3):147–154
167. Hiratsuka K, Wu X, Fukuzawa H, Chua NH (1994) Molecular dissection of GT-1 from *Arabidopsis*. Plant Cell 6(12):1805–1813
168. Villain P, Mache R, Zhou DX (1996) The mechanism of GT element-mediated cell type-specific transcriptional control. J Biol Chem 271(51):32593–32598
169. Ulmasov T, Hagen G, Guilfoyle TJ (1999) Dimerization and DNA binding of auxin response factors. Plant J 19(3):309–319
170. da Costa e Silva O (1994) CG-1, a parsley light-induced DNA-binding protein. Plant Mol Biol 25(5):921–924
171. Bouche N, Scharlat A, Snedden W, Bouchez D, Fromm H (2002) A novel family of calmodulin-binding transcription activators in multicellular organisms. J Biol Chem 277(24):21851–21861
172. Doherty CJ, Van Buskirk HA, Myers SJ, Thomashow MF (2009) Roles for *Arabidopsis* CAMTA transcription factors in cold-regulated gene expression and freezing tolerance. Plant Cell 21(3):972–984
173. Kloks CP, Spronk CA, Lasonder E, Hoffmann A, Vuister GW, Grzesiek S, Hilbers CW (2002) The solution structure and DNA-binding properties of the cold-shock domain of the human Y-box protein YB-1. J Mol Biol 316(2):317–326
174. Kloks CP, Tessari M, Vuister GW, Hilbers CW (2004) Cold shock domain of the human Y-box protein YB-1. Backbone dynamics and equilibrium between the native state and a partially unfolded state. Biochemistry 43(31):10237–10246
175. Kovall RA, Hendrickson WA (2004) Crystal structure of the nuclear effector of Notch signaling, CSL, bound to DNA. Embo J 23(17):3441–3451
176. Chung CN, Hamaguchi Y, Honjo T, Kawaichi M (1994) Site-directed mutagenesis study on DNA binding regions of the mouse homologue of Suppressor of Hairless, RBP-J kappa. Nucleic Acids Res 22(15):2938–2944
177. Tun T, Hamaguchi Y, Matsunami N, Furukawa T, Honjo T, Kawaichi M (1994) Recognition sequence of a highly conserved DNA binding protein RBP-J kappa. Nucleic Acids Res 22(6):965–971
178. Kokoszynska K, Ostrowski J, Rychlewski L, Wyrwicz LS (2008) The fold recognition of CP2 transcription factors gives new insights into the function and evolution of tumor suppressor protein p53. Cell Cycle 7(18):2907–2915
179. Uv AE, Thompson CR, Bray SJ (1994) The *Drosophila* tissue-specific factor Grainyhead contains novel DNA-binding and dimerization domains which are conserved in the human protein CP2. Mol Cell Biol 14(6):4020–4031

180. Shirra MK, Hansen U (1998) LSF and NTF-1 share a conserved DNA recognition motif yet require different oligomerization states to form a stable protein-DNA complex. J Biol Chem 273(30):19260–19268
181. Kim CG, Swendeman SL, Barnhart KM, Sheffery M (1990) Promoter elements and erythroid cell nuclear factors that regulate alpha-globin gene transcription in vitro. Mol Cell Biol 10(11):5958–5966
182. Dynlacht BD, Attardi LD, Admon A, Freeman M, Tjian R (1989) Functional analysis of NTF-1, a developmentally regulated *Drosophila* transcription factor that binds neuronal *cis* elements. Genes Dev 3(11):1677–1688
183. Pellegrini L, Tan S, Richmond TJ (1995) Structure of serum response factor core bound to DNA. Nature 376(6540):490–498
184. Huang K, Louis JM, Donaldson L, Lim FL, Sharrocks AD, Clore GM (2000) Solution structure of the MEF2A-DNA complex: Structural basis for the modulation of DNA bending and specificity by MADS-box transcription factors. Embo J 19(11):2615–2628
185. Kaufmann K, Melzer R, Theissen G (2005) MIKC-type MADS-domain proteins: Structural modularity, protein interactions and network evolution in land plants. Gene 347(2): 183–198
186. Ohki I, Shimotake N, Fujita N, Jee J, Ikegami T, Nakao M, Shirakawa M (2001) Solution structure of the methyl-CpG binding domain of human MBD1 in complex with methylated DNA. Cell 105(4):487–497
187. Fujita N, Shimotake N, Ohki I, Chiba T, Saya H, Shirakawa M, Nakao M (2000) Mechanism of transcriptional regulation by methyl-CpG binding protein MBD1. Mol Cell Biol 20(14):5107–5118
188. Lamoureux JS, Stuart D, Tsang R, Wu C, Glover JN (2002) Structure of the sporulation-specific transcription factor Ndt80 bound to DNA. Embo J 21(21):5721–5732
189. Chu S, Herskowitz I (1998) Gametogenesis in yeast is regulated by a transcriptional cascade dependent on Ndt80. Mol Cell 1(5):685–696
190. McLure KG, Lee PW (1998) How p53 binds DNA as a tetramer. Embo J 17(12):3342–3350
191. el-Deiry WS, Kern SE, Pietenpol JA, Kinzler KW, Vogelstein B (1992) Definition of a consensus binding site for p53. Nat Genet 1(1):45–49
192. Chen FE, Huang DB, Chen YQ, Ghosh G (1998) Crystal structure of p50/p65 heterodimer of transcription factor NF-kappaB bound to DNA. Nature 391(6665):410–413
193. Ghosh G, van Duyne G, Ghosh S, Sigler PB (1995) Structure of NF-kappa B p50 homodimer bound to a kappa B site. Nature 373(6512):303–310
194. Muller CW, Rey FA, Sodeoka M, Verdine GL, Harrison SC (1995) Structure of the NF-kappa B p50 homodimer bound to DNA. Nature 373(6512):311–317
195. Zabel U, Schreck R, Baeuerle PA (1991) DNA binding of purified transcription factor NF-kappa B. Affinity, specificity, Zn2+ dependence, and differential half-site recognition. J Biol Chem 266(1):252–260
196. Nagata T, Gupta V, Sorce D, Kim WY, Sali A, Chait BT, Shigesada K, Ito Y, Werner MH (1999) Immunoglobulin motif DNA recognition and heterodimerization of the PEBP2/CBF Runt domain. Nat Struct Biol 6(7):615–619
197. Wheeler JC, Shigesada K, Gergen JP, Ito Y (2000) Mechanisms of transcriptional regulation by Runt domain proteins. Semin Cell Dev Biol 11(5):369–375
198. Chen X, Vinkemeier U, Zhao Y, Jeruzalmi D, Darnell JE Jr., Kuriyan J (1998) Crystal structure of a tyrosine phosphorylated STAT-1 dimer bound to DNA. Cell 93(5):827–839
199. Horvath CM (2000) STAT proteins and transcriptional responses to extracellular signals. Trends Biochem Sci 25(10):496–502
200. Kim JL, Nikolov DB, Burley SK (1993) Co-crystal structure of TBP recognizing the minor groove of a TATA element. Nature 365(6446):520–527
201. Nikolov DB, Chen H, Halay ED, Hoffman A, Roeder RG, Burley SK (1996) Crystal structure of a human TATA box-binding protein/TATA element complex. Proc Natl Acad Sci U S A 93(10):4862–4867

202. Kim JL, Burley SK (1994) 1.9 A resolution refined structure of TBP recognizing the minor groove of TATAAAAG. Nat Struct Biol 1(9):638–653
203. Kim Y, Geiger JH, Hahn S, Sigler PB (1993) Crystal structure of a yeast TBP/TATA-box complex. Nature 365(6446):512–520
204. Burley SK (1996) The TATA box binding protein. Curr Opin Struct Biol 6(1):69–75
205. Muller CW, Herrmann BG (1997) Crystallographic structure of the T domain-DNA complex of the Brachyury transcription factor. Nature 389(6653):884–888
206. Coll M, Seidman JG, Muller CW (2002) Structure of the DNA-bound T-box domain of human TBX3, a transcription factor responsible for ulnar-mammary syndrome. Structure 10(3):343–356
207. Kispert A, Herrmann BG (1993) The Brachyury gene encodes a novel DNA binding protein. Embo J 12(8):3211–3220
208. Wilson V, Conlon FL (2002) The T-box family. Genome Biol 3(6):REVIEWS3008
209. Desveaux D, Despres C, Joyeux A, Subramaniam R, Brisson N (2000) PBF-2 is a novel single-stranded DNA binding factor implicated in PR-10a gene activation in potato. Plant Cell 12(8):1477–1489
210. Xu RM, Koch C, Liu Y, Horton JR, Knapp D, Nasmyth K, Cheng X (1997) Crystal structure of the DNA-binding domain of Mbp1, a transcription factor important in cell-cycle control of DNA synthesis. Structure 5(3):349–358
211. Taylor IA, Treiber MK, Olivi L, Smerdon SJ (1997) The X-ray structure of the DNA-binding domain from the *Saccharomyces cerevisiae* cell-cycle transcription factor Mbp1 at 2.1 A resolution. J Mol Biol 272(1):1–8
212. Nair M, McIntosh PB, Frenkiel TA, Kelly G, Taylor IA, Smerdon SJ, Lane AN (2003) NMR structure of the DNA-binding domain of the cell cycle protein Mbp1 from *Saccharomyces cerevisiae*. Biochemistry 42(5):1266–1273
213. Johnston LH, Lowndes NF (1992) Cell cycle control of DNA synthesis in budding yeast. Nucleic Acids Res 20(10):2403–2410
214. Iwahara J, Iwahara M, Daughdrill GW, Ford J, Clubb RT (2002) The structure of the Dead ringer-DNA complex reveals how AT-rich interaction domains (ARIDs) recognize DNA. Embo J 21(5):1197–1209
215. Cordier F, Hartmann B, Rogowski M, Affolter M, Grzesiek S (2006) DNA recognition by the brinker repressor – an extreme case of coupling between binding and folding. J Mol Biol 361(4):659–672
216. Sivasankaran R, Vigano MA, Muller B, Affolter M, Basler K (2000) Direct transcriptional control of the Dpp target omb by the DNA binding protein Brinker. Embo J 19(22):6162–6172
217. Iwahara J, Kigawa T, Kitagawa K, Masumoto H, Okazaki T, Yokoyama S (1998) A helix-turn-helix structure unit in human centromere protein B (CENP-B). Embo J 17(3):827–837
218. Yoda K, Kitagawa K, Masumoto H, Muro Y, Okazaki T (1992) A human centromere protein, CENP-B, has a DNA binding domain containing four potential alpha helices at the NH2 terminus, which is separable from dimerizing activity. J Cell Biol 119(6):1413–1427
219. Muro Y, Masumoto H, Yoda K, Nozaki N, Ohashi M, Okazaki T (1992) Centromere protein B assembles human centromeric alpha-satellite DNA at the 17-bp sequence, CENP-B box. J Cell Biol 116(3):585–596
220. Yoda K, Nakamura T, Masumoto H, Suzuki N, Kitagawa K, Nakano M, Shinjo A, Okazaki T (1996) Centromere protein B of African green monkey cells: Gene structure, cellular expression, and centromeric localization. Mol Cell Biol 16(9):5169–5177
221. Yamasaki K, Akiba T, Yamasaki T, Harata K (2007) Structural basis for recognition of the matrix attachment region of DNA by transcription factor SATB1. Nucleic Acids Res 35(15):5073–5084
222. Galande S, Dickinson LA, Mian IS, Sikorska M, Kohwi-Shigematsu T (2001) SATB1 cleavage by caspase 6 disrupts PDZ domain-mediated dimerization, causing detachment from chromatin early in T-cell apoptosis. Mol Cell Biol 21(16):5591–5604

223. Zheng N, Fraenkel E, Pabo CO, Pavletich NP (1999) Structural basis of DNA recognition by the heterodimeric cell cycle transcription factor E2F-DP. Genes Dev 13(6):666–674
224. Mo Y, Vaessen B, Johnston K, Marmorstein R (1998) Structures of SAP-1 bound to DNA targets from the E74 and c-fos promoters: Insights into DNA sequence discrimination by Ets proteins. Mol Cell 2(2):201–212
225. Kodandapani R, Pio F, Ni CZ, Piccialli G, Klemsz M, McKercher S, Maki RA, Ely KR (1996) A new pattern for helix-turn-helix recognition revealed by the PU.1 ETS-domain-DNA complex. Nature 380(6573):456–460
226. Wei GH, Badis G, Berger MF, Kivioja T, Palin K, Enge M, Bonke M, Jolma A, Varjosalo M, Gehrke AR, Yan J, Talukder S, Turunen M, Taipale M, Stunnenberg HG, Ukkonen E, Hughes TR, Bulyk ML, Taipale J (2010) Genome-wide analysis of ETS-family DNA-binding in vitro and in vivo. Embo J 29(13):2147–2160
227. Logan N, Delavaine L, Graham A, Reilly C, Wilson J, Brummelkamp TR, Hijmans EM, Bernards R, La Thangue NB (2004) E2F-7: A distinctive E2F family member with an unusual organization of DNA-binding domains. Oncogene 23(30):5138–5150
228. Di Stefano L, Jensen MR, Helin K (2003) E2F7, a novel E2F featuring DP-independent repression of a subset of E2F-regulated genes. Embo J 22(23):6289–6298
229. Li J, Ran C, Li E, Gordon F, Comstock G, Siddiqui H, Cleghorn W, Chen HZ, Kornacker K, Liu CG, Pandit SK, Khanizadeh M, Weinstein M, Leone G, de Bruin A (2008) Synergistic function of E2F7 and E2F8 is essential for cell survival and embryonic development. Dev Cell 14(1):62–75
230. Kosugi S, Ohashi Y (2002) E2Ls, E2F-like repressors of *Arabidopsis* that bind to E2F sites in a monomeric form. J Biol Chem 277(19):16553–16558
231. Jin C, Marsden I, Chen X, Liao X (1999) Dynamic DNA contacts observed in the NMR structure of winged helix protein-DNA complex. J Mol Biol 289(4):683–690
232. Kaufmann E, Müller D, Knöchel W (1995) DNA recognition site analysis of *Xenopus* winged helix proteins. J Mol Biol 248(2):239–254
233. Pierrou S, Hellqvist M, Samuelsson L, Enerback S, Carlsson P (1994) Cloning and characterization of seven human forkhead proteins: Binding site specificity and DNA bending. Embo J 13(20):5002–5012
234. Biggs WH 3rd, Cavenee WK, Arden KC (2001) Identification and characterization of members of the FKHR (FOX O) subclass of winged-helix transcription factors in the mouse. Mamm Genome 12(6):416–425
235. Windhovel A, Hein I, Dabrowa R, Stockhaus J (2001) Characterization of a novel class of plant homeodomain proteins that bind to the C4 phosphoenolpyruvate carboxylase gene of Flaveria trinervia. Plant Mol Biol 45(2):201–214
236. Park HC, Kim ML, Lee SM, Bahk JD, Yun DJ, Lim CO, Hong JC, Lee SY, Cho MJ, Chung WS (2007) Pathogen-induced binding of the soybean zinc finger homeodomain proteins GmZF-HD1 and GmZF-HD2 to two repeats of ATTA homeodomain binding site in the calmodulin isoform 4 (GmCaM4) promoter. Nucleic Acids Res 35(11):3612–3623
237. Tan QK, Irish VF (2006) The *Arabidopsis* zinc finger-homeodomain genes encode proteins with unique biochemical properties that are coordinately expressed during floral development. Plant Physiol 140(3):1095–1108
238. Vuister GW, Kim SJ, Orosz A, Marquardt J, Wu C, Bax A (1994) Solution structure of the DNA-binding domain of *Drosophila* heat shock transcription factor. Nat Struct Biol 1(9):605–614
239. Perisic O, Xiao H, Lis JT (1989) Stable binding of Drosophila heat shock factor to head-to-head and tail-to-tail repeats of a conserved 5 bp recognition unit. Cell 59(5):797–806
240. LaRonde-LeBlanc NA, Wolberger C (2003) Structure of HoxA9 and Pbx1 bound to DNA: Hox hexapeptide and DNA recognition anterior to posterior. Genes Dev 17(16):2060–2072
241. Kissinger CR, Liu BS, Martin-Blanco E, Kornberg TB, Pabo CO (1990) Crystal structure of an engrailed homeodomain-DNA complex at 2.8 Å resolution: A framework for understanding homeodomain-DNA interactions. Cell 63(3):579–590

242. Berger MF, Badis G, Gehrke AR, Talukder S, Philippakis AA, Pena-Castillo L, Alleyne TM, Mnaimneh S, Botvinnik OB, Chan ET, Khalid F, Zhang W, Newburger D, Jaeger SA, Morris QD, Bulyk ML, Hughes TR (2008) Variation in homeodomain DNA binding revealed by high-resolution analysis of sequence preferences. Cell 133(7):1266–1276
243. Escalante CR, Yie J, Thanos D, Aggarwal AK (1998) Structure of IRF-1 with bound DNA reveals determinants of interferon regulation. Nature 391(6662):103–106
244. Fujii Y, Shimizu T, Kusumoto M, Kyogoku Y, Taniguchi T, Hakoshima T (1999) Crystal structure of an IRF-DNA complex reveals novel DNA recognition and cooperative binding to a tandem repeat of core sequences. Embo J 18(18):5028–5041
245. Hames C, Ptchelkine D, Grimm C, Thevenon E, Moyroud E, Gerard F, Martiel JL, Benlloch R, Parcy F, Muller CW (2008) Structural basis for LEAFY floral switch function and similarity with helix-turn-helix proteins. Embo J 27(19):2628–2637
246. William DA, Su Y, Smith MR, Lu M, Baldwin DA, Wagner D (2004) Genomic identification of direct target genes of LEAFY. Proc Natl Acad Sci U S A 101(6):1775–1780
247. Cutler G, Perry KM, Tjian R (1998) Adf-1 is a nonmodular transcription factor that contains a TAF-binding Myb-like motif. Mol Cell Biol 18(4):2252–2261
248. England BP, Heberlein U, Tjian R (1990) Purified *Drosophila* transcription factor, *Adh* distal factor-1 (Adf-1), binds to sites in several *Drosophila* promoters and activates transcription. J Biol Chem 265(9):5086–5094
249. Bhaskar V, Courey AJ (2002) The MADF-BESS domain factor Dip3 potentiates synergistic activation by Dorsal and Twist. Gene 299(1–2):173–184
250. Bender A, Sprague GF Jr. (1987) MAT alpha 1 protein, a yeast transcription activator, binds synergistically with a second protein to a set of cell-type-specific genes. Cell 50(5): 681–691
251. Biedenkapp H, Borgmeyer U, Sippel AE, Klempnauer KH (1988) Viral myb oncogene encodes a sequence-specific DNA-binding activity. Nature 335(6193):835–837
252. Romero I, Fuertes A, Benito MJ, Malpica JM, Leyva A, Paz-Ares J (1998) More than 80 R2R3-MYB regulatory genes in the genome of *Arabidopsis thaliana*. Plant J 14(3):273–284
253. Xu W, Rould MA, Jun S, Desplan C, Pabo CO (1995) Crystal structure of a paired domain-DNA complex at 2.5 A resolution reveals structural basis for Pax developmental mutations. Cell 80(4):639–650
254. Lehmann M, Siegmund T, Lintermann KG, Korge G (1998) The pipsqueak protein of *Drosophila melanogaster* binds to GAGA sequences through a novel DNA-binding domain. J Biol Chem 273(43):28504–28509
255. Siegmund T, Lehmann M (2002) The Drosophila Pipsqueak protein defines a new family of helix-turn-helix DNA-binding proteins. Dev Genes Evol 212(3):152–157
256. Klemm JD, Rould MA, Aurora R, Herr W, Pabo CO (1994) Crystal structure of the Oct-1 POU domain bound to an octamer site: DNA recognition with tethered DNA-binding modules. Cell 77(1):21–32
257. Cook AL, Sturm RA (2008) POU domain transcription factors: BRN2 as a regulator of melanocytic growth and tumourigenesis. Pigment Cell Melanoma Res 21(6):611–626
258. Yousef MS, Matthews BW (2005) Structural basis of Prospero-DNA interaction: Implications for transcription regulation in developing cells. Structure 13(4):601–607
259. Hassan B, Li L, Bremer KA, Chang W, Pinsonneault J, Vaessin H (1997) Prospero is a panneural transcription factor that modulates homeodomain protein activity. Proc Natl Acad Sci U S A 94(20):10991–10996
260. Cook T, Pichaud F, Sonneville R, Papatsenko D, Desplan C (2003) Distinction between color photoreceptor cell fates is controlled by Prospero in *Drosophila*. Dev Cell 4(6):853–864
261. Konig P, Giraldo R, Chapman L, Rhodes D (1996) The crystal structure of the DNA-binding domain of yeast RAP1 in complex with telomeric DNA. Cell 85(1):125–136
262. Henry YA, Chambers A, Tsang JS, Kingsman AJ, Kingsman SM (1990) Characterisation of the DNA binding domain of the yeast RAP1 protein. Nucleic Acids Res 18(9):2617–2623
263. Vignais ML, Huet J, Buhler JM, Sentenac A (1990) Contacts between the factor TUF and RPG sequences. J Biol Chem 265(24):14669–14674

264. Gajiwala KS, Chen H, Cornille F, Roques BP, Reith W, Mach B, Burley SK (2000) Structure of the winged-helix protein hRFX1 reveals a new mode of DNA binding. Nature 403(6772):916–921
265. Siegrist CA, Durand B, Emery P, David E, Hearing P, Mach B, Reith W (1993) RFX1 is identical to enhancer factor C and functions as a transactivator of the hepatitis B virus enhancer. Mol Cell Biol 13(10):6375–6384
266. Emery P, Strubin M, Hofmann K, Bucher P, Mach B, Reith W (1996) A consensus motif in the RFX DNA binding domain and binding domain mutants with altered specificity. Mol Cell Biol 16(8):4486–4494
267. Lysenko EA (2007) Plant sigma factors and their role in plastid transcription. Plant Cell Rep 26(7):845–859
268. Hakimi MA, Privat I, Valay JG, Lerbs-Mache S (2000) Evolutionary conservation of C-terminal domains of primary sigma(70)-type transcription factors between plants and bacteria. J Biol Chem 275(13):9215–9221
269. Campbell EA, Muzzin O, Chlenov M, Sun JL, Olson CA, Weinman O, Trester-Zedlitz ML, Darst SA (2002) Structure of the bacterial RNA polymerase promoter specificity sigma subunit. Mol Cell 9(3):527–539
270. Anbanandam A, Albarado DC, Nguyen CT, Halder G, Gao X, Veeraraghavan S (2006) Insights into transcription enhancer factor 1 (TEF-1) activity from the solution structure of the TEA domain. Proc Natl Acad Sci U S A 103(46):17225–17230
271. Hwang JJ, Chambon P, Davidson I (1993) Characterization of the transcription activation function and the DNA binding domain of transcriptional enhancer factor-1. Embo J 12(6):2337–2348
272. Weider M, Machnik A, Klebl F, Sauer N (2006) Vhr1p, a new transcription factor from budding yeast, regulates biotin-dependent expression of VHT1 and BIO5. J Biol Chem 281(19):13513–13524
273. Yamaguchi-Iwai Y, Stearman R, Dancis A, Klausner RD (1996) Iron-regulated DNA binding by the AFT1 protein controls the iron regulon in yeast. Embo J 15(13): 3377–3384
274. Allen MD, Yamasaki K, Ohme-Takagi M, Tateno M, Suzuki M (1998) A novel mode of DNA recognition by a beta-sheet revealed by the solution structure of the GCC-box binding domain in complex with DNA. Embo J 17(18):5484–5496
275. Lindner SE, De Silva EK, Keck JL, Llinas M (2010) Structural determinants of DNA binding by a P. falciparum ApiAP2 transcriptional regulator. J Mol Biol 395(3): 558–567
276. Ohme-Takagi M, Shinshi H (1995) Ethylene-inducible DNA binding proteins that interact with an ethylene-responsive element. Plant Cell 7(2):173–182
277. De Silva EK, Gehrke AR, Olszewski K, Leon I, Chahal JS, Bulyk ML, Llinas M (2008) Specific DNA-binding by apicomplexan AP2 transcription factors. Proc Natl Acad Sci U S A 105(24):8393–8398
278. Huth JR, Bewley CA, Nissen MS, Evans JN, Reeves R, Gronenborn AM, Clore GM (1997) The solution structure of an HMG-I(Y)-DNA complex defines a new architectural minor groove binding motif. Nat Struct Biol 4(8):657–665
279. Xing Y, Fikes JD, Guarente L (1993) Mutations in yeast HAP2/HAP3 define a hybrid CCAAT box binding domain. Embo J 12(12):4647–4655
280. Olesen JT, Guarente L (1990) The HAP2 subunit of yeast CCAAT transcriptional activator contains adjacent domains for subunit association and DNA recognition: Model for the HAP2/3/4 complex. Genes Dev 4(10):1714–1729
281. Romier C, Cocchiarella F, Mantovani R, Moras D (2003) The NF-YB/NF-YC structure gives insight into DNA binding and transcription regulation by CCAAT factor NF-Y. J Biol Chem 278(2):1336–1345
282. Bi W, Wu L, Coustry F, de Crombrugghe B, Maity SN (1997) DNA binding specificity of the CCAAT-binding factor CBF/NF-Y. J Biol Chem 272(42):26562–26572

283. Scott S, Dorrington R, Svetlov V, Beeser AE, Distler M, Cooper TG (2000) Functional domain mapping and subcellular distribution of Dal82p in *Saccharomyces cerevisiae*. J Biol Chem 275(10):7198–7204
284. Dorrington RA, Cooper TG (1993) The DAL82 protein of *Saccharomyces cerevisiae* binds to the DAL upstream induction sequence (UIS). Nucleic Acids Res 21(16):3777–3784
285. Carrasco JL, Ancillo G, Castello MJ, Vera P (2005) A novel DNA-binding motif, hallmark of a new family of plant transcription factors. Plant Physiol 137(2):602–606
286. Carrasco JL, Ancillo G, Mayda E, Vera P (2003) A novel transcription factor involved in plant defense endowed with protein phosphatase activity. Embo J 22(13):3376–3384
287. Yamasaki K, Kigawa T, Inoue M, Yamasaki T, Yabuki T, Aoki M, Seki E, Matsuda T, Tomo Y, Terada T, Shirouzu M, Tanaka A, Seki M, Shinozaki K, Yokoyama S (2005) Solution structure of the major DNA-binding domain of *Arabidopsis thaliana* ethylene-insensitive3-like3. J Mol Biol 348(2):253–264
288. Solano R, Stepanova A, Chao Q, Ecker JR (1998) Nuclear events in ethylene signaling: A transcriptional cascade mediated by ETHYLENE-INSENSITIVE3 and ETHYLENE-RESPONSE-FACTOR1. Genes Dev 12(23):3703–3714
289. Kosugi S, Ohashi Y (2000) Cloning and DNA-binding properties of a tobacco Ethylene-insensitive3 (EIN3) homolog. Nucleic Acids Res 28(4):960–967
290. Baker HV (1991) GCR1 of *Saccharomyces cerevisiae* encodes a DNA binding protein whose binding is abolished by mutations in the CTTCC sequence motif. Proc Natl Acad Sci U S A 88(21):9443–9447
291. Uemura H, Koshio M, Inoue Y, Lopez MC, Baker HV (1997) The role of Gcr1p in the transcriptional activation of glycolytic genes in yeast *Saccharomyces cerevisiae*. Genetics 147(2):521–532
292. Huie MA, Baker HV (1996) DNA-binding properties of the yeast transcriptional activator, Gcr1p. Yeast 12(4):307–317
293. Huie MA, Scott EW, Drazinic CM, Lopez MC, Hornstra IK, Yang TP, Baker HV (1992) Characterization of the DNA-binding activity of GCR1: In vivo evidence for two GCR1-binding sites in the upstream activating sequence of TPI of *Saccharomyces cerevisiae*. Mol Cell Biol 12(6):2690–2700
294. Richards DE, Peng J, Harberd NP (2000) Plant GRAS and metazoan STATs: One family? Bioessays 22(6):573–577
295. Hirsch S, Kim J, Munoz A, Heckmann AB, Downie JA, Oldroyd GE (2009) GRAS proteins form a DNA binding complex to induce gene expression during nodulation signaling in Medicago truncatula. Plant Cell 21(2):545–557
296. Olsen AN, Ernst HA, Leggio LL, Skriver K (2005) NAC transcription factors: Structurally distinct, functionally diverse. Trends Plant Sci 10(2):79–87
297. Tran LS, Nakashima K, Sakuma Y, Simpson SD, Fujita Y, Maruyama K, Fujita M, Seki M, Shinozaki K, Yamaguchi-Shinozaki K (2004) Isolation and functional analysis of *Arabidopsis* stress-inducible NAC transcription factors that bind to a drought-responsive cis-element in the early responsive to dehydration stress 1 promoter. Plant Cell 16(9):2481–2498
298. Surdo PL, Bottomley MJ, Sattler M, Scheffzek K (2003) Crystal structure and nuclear magnetic resonance analyses of the SAND domain from glucocorticoid modulatory element binding protein-1 reveals deoxyribonucleic acid and zinc binding regions. Mol Endocrinol 17(7):1283–1295
299. Huggenvik JI, Michelson RJ, Collard MW, Ziemba AJ, Gurley P, Mowen KA (1998) Characterization of a nuclear deformed epidermal autoregulatory factor-1 (DEAF-1)-related (NUDR) transcriptional regulator protein. Mol Endocrinol 12(10):1619–1639
300. Murphy EC, Zhurkin VB, Louis JM, Cornilescu G, Clore GM (2001) Structural basis for SRY-dependent 46-X,Y sex reversal: Modulation of DNA bending by a naturally occurring point mutation. J Mol Biol 312(3):481–499
301. Harley VR, Lovell-Badge R, Goodfellow PN (1994) Definition of a consensus DNA binding site for SRY. Nucleic Acids Res 22(8):1500–1501

302. Zhou DX, Bisanz-Seyer C, Mache R (1995) Molecular cloning of a small DNA binding protein with specificity for a tissue-specific negative element within the rps1 promoter. Nucleic Acids Res 23(7):1165–1169
303. Cho G, Kim J, Rho HM, Jung G (1995) Structure-function analysis of the DNA binding domain of *Saccharomyces cerevisiae* ABF1. Nucleic Acids Res 23(15):2980–2987
304. Dorsman JC, van Heeswijk WC, Grivell LA (1990) Yeast general transcription factor GFI: Sequence requirements for binding to DNA and evolutionary conservation. Nucleic Acids Res 18(9):2769–2776
305. Bastola DR, Pethe VV, Winicov I (1998) Alfin1, a novel zinc-finger protein in alfalfa roots that binds to promoter elements in the salt-inducible MsPRP2 gene. Plant Mol Biol 38(6):1123–1135
306. Santi L, Wang Y, Stile MR, Berendzen K, Wanke D, Roig C, Pozzi C, Muller K, Muller J, Rohde W, Salamini F (2003) The GA octodinucleotide repeat binding factor BBR participates in the transcriptional regulation of the homeobox gene Bkn3. Plant J 34(6): 813–826
307. Kooiker M, Airoldi CA, Losa A, Manzotti PS, Finzi L, Kater MM, Colombo L (2005) BASIC PENTACYSTEINE1, a GA binding protein that induces conformational changes in the regulatory region of the homeotic *Arabidopsis* gene SEEDSTICK. Plant Cell 17(3):722–729
308. Hagman J, Gutch MJ, Lin H, Grosschedl R (1995) EBF contains a novel zinc coordination motif and multiple dimerization and transcriptional activation domains. Embo J 14(12):2907–2916
309. Hagman J, Belanger C, Travis A, Turck CW, Grosschedl R (1993) Cloning and functional characterization of early B-cell factor, a regulator of lymphocyte-specific gene expression. Genes Dev 7(5):760–773
310. Travis A, Hagman J, Hwang L, Grosschedl R (1993) Purification of early-B-cell factor and characterization of its DNA-binding specificity. Mol Cell Biol 13(6):3392–3400
311. Turner RB, Smith DL, Zawrotny ME, Summers MF, Posewitz MC, Winge DR (1998) Solution structure of a zinc domain conserved in yeast copper-regulated transcription factors. Nat Struct Biol 5(7):551–555
312. Buchman C, Skroch P, Dixon W, Tullius TD, Karin M (1990) A single amino acid change in CUP2 alters its mode of DNA binding. Mol Cell Biol 10(9):4778–4787
313. Dobi A, Dameron CT, Hu S, Hamer D, Winge DR (1995) Distinct regions of Cu(I).ACE1 contact two spatially resolved DNA major groove sites. J Biol Chem 270(17): 10171–10178
314. Cvitanich C, Pallisgaard N, Nielsen KA, Hansen AC, Larsen K, Pihakaski-Maunsbach K, Marcker KA, Jensen EO (2000) CPP1, a DNA-binding protein involved in the expression of a soybean leghemoglobin c3 gene. Proc Natl Acad Sci U S A 97(14):8163–8168
315. Fauth T, Muller-Planitz F, Konig C, Straub T, Becker PB (2010) The DNA binding CXC domain of MSL2 is required for faithful targeting the Dosage Compensation Complex to the X chromosome. Nucleic Acids Res 38(10):3209–3221
316. Allen MD, Grummitt CG, Hilcenko C, Min SY, Tonkin LM, Johnson CM, Freund SM, Bycroft M, Warren AJ (2006) Solution structure of the nonmethyl-CpG-binding CXXC domain of the leukaemia-associated MLL histone methyltransferase. Embo J 25(19): 4503–4512
317. Lee JH, Voo KS, Skalnik DG (2001) Identification and characterization of the DNA binding domain of CpG-binding protein. J Biol Chem 276(48):44669–44676
318. Jorgensen HF, Ben-Porath I, Bird AP (2004) Mbd1 is recruited to both methylated and nonmethylated CpGs via distinct DNA binding domains. Mol Cell Biol 24(8):3387–3395
319. Kim JG, Hudson LD (1992) Novel member of the zinc finger superfamily: A C2-HC finger that recognizes a glia-specific gene. Mol Cell Biol 12(12):5632–5639
320. Kim JG, Armstrong RC, v Agoston D, Robinsky A, Wiese C, Nagle J, Hudson LD (1997) Myelin transcription factor 1 (Myt1) of the oligodendrocyte lineage, along with a closely

related CCHC zinc finger, is expressed in developing neurons in the mammalian central nervous system. J Neurosci Res 50(2):272–290
321. Elrod-Erickson M, Benson TE, Pabo CO (1998) High-resolution structures of variant Zif268-DNA complexes: Implications for understanding zinc finger-DNA recognition. Structure 6(4):451–464
322. Zhu L, Wilken J, Phillips NB, Narendra U, Chan G, Stratton SM, Kent SB, Weiss MA (2000) Sexual dimorphism in diverse metazoans is regulated by a novel class of intertwined zinc fingers. Genes Dev 14(14):1750–1764
323. Erdman SE, Chen HJ, Burtis KC (1996) Functional and genetic characterization of the oligomerization and DNA binding properties of the Drosophila doublesex proteins. Genetics 144(4):1639–1652
324. Shimofurutani N, Kisu Y, Suzuki M, Esaka M (1998) Functional analyses of the Dof domain, a zinc finger DNA-binding domain, in a pumpkin DNA-binding protein AOBP. FEBS Lett 430(3):251–256
325. Lin R, Ding L, Casola C, Ripoll DR, Feschotte C, Wang H (2007) Transposase-derived transcription factors regulate light signaling in *Arabidopsis*. Science 318(5854): 1302–1305
326. Omichinski JG, Clore GM, Schaad O, Felsenfeld G, Trainor C, Appella E, Stahl SJ, Gronenborn AM (1993) NMR structure of a specific DNA complex of Zn-containing DNA binding domain of GATA-1. Science 261(5120):438–446
327. Bates DL, Chen Y, Kim G, Guo L, Chen L (2008) Crystal structures of multiple GATA zinc fingers bound to DNA reveal new insights into DNA recognition and self-association by GATA. J Mol Biol 381(5):1292–1306
328. Svetlov VV, Cooper TG (1998) The *Saccharomyces cerevisiae* GATA factors Dal80p and Deh1p can form homo- and heterodimeric complexes. J Bacteriol 180(21):5682–5688
329. Schnitzler GR, Fischer WH, Firtel RA (1994) Cloning and characterization of the G-box binding factor, an essential component of the developmental switch between early and late development in *Dictyostelium*. Genes Dev 8(4):502–514
330. Brown JM, Firtel RA (2001) Functional and regulatory analysis of the *Dictyostelium* G-box binding factor. Dev Biol 234(2):521–534
331. Hjorth AL, Pears C, Williams JG, Firtel RA (1990) A developmentally regulated trans-acting factor recognizes dissimilar G/C-rich elements controlling a class of cAMP-inducible *Dictyostelium* genes. Genes Dev 4(3):419–432
332. Cohen SX, Moulin M, Hashemolhosseini S, Kilian K, Wegner M, Muller CW (2003) Structure of the GCM domain-DNA complex: A DNA-binding domain with a novel fold and mode of target site recognition. Embo J 22(8):1835–1845
333. Akiyama Y, Hosoya T, Poole AM, Hotta Y (1996) The gcm-motif: A novel DNA-binding motif conserved in *Drosophila* and mammals. Proc Natl Acad Sci U S A 93(25): 14912–14916
334. Schreiber J, Sock E, Wegner M (1997) The regulator of early gliogenesis glial cells missing is a transcription factor with a novel type of DNA-binding domain. Proc Natl Acad Sci U S A 94(9):4739–4744
335. Raventos D, Skriver K, Schlein M, Karnahl K, Rogers SW, Rogers JC, Mundy J (1998) HRT, a novel zinc finger, transcriptional repressor from barley. J Biol Chem 273(36): 23313–23320
336. Husbands A, Bell EM, Shuai B, Smith HM, Springer PS (2007) LATERAL ORGAN BOUNDARIES defines a new family of DNA-binding transcription factors and can interact with specific BHLH proteins. Nucleic Acids Res 35(19):6663–6671
337. Spahr H, Samuelsson T, Hallberg BM, Gustafsson CM (2010) Structure of mitochondrial transcription termination factor 3 reveals a novel nucleic acid-binding domain. Biochem Biophys Res Commun 397(3):386–390
338. Fernandez-Silva P, Martinez-Azorin F, Micol V, Attardi G (1997) The human mitochondrial transcription termination factor (mTERF) is a multizipper protein but binds to DNA

as a monomer, with evidence pointing to intramolecular leucine zipper interactions. Embo J 16(5):1066–1079

339. Hyvarinen AK, Pohjoismaki JL, Reyes A, Wanrooij S, Yasukawa T, Karhunen PJ, Spelbrink JN, Holt IJ, Jacobs HT (2007) The mitochondrial transcription termination factor mTERF modulates replication pausing in human mitochondrial DNA. Nucleic Acids Res 35(19):6458–6474
340. Song Z, Krishna S, Thanos D, Strominger JL, Ono SJ (1994) A novel cysteine-rich sequence-specific DNA-binding protein interacts with the conserved X-box motif of the human major histocompatibility complex class ii genes via a repeated Cys-His domain and functions as a transcriptional repressor. J Exp Med 180(5):1763–1774
341. Rastinejad F, Wagner T, Zhao Q, Khorasanizadeh S (2000) Structure of the RXR-RAR DNA-binding complex on the retinoic acid response element DR1. Embo J 19(5):1045–1054
342. Cheskis B, Freedman LP (1994) Ligand modulates the conversion of DNA-bound vitamin D3 receptor (VDR) homodimers into VDR-retinoid X receptor heterodimers. Mol Cell Biol 14(5):3329–3338
343. Ribeiro RC, Kushner PJ, Baxter JD (1995) The nuclear hormone receptor gene superfamily. Annu Rev Med 46:443–453
344. Nagano Y, Furuhashi H, Inaba T, Sasaki Y (2001) A novel class of plant-specific zinc-dependent DNA-binding protein that binds to A/T-rich DNA sequences. Nucleic Acids Res 29(20):4097–4105
345. Shi Y, Wang YF, Jayaraman L, Yang H, Massague J, Pavletich NP (1998) Crystal structure of a Smad MH1 domain bound to DNA: Insights on DNA binding in TGF-beta signaling. Cell 94(5):585–594
346. Chai J, Wu JW, Yan N, Massague J, Pavletich NP, Shi Y (2003) Features of a Smad3 MH1-DNA complex. Roles of water and zinc in DNA binding. J Biol Chem 278(22): 20327–20331
347. Massague J, Seoane J, Wotton D (2005) Smad transcription factors. Genes Dev 19(23): 2783–2810
348. Sabogal A, Lyubimov AY, Corn JE, Berger JM, Rio DC (2010) THAP proteins target specific DNA sites through bipartite recognition of adjacent major and minor grooves. Nat Struct Mol Biol 17(1):117–123
349. Mitsuda N, Hisabori T, Takeyasu K, Sato MH (2004) VOZ; isolation and characterization of novel vascular plant transcription factors with a one-zinc finger from *Arabidopsis thaliana*. Plant Cell Physiol 45(7):845–854
350. Rushton PJ, Somssich IE, Ringler P, Shen QJ (2010) WRKY transcription factors. Trends Plant Sci 15(5):247–258
351. Marmorstein R, Carey M, Ptashne M, Harrison SC (1992) DNA recognition by GAL4: Structure of a protein-DNA complex. Nature 356(6368):408–414
352. Cahuzac B, Cerdan R, Felenbok B, Guittet E (2001) The solution structure of an AlcR-DNA complex sheds light onto the unique tight and monomeric DNA binding of a Zn(2)Cys(6) protein. Structure 9(9):827–836
353. Mamnun YM, Pandjaitan R, Mahe Y, Delahodde A, Kuchler K (2002) The yeast zinc finger regulators Pdr1p and Pdr3p control pleiotropic drug resistance (PDR) as homo- and heterodimers in vivo. Mol Microbiol 46(5):1429–1440
354. Finn RD, Mistry J, Tate J, Coggill P, Heger A, Pollington JE, Gavin OL, Gunasekaran P, Ceric G, Forslund K, Holm L, Sonnhammer EL, Eddy SR, Bateman A (2010) The Pfam protein families database. Nucleic Acids Res 38(Database issue):D211–222
355. Hunter S, Apweiler R, Attwood TK, Bairoch A, Bateman A, Binns D, Bork P, Das U, Daugherty L, Duquenne L, Finn RD, Gough J, Haft D, Hulo N, Kahn D, Kelly E, Laugraud A, Letunic I, Lonsdale D, Lopez R, Madera M, Maslen J, McAnulla C, McDowall J, Mistry J, Mitchell A, Mulder N, Natale D, Orengo C, Quinn AF, Selengut JD, Sigrist CJ, Thimma M, Thomas PD, Valentin F, Wilson D, Wu CH, Yeats C (2009) InterPro: The integrative protein signature database. Nucleic Acids Res 37(Database issue):D211–215

356. Gruschus JM, Tsao DH, Wang LH, Nirenberg M, Ferretti JA (1997) Interactions of the vnd/NK-2 homeodomain with DNA by nuclear magnetic resonance spectroscopy: basis of binding specificity. Biochemistry 36:5372–5380
357. Remenyi A, Tomilin A, Pohl E, Lins K, Philippsen A, Reinbold R, Scholer HR, Wilmanns M (2001) Differential dimer activities of the transcription factor Oct-1 by DNA-induced interface swapping. Mol Cell 8:569–580
358. Rastinejad F, Wagner T, Zhao Q, Khorasanizadeh S (2000) Structure of the RXR-RAR DNA-binding complex on the retinoic acid response element DR1. Embo J 19:1045–1054
359. Lamoureux JS, Stuart D, Tsang R, Wu C, Glover JN (2002) Structure of the sporulation-specific transcription factor Ndt80 bound to DNA. Embo J 21:5721–5732
360. Berman HM, Westbrook J, Feng Z, Gilliland G, Bhat TN, Weissig H, Shindyalov IN, Bourne PE (2000) The Protein Data Bank. Nucleic Acids Res 28(1):235–242
361. Durbin R (1998) Biological sequence analysis: Probabilistic models of proteins and nucleic acids. Cambridge University Press, Cambridge, UK, New York

Chapter 4
Function and Evolution of C2H2 Zinc Finger Arrays

Lisa Stubbs, Younguk Sun, and Derek Caetano-Anolles

Abstract *Krüppel*-type or C2H2 zinc fingers represent a dominant DNA-binding motif in eukaryotic transcription factor (TF) proteins. In *Krüppel*-type (KZNF) TFs, KZNF motifs are arranged in arrays of three to as many as 40 tandem units, which cooperate to define the unique DNA recognition properties of the protein. Each finger contains four amino acids located at specific positions, which are brought into direct contact with adjacent nucleotides in the DNA sequence as the KZNF array winds around the major groove of the alpha helix. This arrangement creates an intimate and potentially predictable relationship between the amino acid sequence of KZNF arrays and the nucleotide sequence of target binding sites. The large number of possible combinations and arrangements of modular KZNF motifs, and the increasing lengths of KZNF arrays in vertebrate species, has created huge repertoires of functionally unique TF proteins. The properties of this versatile DNA-binding motif have been exploited independently many times over the course of evolution, through attachment to effector motifs that confer activating, repressing or other activities to the proteins. Once created, some of these novel inventions have expanded in specific evolutionary clades, creating large families of TFs that are lineage- or species-unique. This chapter reviews the properties and their remarkable evolutionary history of eukaryotic KZNF TF proteins, with special focus on large families that dominate the TF landscapes in different metazoan species.

4.1 Introduction

The C2H2 zinc finger motif, first identified in studies of the *Xenopus* TF TIFIIA [1] is by far the most common protein domain in metazoan TFs (see Chapter 3). Most versions of this abundant motif correspond to a subtype called the "*Krüppel*-type", named for the *Drosophila Krüppel* protein, a developmentally active TF that bears the canonical C2H2 zinc-binding structure [1, 2]. The C2H2 zinc finger motif was

L. Stubbs (✉)
Department of Cell and Developmental Biology, Institute for Genomic Biology, University of Illinois, Urbana, IL 61801, USA
e-mail: ljstubbs@illinois.edu

T.R. Hughes (ed.), *A Handbook of Transcription Factors*, Subcellular Biochemistry 52,
DOI 10.1007/978-90-481-9069-0_4,

originally thought to be specific to eukaryotes, but a very similar structural domain has been identified in some bacterial TF genes, hinting at more ancient origins [3]. The most striking and characteristic feature of these 28 amino acid motifs is a secondary structure that is dependent upon the coordination of a single zinc atom by paired cysteine (C) and histidine (H) residues (Fig. 4.1). This zinc-dependent structure is required for the interaction between the finger motif and nucleic acids; in the absence of zinc, or if elements of the conserved C2H2 structure are abolished through mutation, zinc fingers lose their ability to fold properly and to bind DNA [1, 4–6].

In addition to the paired cysteine and histidine residues, *Krüppel*-type zinc finger (KZNF) motifs contain a highly conserved inter-finger "spacer", or H/C link sequence, a seven amino acid segment with the consensus sequence TGEKP(Y/F) (Fig. 4.1). KZNF proteins carry out many different kinds of molecular functions, including protein-protein interactions, RNA binding, and sequence-specific binding to DNA. Some DNA-binding KZNFs are now known to carry out functions related to meiotic recombination and chromosome segregation [7–11], or maintenance of DNA methylation marks [12, 13]. Additional functions related to chromosome structure and maintenance may be found as new research is completed. However, most KZNFs with specific DNA recognition capabilities are thought to function as TFs, and this latter class of proteins is the primary focus of this chapter.

Typically, DNA binding KZNF proteins contain 3 or more zinc-finger motifs, which are arranged in tandem within the protein (Figs. 4.1 and 4.2). These multi-fingered, or "polydactyl" KZNF proteins include many of the best-known TFs in eukaryotes, including yeast, plants, invertebrate and vertebrate species. TF proteins with as many as 40 tandem KZNF motifs can be found in most vertebrate genomes and long polydactyl KZNF proteins are also found in plants [14]. The tandem

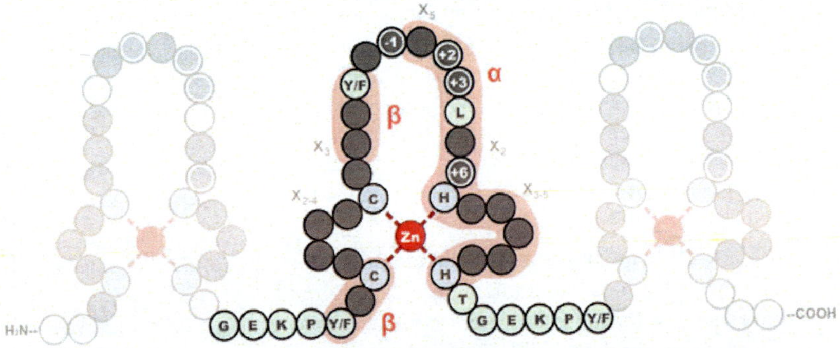

Fig. 4.1 Tandem *Krüppel*-type zinc finger structure displaying the C2H2 motif. Individual zinc ions interact with paired cysteine (C) and histidine (H) residues, stabilizing protein fold structure within zinc fingers. Each finger consists of two β-sheets and one α-helix, the latter of which contains residues that make up the DNA-binding interface (at positions –1, 2, 3, and 6 relative to the helix) as indicated in the figure. The common structure of a finger sequence motif is represented, with *X* denoting an amino acid residue of any type with the subscript representing the number (X_{2-4} represents between 2 and 4 non-specified amino acid residues). The consensus sequence, TGEKP(Y/F), is a highly conserved "H/C link" region between consecutive fingers

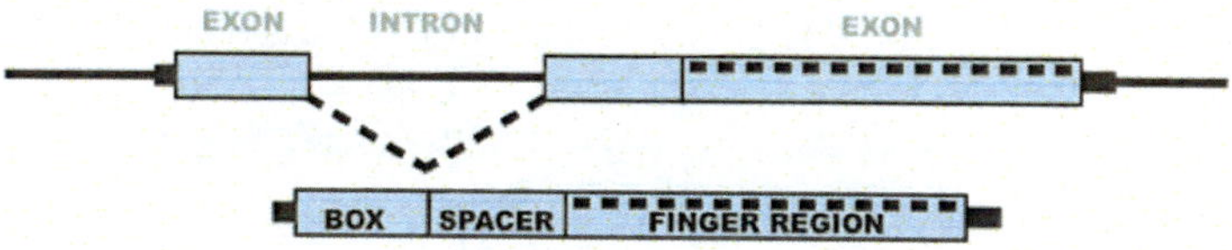

Fig. 4.2 Exon and protein structure of a typical KZNF TF protein. Many KZNF genes, including those of the KRAB, SCAN, BTB/POZ, ZAD and other families, contain one or more exons encoding a specific N-terminal effector domain, and a second exon encoding a "spacer" or "tether" region and an array of 3–40 zinc finger motifs. The tandem arrangement of KZNF-encoding sequences, which contain highly conserved structural elements, has enabled rapid evolution of proteins of this type

arrangement of KZNF motifs permits the adjacent fingers to interact and stabilize DNA binding of the protein at specific sites, as will be discussed in more detail in the following sections. While zinc-fingers define binding site specificity and stability for KZNF proteins, most TFs of this type also require one or more "effector" motifs to translate site-specific DNA binding into gene regulatory activities impacting neighboring genes.

Over the course of evolution, exons encoding tandem KZNF arrays have become associated with coding sequences for a wide variety of different effector domains, to generate proteins with novel structures and activities. Many of these novel KZNF proteins have arisen in, and remain exclusive to, particular evolutionary lineages; some of these species-specific genes have expanded through repeated duplication events to form large families of lineage-specific genes. While this same process has occurred for many gene types, the lineage-specific expansion of KZNF genes is a striking and extraordinary story. This chapter will focus on basic functions of the KZNF motifs, the types of TFs that rely on their highly specific targeting abilities, and their remarkable evolutionary trajectory.

4.2 Zinc Finger–DNA Interactions

The structural elements that control the interaction between KZNF motifs and DNA "target sites" include the paired cysteine and histidine residues, as well as the amino acids surrounding them. The arrangement and spacing of elements within the finger motif, including the H/C link, are critical to maintaining the zinc-finger structure, and are therefore very highly conserved [1]. Most importantly for DNA binding, residues near the C-terminal end of each finger fold into an alpha helix, positioning specific amino acids within the helix to interact directly with DNA (Figs. 4.1 and 4.3). In particular, positions –1, 2, 3 and 6 (relative to the alpha helix) play a critical role in DNA interaction: together, the amino acids at these four positions in each finger are thought largely to determine DNA binding specificity of the protein [15, 16].

The array of multiple, adjacent fingers in these proteins winds around the DNA double strand within the major groove, wrapping around the DNA molecule in an intimate spatial relationship that places the DNA-contacting residues of each finger in register with nucleotides within a turn of the helix. The interaction between

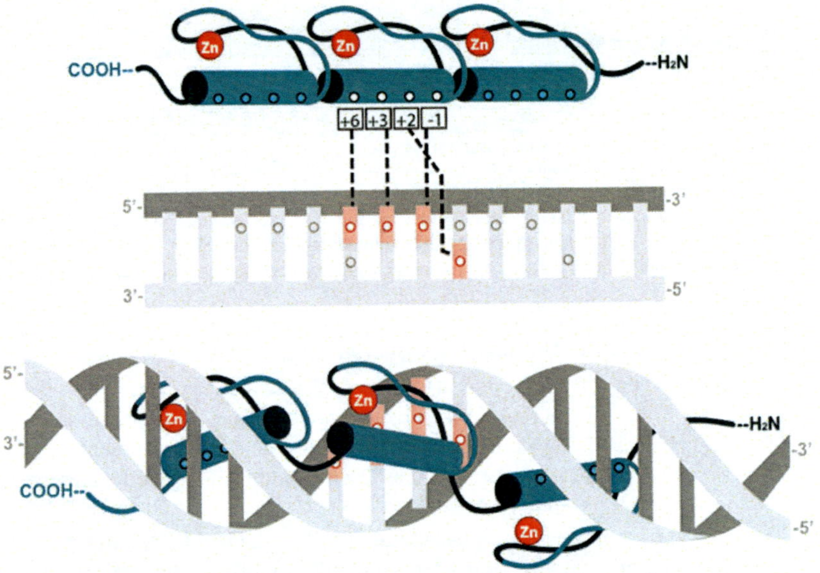

Fig. 4.3 KZNF motif DNA-binding interactions. The alpha helices of KZNF motifs contain amino acid residues that bind to DNA nucleotides (at the –1, 2, 3, and 6 sites as shown at *top*). The relationship between fingers and nucleotides is not one-to-one, as the amino acid at the +2 position will interact with the nucleotide complementary to the neighboring finger's +6 binding site. In this fashion, fingers wind around the major groove of the DNA molecule (illustrated in the *lower panel* of the figure)

the four DNA contacting amino-acid residues in each finger and nucleotides at the DNA binding site is not a simple 1:1 relationship, as there is some overlap between nucleotides bound by adjacent fingers [6, 17–19]. However, the arrangement is such that each finger defines binding specificity at a net of 3 adjacent nucleotides, while exerting some influence over the binding specificity of neighboring KZNF motifs (Fig. 4.3).

4.2.1 Predicting a Zinc-Finger Code

This precisely structured relationship between nucleotides in a binding site and specific amino acids in the DNA-contacting portion of each C2H2 finger implies the existence of a zinc-finger DNA binding "code", and the possibility that a KZNF protein's binding preferences might be predicted de novo from its amino acid sequence. In fact, several different groups have designed mathematical formulas and informatics tools that predict KZNF binding codes [19–21]. These programs are built upon knowledge derived from in vitro DNA binding experiments and structural data, together with calculations of predicted energies of interaction between specific amino acids and nucleotides. Although these methods have proved successful in designing custom KZNF proteins to bind with maximum efficiency to unique sites both in vitro and in vivo [22] it is still unclear if they can accurately predict the binding preferences of KZNF proteins as they exist in normal cellular contexts.

There are several reasons why KZNF arrays, and especially long polydactyl proteins, might not behave in living systems as in vitro models would predict. Firstly, most in vitro studies have focused on predetermined libraries of zinc-finger triplets that are selected for maximal binding to naked DNA under non-biological conditions; by contrast, natural selection in living systems operates in a much more complex milieu, and has taken full advantage of the combinatorial possibilities to produce KZNF protein repertoires of remarkable diversity. Since zinc-finger DNA binding is known to be context-dependent, extrapolations from in vitro experiments with specific KZNF triplets to the behavior of the highly diverse KZNF proteins in metazoan cells are fraught with uncertainties.

Secondly, there is some evidence to suggest that some KZNF proteins might be modified post-translationally in a cell-type specific way that could alter their DNA recognition specificity. For example, phosphorylation of the DNA binding domains in the KZNF protein, *Yin yang 1* (YY1) has been shown to affect the protein's ability to bind DNA targets [23]. The extent to which most KZNF proteins are modified in vivo is unknown.

Thirdly, it is not clear that all fingers in polydactyl proteins need be necessarily engaged simultaneously, or ever engaged at all, in DNA binding. Indeed, several proteins have been described in which the same KZNF motifs can act as DNA binding elements in some instances, and serve alternative, unrelated functions in other circumstances. For example, in the yeast KZNF protein, ZAP1, two of the five zinc-fingers can serve alternatively as DNA-binding or zinc-response elements [24]. In mammals, two C-terminal fingers in the KZNF protein, ZAC, can either participate in DNA binding or be sequestered for interactions with protein partner, p300 [25]. Through differential use of specific subsets of its seven KZNF motifs, ZAC can recognize two distinct sets of high-affinity binding sites [26]. Similarly, a 30-fingered protein, OAZ, can use subsets of fingers to recognize more than one DNA binding site, and use others to mediate dimer formation or interactions with protein co-factors [27]. Similar dual-purpose activities have been implicated for a large number of KZNF proteins in many species [28]. There is therefore good reason to suspect that many polydactyl proteins will act in this way, utilizing subsets of fingers alternatively for various functions in a range of biological contexts.

4.2.2 Experimental Data on KZNF–DNA Interactions

Much current knowledge regarding the interactions between polydactyl KZNF proteins and DNA binding sites is based on in vitro experiments; the in vivo functions of most members of this abundant protein class remain a mystery. The picture should be clarified significantly in the next several years, as in vivo DNA binding sites for more polydactyl KZNF proteins are mapped through unbiased methods, in particular, through chromatin-immunoprecipitation followed by high-throughput sequencing ("ChIP-seq"). To date, only a small number of proteins have been examined using these unbiased methods. The conserved polydactyl KZNF protein, RE1-silencing TF (REST), was one of the first TFs to be mapped using ChIP-seq methods [29]. REST in vivo binding sites had been studied extensively on a gene-by-gene level, and the results of ChIP-seq studies, while fascinating, largely confirmed what was known about the binding site and types of preferred gene targets for this regulatory protein.

However, the analysis revealed significant levels of protein binding to REST "half sites", representing 5′ or 3′ segments of the strong, well characterized 21 base-pair consensus sequence, referred to as "NRSE" (for neuron-restrictive silencer element) that correlates well with the predicted binding site for the 8-fingered REST protein (Table 4.1). The levels of half-site binding indicate that in some contexts, REST may use only a portion of its fingers to recognize DNA, thereby significantly increasing the potential regulatory repertoire of this abundant transcriptional regulator [29]. Binding sites for a second polydactyl protein, the SCAN-KRAB protein ZNF263, have also recently been identified using ChIP-seq; the single 24 nucleotide consensus binding site predicted in these studies suggests that this protein uses most of its 9 zinc-fingers for DNA binding [30]. The binding site predicted for ZNF263 bears some similarity to the site that would be computationally predicted from the protein's amino acid sequence computationally, as well as some striking differences.

Additional insights have also been provided through earlier ChIP experiments coupled to microarrays ("ChIP-chip") for a small number of additional KZNF proteins, including the multifunctional protein, CTCF [31, 32]. However, despite this progress, a remarkably tiny fraction of this exceptionally large and versatile protein family has known regulatory functions, gene targets, or DNA binding sites. For that reason, most of what we know about their functional properties comes from "special case" stories focused on the products of single, possibly unrepresentative, KZNF genes. This picture should change dramatically, with the advent of "next-generation" sequencing technologies and their coupling to chromatin-binding assays, in the next few years.

4.3 Evolutionary History: The Rise and Fall of Lineage-Specific KZNF Families

The polydactyl KZNF TF family includes hundreds of members in many eukaryotic species (Fig. 4.4), many of which have highly been conserved over the course of evolution [33]. An example includes the mammalian *Krüppel*-like factor (KLF) family, a group of 17 three-fingered genes related distantly to the ancient TF, SP1 [34]. The *Drosophila* genome contains 4 related *Klf* genes that share many properties, including developmental expression and key roles in differentiation and development, with the mammalian proteins [33, 34]. The KLF family exemplifies the features typical

Table 4.1 Binding sites and functions for 10 well known polydactyl KZNF proteins

Gene name[a]	Number of C2H2 domains	Binding site sequence motif logo[b]	Method[b]	Known functions[c, d]
YY1 [1]	4	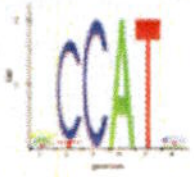	Compiled	■ Positive and negative control of cellular and viral genes via histone modification
REST [2]	9		ChIP-seq	■ Represses neuronal gene transcription in non-neuronal cells
RREB1 [3]	15		SELEX	■ May be involved in Ras/Raf-mediated cell differentiation
Klf4 [4]	3		ChIP-seq	■ Required for establishing the barrier function of the skin and for postnatal maturation and maintenance of the ocular surface
Evi1 [5]	10		SELEX	■ Plays a role in development, cell proliferation and differentiation
CTCF [6]	11		ChIP-seq	■ Transcriptional regulation by preventing interaction between promoter and nearby enhancers and silencers

Table 4.1 (continued)

Gene name[a]	Number of C2H2 domains	Binding site sequence motif logo[b]	Method[b]	Known functions[c, d]
Zfx [4]	13		ChIP-seq	■ Required for self-renewal of embryonic and adult hematopoietic stem cell
Zfp423 [7]	30		SELEX	■ Plays a central role in BMP signaling and olfactory neurogenesis
SP1 [8]	3		Compiled	■ Regulates the expression of genes involved in cell growth, apoptosis, differentiation and immune responses
ZNF143 [9]	7		Compiled	■ Activates the gene for selenocysteine tRNA

[a]Number in bracket refers to literature reporting binding sites [1–9, 29, 73–75, 78, 80–83].
[b]Taken from JASPAR database (http://jaspar.genereg.net/); In Method column, "compiled" refers to binding sites determined by multiple methods.
[c]Taken from the UniProt database (http://www.uniprot.org/).
[d]Adapted from the NCBI Entrez database (http://www.ncbi.nlm.nih.gov/Entrez/).

	Mouse-ear Cress *A. thaliana*	Budding Yeast *S. cerevisiae*	Roundworm Nematode *C. elegans*	Common Fruit Fly *D. melanogaster*	California Purple Sea Urchin *S. purpuratus*	Zebrafish *D. rerio*	Western Clawed Frog *X. tropicalis*	Domestic Chicken *G. gallus*	Mouse *M. musculus*	Human *H. sapiens*
SCAN	0	0	0	0	0	0	0	0	48 [4]	57 [4]
KRAB	0	0	0	0	1	0	21 [5]	33 [5]	502 [4]	381 [4]
ZAD	0	0	0	98 [3]	0	1 [3]	1 [3]	1 [3]	1 [3]	1 [3]
BTB/POZ	110	1	1 [3]	11 [3]	0	46 [5]	30 [5]	26 [5]	44 [5]	50 [5]
Poly-C2H2	105 [1]	53 [1]	139 [1]	291 [1]	377 [2]	405 [5]	347 [5]	224 [5]	583 [5]	712 [5]

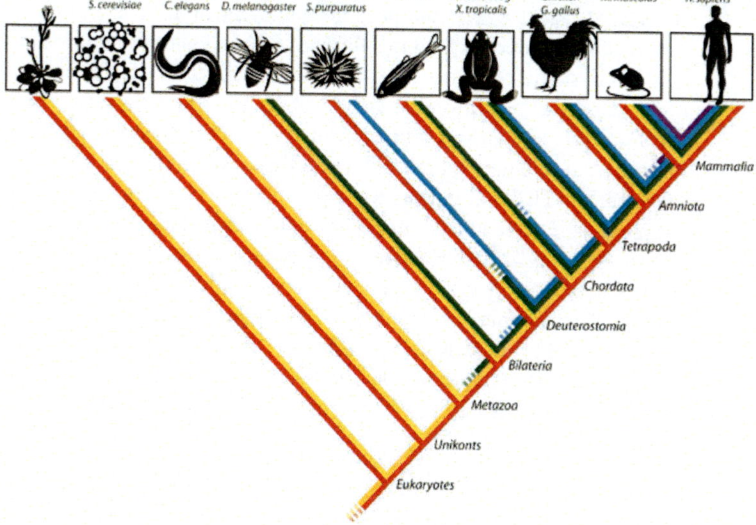

Fig. 4.4 Phylogenetic distribution of polydactyl KZNF protein families. The distribution and number of polydactyl KZNF proteins in different families is shown in this phylogeny of all eukaryotic model systems. The gain and loss of genes over evolution can be seen along the tree for all polydactyl C2H2 (*red*) BTB/POZ (*orange*), ZAD (*green*), KRAB (*blue*), and SCAN (*purple*) KZNF families. The numbers of genes in each family are shown in the accompanying table. Information was compiled from the PFAM Database [76], unless otherwise noted as coming from [1][79], [2][77], [3][39], [4][42] or [5][41]

of many KZNF family groups: most of these proteins contain short KZNF arrays, and have been well conserved in metazoan species.

However, most genomes contain subfamilies of KZNF genes with very different evolutionary histories and fates. Over the course of evolution, distinct KZNF families have emerged independently in different lineages, through exon shuffling events that bring DNA sequences encoding polydactyl KZNF arrays together with different types of protein-interaction or chromatin-modifying "effector" domains (see *Chapter 12* for a general overview of TF effector domains). New versions of KZNF proteins, coupling long polydactyl arrays with different types of activation, repression, or protein-interaction effectors, have arisen in different evolutionary lineages. Some of the genes encoding these novel constructs have subsequently expanded by repeated duplication events into large gene families; these in turn have either been integrated into key regulatory networks and conserved, or lost and replaced by other KZNF families in subsequent lineages.

4.3.1 An Ancient Family: BTB/POZ

One of the most ancient families of this type, in which arrays of KZNF fingers are attached to an N-terminal BTB/POZ motif, is represented in most eukaryotic species. As with many families, the numbers of BTB/POZ proteins has varied throughout evolution, changing through whole genome duplications, single-gene duplications, and gene loss (Fig. 4.4). The BTB/POZ domain (named BTB because of its presence in *Drosophila Broad Complex, tramtrack* and *bric a brac* genes, and POZ for "poxviruses and zinc finger") is found associated with several types of proteins, including but not limited to those containing KZNF array. The primary function of BTB/POZ appears to be protein dimerization, although the activities of the proteins in which this domain are found suggest a more specific functional role. Several BTB/POZ-KZNF proteins are found in *Drosophila,* where they play key roles in both local gene regulation and higher-order chromatin structure, often in the context of embryonic development [35]. Similar developmental functions have been attributed to BTB/POZ-KZNF proteins in humans and mice. Whereas the originally discovered BTB/POZ genes function mainly as transcriptional repressors, these proteins can operate as agents of chromosome decondensation and gene activation as well [36].

4.3.2 Lineage-Specific Inventions

In addition to this older family of genes, most metazoan genomes appear to carry surprisingly large numbers of lineage-specific KZNF genes. Typically, these genes represent novel constructs, in which exons encoding specific N-terminal effector domains are spliced to one or more exons encoding adjacent elements of a C-terminal KZNF array (Fig. 4.2). Most of these proteins also contain a region between the effector and KZNF motifs, usually referred to as a tether or spacer sequence. This typical structure is found in KZNF proteins of several different types, which are restricted to certain evolutionary lineages and have expanded by duplication into large TF families.

In *Drosophila*, 98 genes are found that encode KZNF attached to an N-terminal repressive motif called ZAD, whose function is as-yet poorly characterized [37]. Like BTB/POZ, the ZAD domain facilitates protein-protein interactions. A single ZAD-like gene, *ZNF276*, exists in vertebrate species, but an expanded ZAD family is found only in insects with the largest numbers found in higher homometabolous species (i.e. those that go through metamorphosis) (Fig. 4.4) [38]. Like the largest KZNF families in other species, ZAD-KZNF genes are found clustered together on insect chromosomes, reflecting the fact that the families arose through repeated rounds of tandem segmental duplications [39]. Although most ZAD-KZNF genes are of unknown function, most are expressed in the female germline and a few have been linked to developmental mutations in *Drosophila* [38]. The lineage-specific expansion of this class of KZNF proteins, phenotypes associated with certain family members, and their developmental expression make it likely that the ZAD-KZNF proteins play a role in species-specific developmental processes. A role for these genes in regulating developmental pathways could explain the dramatic expansion of the ZAD-ZNF family particularly in metamorphic species.

In vertebrate genomes, two other KZNF families have expanded into large gene families that are limited to certain evolutionary clades. In proteins of the SCAN-KZNF family, a C-terminal protein-interacting SCAN domain is attached through a tether sequence of varying length to N-terminal KZNF arrays [40]. SCAN domains are found in most vertebrates and are associated with a variety of other protein motifs, but the combination of SCAN with KZNF arrays has only been detected in mammals [40, 41]. After SCAN-KZNF genes arose, they expanded into a small family in most mammalian species, with a total of 57 protein-coding genes in the human genome (Fig. 4.4). Like the ZAD-KZNF genes in insects, SCAN-KZNF coding genes are frequently found in clusters, with related family members located adjacent to each other at specific chromosomal sites. The primary expansion of this family through segmental duplications must have occurred relatively early in mammalian evolution, since most SCAN-KZNF gene clusters, and the genes that are resident within them, are represented by orthologs in the different mammalian species. Nevertheless, a small number of lineage specific SCAN-KZNF gene duplicates have also been identified in comparisons between the gene sets of human, dog and mouse [40, 42]. A small number of mammalian SCAN-containing KZNF proteins also include a KRAB motif (see below), and SCAN- and KRAB-containing KZNF genes are sometimes found together in chromosomal clusters [41, 42]. These data indicate some intermingling of genes of these two types over the course of evolutionary history.

4.3.3 A Case Study: The KRAB-ZNF Family

A second major KZNF family has diverged rapidly and dramatically in gene copy number in different mammalian lineages. The KRAB-A, or *Krüppel*-associated box, type A domain is a 41-residue element that interacts with a ubiquitous co-factor, called KAP-1, to attract histone deacetylase complexes to specific DNA sites [43–47] (also see *Chapter 12*). A single gene, called *Meisetz* or *Prdm9,* was formed through association of an exon encoding a KRAB domain, together with sequences encoding a second effector, the SET domain, to an exon encoding tether sequence and polydactyl KZNF array, in early metazoan history [48]. A recognizable *Prdm9* ortholog can be found in echinoderms, protochordates, and vertebrate species. However both KZNF-motif number and sequence of the DNA-binding amino acids in the PRDM9 protein vary widely between species, exhibiting signs of strong positive selection [49]. In addition to its predicted role in transcriptional regulation, *Prdm9* has recently been shown to play a key role in marking hotspots for meiotic recombination in mammals [9, 10, 50].

Whereas *Prdm9* and its close relatives form a very small family in most vertebrates, a revised version of this protein type, containing one or more KRAB domains and a KZNF array but lacking the SET domain, has undergone dramatic expansion especially in mammalian lineages. Over 400 KRAB-KZNF genes exist in the human genome, and similar numbers are found in all mammalian genomes that have been examined [42]. By their sheer numbers, this single family of KZNF proteins dominates the mammalian transcription-factor landscape, comprising up to one-fourth of that total number of predicted human TF genes [51]. Most intriguingly, although all mammals have roughly equal numbers of proteins of this type, the number of 1:1 orthologous pairs is remarkably small. For example, although both human and mouse possess hundreds of KRAB-ZNF genes, only 112 genes represent convincing orthologs that are shared by these two species [42]. About one-third of human KRAB-ZNF genes are primate-specific, and a similar number of mouse genes can be found only in other rodents. For example, a cluster of mouse genes on chromosome 13 (chr13), including genes involved in regulating the sex-limited expression of target genes, contains many KRAB-ZNF coding sequences that are restricted to the *Mus* lineage [52]. Similarly, about 30 human genes of this type have arisen through segmental duplication since the divergence of old world monkeys, creating novel transcriptional regulators that exist only in higher primates [53].

The tendency toward tandem segmental duplication may help explain why KRAB-KZNF genes have been gained and lost so frequently over the course of vertebrate evolution. Tandem segmental duplications, like those found in the KZNF gene clusters, are known to be hotspots of copy number variation (CNV) both between and within species, driving duplications and deletions through illegitimate recombination events [54, 55]. If the duplication units include a full-length gene, each recombination event can give rise to versions of the chromosome with one less or one additional gene, respectively. Recent studies have confirmed that many protein-coding genes are copy-number variant in the human population, and genes located in segmental duplications rank among those most likely to lost or gained in certain human individuals. Not surprisingly, many KRAB-KZNF loci are found among recently generated segmental duplications [53] and among these copy-number-variant genes.

As these data show, the KRAB-KZNF gene family has evolved rapidly, and still is evolving, with novel genes created through duplication, and even some conserved genes displaying sequence changes that reflect the influence of positive selection. Recent studies show that as new duplicates arise, they can change rapidly in function through two different routes. First, the newly duplicated genes can change in expression pattern, diverging from the parental gene copy in tissue-specific sites and levels of gene expression [53]. Although KRAB-KZNF genes reside in closely packed gene clusters, neighboring genes do not often share similar expression patterns, even when the two genes are closely related [42, 53, 56–58]. These data suggest that (1) the genes are typically duplicated along with the regulatory elements needed to drive their tissue-specific expression patterns, and (2) that neighbors are probably shielded from the influence of enhancers or repressive elements controlling the surrounding genes. Whatever the mechanism, the ability to quickly adapt unique expression patterns after duplication has provided a rapid path to functional divergence for KRAB-KZNF genes.

The second route through which new KRAB-KZNF paralogs can diverge rapidly from parental genes is through sequence changes that affect the DNA binding properties of the encoded proteins. This divergence occurs through two different mechanisms. First, paralogous gene copies can acquire non-synonymous mutations in the DNA-binding amino acids in the finger motifs; for many KRAB-KZNF gene paralogs, the acquisition of novel DNA-binding sequences has occurred under the influence of positive selection [41, 53, 56, 59, 60]. An alternative path to paralog divergence involves a mechanism that is unique to proteins like the polydactyl KZNFs, which contain multiple, similar motifs that are encoded in a single exon (Fig. 4.3). The sequences encoding these protein motifs are essentially tandem repeat sequences, and are prone to the same types of duplications and deletions observed for microsatellites and other simple genomic repeats. As a result, paralogous KRAB-KZNF proteins often differ from each other in KZNF motif number, often due to the deletion or duplication of one or more zinc-fingers from the middle of the KZNF array [59, 60]. This process can occur rapidly, giving rise to proteins that are otherwise nearly identical, but contain different numbers and arrangements of tandem KZNF motifs [53]. Because of the intimate relationship that exists between an ordered array of amino acids in the KZNF alpha-helical region and the nucleotide sequence at target sites, deletion or duplication of fingers from within an array is expected to have significant impact on DNA binding, target-site preference, and stability of KZNF association with DNA.

4.3.4 A General Path to Rapid Divergence for Polydactyl KZNF Genes

Although these paths to paralog divergence are best described for mammalian KRAB-ZNF genes, the pattern of divergence also follows for SCAN-KZNF subfamily [42] and our recent studies indicate a similar pattern for primate BTB/POZ-KZNF proteins [53]. There is no reason to believe that similar patterns of divergence would not have defined the growth of KZNF gene families of other types and in other species as well. In fact, a recent survey of KZNF genes in multiple species detected lineage-specific family members in virtually every genome analyzed, and showed that positive selection acting to diversify DNA-binding capabilities of KZNF proteins of many different types [41]. The key feature that drives duplication and deletion of KZNF motifs in KRAB-KZNF genes is the occurrence of multiple, tandemly arranged finger-encoding repeats in a single exon; this kind of structure is present in a large fraction of KZNF genes in every species (Fig. 4.3). Whether finger deletion and duplication are driven by illegitimate recombination between the adjacent repeats, or a mechanism such as replication slippage, remains to be determined. However, the high frequency of these events and the relatively rapid pace in which they have occurred in divergence of recent primate duplicates argue for the latter mechanism, which is known to drive a similar pace of genomic divergence at microsatellites and other simple sequence repeats [61].

The ability to create new DNA binding capabilities through binding-sequence divergence or zinc finger number and arrangement is likely to underlie much of this gene family's remarkable growth and success. However, despite similarities in structure, with N-terminal effectors and KZNF arrays encoded intact on separate exons, the different families of polydactyl KZNF genes display very different evolutionary histories. Why have the older BTB/POZ-KZNF genes and the vertebrate-specific SCAN-KZNF families not exploded in numbers, as the KRAB-KZNFs have done? What drove the expansion of ZAD-KZNF genes in insects, and the expansion of a less-characterized family, the FAX/FAD-KZNFs [62] in amphibian genomes?

In considering the functions of the major mammalian KZNF effectors, we may find a clue to this mystery. Whereas BTB and SCAN appear to be concerned primarily with protein homo- and hetero-dimerization, KRAB is thought to play a very different role. Whereas future studies of KRAB domain function may still hold some surprises, it is thought primarily to interact directly with a single, abundant, and ubiquitous co-factor, KAP-1 [63, 64]. Because KAP-1 is so abundant, and serves as a "universal" KRAB co-factor, new KRAB proteins can arise with little effect on other interaction partners. KRAB does not mediate dimer formation, and KRAB-KZNF proteins appear to bind to target sites without the need for partners to stabilize their interaction with DNA. The long polydactyl KZNF arrays that are found in most mammalian proteins of this type probably underlie the independence of proteins of this type. Human KRAB-KZNF proteins contain an average of 12 KZNF motifs; an array of this length could theoretically specific a binding site of 36 bp, an extraordinary length compared to the binding sites of most known TFs. In reality, most binding sites that have been determined for polydactyl KZNFs range from 6 to 27 bp; some examples of well established binding sites are shown as "motif logos" (illustrations that represent the probability of finding a particular nucleotide at a position within the binding site) in Table 4.1. For proteins with binding sites on the longer end of this range, the binding between DNA sequence and the KZNF protein, wound with precision around the double-stranded DNA, would be predicted to be unusually specific and stable.

In contrast to KRAB-KZNFs, the average SCAN-KZNF and BTB/POZ-KZNF proteins contain a smaller number of zinc fingers [42], consistent with the idea that these proteins need to dimerize with other, similar proteins for secure binding to DNA. The potential combinatorial action of these dimerizing proteins provides a way to achieve functional diversity far beyond that implied by the numbers of individual genes. However, their predicted dependence on other proteins for activity may also constrain the ability of new genes to evolve, and established genes to be lost in these gene families. These concepts may help explain why genes encoding BTB/POZ and SCAN-containing KZNF genes have been more restrained than their ZAD-KZNF and KRAB-KZNF cousins in their tendency to gain and lose members over evolutionary time. Because they cannot diverge without affecting the functions of interacting proteins, TFs that require partners for activity tend to be more conserved, and more likely to be locked in to larger regulatory pathways, than independently acting proteins might be.

These basic tenets of protein evolution allow some predictions for the functions of effectors for non-mammalian KZNF effectors, like the insect effector, ZAD. Although the exact functions of the ZAD are not yet know, the prolific expansion of ZAD-KZNF genes in insect genomes, and the differences in ZAD-KZNF repertoires observed in comparisons of different insect genomes [38, 65], suggest that, like KRAB, this effector might function in concert with a ubiquitous co-factor; in analogy to KAP-1, this co-factor might be predicted to correspond to a protein or complex that interfaces with the chromatin remodeling machinery.

4.4 Challenges and Future Directions

Although the KRAB-KZNF family is by far the most dynamic group in mammalian genomes, the larger KZNF family has clearly played a significant role in the evolution of all eukaryotic clades. The versatile building block provided by the ancient C2H2 motif, its ability to assemble into long arrays for stable DNA binding, and the sheer diversity in DNA recognition capabilities that can be achieved by their combinatorial action, have made them a mainstay of TF repertoires and a dominant component of all eukaryotic genomes.

Despite their dominance, and the molecular "recognition code" that is believed to underlie their DNA binding capabilities, the functions of only a tiny fraction of KZNF proteins in any genome is known, and indeed it is not known whether predicted sequence specificities are generally correct. This lack of functional knowledge is especially acute for the polydactyl KZNF genes, due in part to their lack of interspecies conservation and their duplicative histories, which ensure some degree of functional redundancy. In vitro studies of purified polydactyl KZNF proteins are hampered by their low solubility, due to the high cysteine content of the proteins; in vivo studies are complicated by the high degree of similarity between paralogous proteins. And because of the extreme evolutionary diversity of KRAB-KZNFs in vertebrates and similar lineage-specific families, repertoires of such proteins have been fully counted only a small number of completely sequenced genomes [41].

However, the emergence of new technologies is beginning to shine new light on this shadowy component of the metazoan regulatory machinery. Microarrays bearing double-stranded oligonucleotides are currently being used to map binding-site preferences for a large number of GST-tagged TF proteins, including some KZNFs [66]; this method offers a significant advantage in terms of effort and time required to map binding sites compared to previous in vitro methods. However, polydactyl KZNFs have presented unique challenges to methods such as this, which depend on availability of soluble tagged proteins.

Some progress has been made through strategies such as tagging short peptides containing overlapping subsets of fingers from a longer KZNF array (T.R. Hughes, personal communication). Binding sites for other proteins have been successfully determined using established methods, such as "Systematic Evolution of Ligands by Exponential Enrichment" (or SELEX, [67–69] (Table 4.1)), and more recently through the application of bacterial one-hybrid selection systems [70]. However, ultimately, an understanding of the binding properties of long polydactyl KZNF proteins, of the prevalence of finger "multitasking", and of the functional consequences of their unique patterns of evolutionary divergence, will require methods that fully report their binding-site occupancy in living cells. Because of paralog sequence similarity and other factors, mapping binding sites of KRAB-KZNF proteins and the member of other, similar lineage-specific protein families presents a special challenge. However, new strategies including the use of "designer" KZNF recombinases [71, 72] to facilitate in vivo TF tagging, in combination with high-throughput sequencing, hold significant promise to unlock the long-standing mysteries regarding the functions of these abundant eukaryotic TFs. The true impact of the KZNF family's dynamic evolutionary history on speciation, interspecies divergence, and individual differences in gene regulation eventually will only be deciphered when their binding sites, regulatory activities, and interactions with other chromatin proteins are known.

References

1. Miller J, McLachlan AD, Klug A (1985) Repetitive zinc-binding domains in the protein transcription factor IIIA from *Xenopus* oocytes. EMBO J 4:1609–1614
2. Ollo R, Maniatis T (1987) *Drosophila* Krüppel gene product produced in a baculovirus expression system is a nuclear phosphoprotein that binds to DNA. Proc Natl Acad Sci USA 84:5700–5704
3. Bouhouche N, Syvanen M, Kado CI (2000) The origin of prokaryotic C2H2 zinc finger regulators. Trends Microbiol 8:77–81
4. Frankel AD, Berg JM, Pabo CO (1987) Metal-dependent folding of a single zinc finger from transcription factor IIIA. Proc Natl Acad Sci USA 84:4841–4845
5. Lee MS, Gippert GP, Soman KV, Case DA, Wright PE (1989) Three-dimensional solution structure of a single zinc finger DNA-binding domain. Science 245:635–637
6. Pavletich NP, Pabo CO (1991) Zinc finger-DNA recognition: crystal structure of a Zif268-DNA complex at 2.1 A. Science 252:809–817
7. Arya GH, Lodico MJ, Ahmad OI, Amin R, Tomkiel JE (2006) Molecular characterization of teflon, a gene required for meiotic autosome segregation in male *Drosophila melanogaster*. Genetics 174:125–134
8. Baudat F, Buard J, Grey C, Fledel-Alon A, Ober C, Przeworski M, Coop G, de Massy B (2010) PRDM9 is a major determinant of meiotic recombination hotspots in humans and mice. Science 327:836–840
9. Myers S, Bowden R, Tumian A, Bontrop RE, Freeman C, MacFie TS, McVean G, Donnelly P (2010) Drive against hotspot motifs in primates implicates the PRDM9 gene in meiotic recombination. Science 327:876–879
10. Parvanov ED, Petkov PM, Paigen K (2010) Prdm9 controls activation of mammalian recombination hotspots. Science 327:835
11. Phillips CM, Dernburg AF (2006) A family of zinc-finger proteins is required for chromosome-specific pairing and synapsis during meiosis in *C. elegans*. Dev Cell 11:817–829
12. Dickson J, Gowher H, Strogantsev R, Gaszner M, Hair A, Felsenfeld G, West AG (2010) VEZF1 elements mediate protection from DNA methylation. PLoS Genet 6:e1000804
13. Li X, Ito M, Zhou F, Youngson N, Zuo X, Leder P, Ferguson-Smith AC (2008) A maternal-zygotic effect gene, Zfp57, maintains both maternal and paternal imprints. Dev Cell 15: 547–557

14. Englbrecht CC, Schoof H, Bohm S (2004) Conservation, diversification and expansion of C2H2 zinc finger proteins in the *Arabidopsis thaliana* genome. BMC Genomics 5:39
15. Choo Y, Klug A (1994) Toward a code for the interactions of zinc fingers with DNA: selection of randomized fingers displayed on phage. Proc Natl Acad Sci USA 91:11163–11167
16. Wuttke DS, Foster MP, Case DA, Gottesfeld JM, Wright PE (1997) Solution structure of the first three zinc fingers of TFIIIA bound to the cognate DNA sequence: determinants of affinity and sequence specificity. J Mol Biol 273:183–206
17. Elrod-Erickson M, Pabo CO (1999) Binding studies with mutants of Zif268. Contribution of individual side chains to binding affinity and specificity in the Zif268 zinc finger-DNA complex. J Biol Chem 274:19281–19285
18. Elrod-Erickson M, Rould MA, Nekludova L, Pabo CO (1996) Zif268 protein-DNA complex refined at 1.6 A: a model system for understanding zinc finger-DNA interactions. Structure 4:1171–1180
19. Kaplan T, Friedman N, Margalit H (2005) Ab initio prediction of transcription factor targets using structural knowledge. PLoS Comput Biol 1:e1
20. Liu J, Stormo GD (2008) Context-dependent DNA recognition code for C2H2 zinc-finger transcription factors. Bioinformatics 24:1850–1857
21. Persikov AV, Osada R, Singh M (2009) Predicting DNA recognition by Cys2His2 zinc finger proteins. Bioinformatics 25:22–29
22. Klug A (2010) The discovery of zinc fingers and their applications in gene regulation and genome manipulation. Annu Rev Biochem 79:213–231
23. Rizkallah R, Hurt MM (2009) Regulation of the transcription factor YY1 in mitosis through phosphorylation of its DNA-binding domain. Mol Biol Cell 20:4766–4776
24. Bird AJ, Zhao H, Luo H, Jensen LT, Srinivasan C, Evans-Galea M, Winge DR, Eide DJ (2000) A dual role for zinc fingers in both DNA binding and zinc sensing by the Zap1 transcriptional activator. EMBO J 19:3704–3713
25. Hoffmann A, Barz T, Spengler D (2006) Multitasking C2H2 zinc fingers link Zac DNA binding to coordinated regulation of p300-histone acetyltransferase activity. Mol Cell Biol 26:5544–5557
26. Hoffmann A, Ciani E, Boeckardt J, Holsboer F, Journot L, Spengler D (2003) Transcriptional activities of the zinc finger protein Zac are differentially controlled by DNA binding. Mol Cell Biol 23:988–1003
27. Hata A, Seoane J, Lagna G, Montalvo E, Hemmati-Brivanlou A, Massague J (2000) OAZ uses distinct DNA- and protein-binding zinc fingers in separate BMP-Smad and Olf signaling pathways. Cell 100:229–240
28. Brayer KJ, Segal DJ (2008) Keep your fingers off my DNA: protein-protein interactions mediated by C2H2 zinc finger domains. Cell Biochem Biophys 50:111–131
29. Johnson DS, Mortazavi A, Myers RM, Wold B (2007) Genome-wide mapping of in vivo protein–DNA interactions. Science 316:1497–1502
30. Frietze S, Lan X, Jin VX, Farnham PJ (2010) Genomic targets of the KRAB and SCAN domain-containing zinc finger protein 263. J Biol Chem 285:1393–1403
31. Kim TH, Abdullaev ZK, Smith AD, Ching KA, Loukinov DI, Green RD, Zhang MQ, Lobanenkov VV, Ren B (2007) Analysis of the vertebrate insulator protein CTCF-binding sites in the human genome. Cell 128:1231–1245
32. Smith ST, Wickramasinghe P, Olson A, et al. (2009) Genome wide ChIP-chip analyses reveal important roles for CTCF in *Drosophila* genome organization. Dev Biol 328:518–528
33. Knight RD, Shimeld SM (2001) Identification of conserved C2H2 zinc-finger gene families in the Bilateria. Genome Biol 2:RESEARCH0016
34. Pearson R, Fleetwood J, Eaton S, Crossley M, Bao S (2008) Kruppel-like transcription factors: a functional family. Int J Biochem Cell Biol 40:1996–2001
35. Albagli O, Dhordain P, Deweindt C, Lecocq G, Leprince D (1995) The BTB/POZ domain: a new protein-protein interaction motif common to DNA- and actin-binding proteins. Cell Growth Differ 6:1193–1198

36. Kelly KF, Daniel JM (2006) POZ for effect – POZ-ZF transcription factors in cancer and development. Trends Cell Biol 16:578–587
37. Jauch R, Bourenkov GP, Chung HR, Urlaub H, Reidt U, Jäckle H, Wahl MC (2003) The zinc finger-associated domain of the *Drosophila* transcription factor grauzone is a novel zinc-coordinating protein-protein interaction module. Structure 11:1393–1402
38. Chung HR, Löhr U, Jäckle H (2007) Lineage-specific expansion of the zinc finger associated domain ZAD. Mol Biol Evol 24:1934–1943
39. Chung HR, Schafer U, Jäckle H, Böhm S (2002) Genomic expansion and clustering of ZAD-containing C2H2 zinc-finger genes in *Drosophila*. EMBO Rep 3:1158–1162
40. Edelstein LC, Collins T (2005) The SCAN domain family of zinc finger transcription factors. Gene 359:1–17
41. Emerson RO, Thomas JH (2009) Adaptive evolution in zinc finger transcription factors. PLoS Genet 5:e1000325
42. Huntley S, Baggott DM, Hamilton AT, Tran-Gyamfi M, Yang S, Kim J, Gordon L, Branscomb E, Stubbs L (2006) A comprehensive catalog of human KRAB-associated zinc finger genes: insights into the evolutionary history of a large family of transcriptional repressors. Genome Res 16:669–677
43. Abrink M, Ortiz JA, Mark C, Sanchez C, Looman C, Hellman L, Chambon P, Losson R (2001) Conserved interaction between distinct Krüppel-associated box domains and the transcriptional intermediary factor 1 beta. Proc Natl Acad Sci USA 98:1422–1426
44. Friedman JR, Fredericks WJ, Jensen DE, Speicher DW, Huang XP, Neilson EG, Rauscher FJ 3rd. (1996) KAP-1, a novel corepressor for the highly conserved KRAB repression domain. Genes Dev 10:2067–2078
45. Lorenz P, Koczan D, Thiesen HJ (2001) Transcriptional repression mediated by the KRAB domain of the human C2H2 zinc finger protein Kox1/ZNF10 does not require histone deacetylation. Biol Chem 382:637–644
46. Margolin JF, Friedman JR, Meyer WK, Vissing H, Thiesen HJ, Rauscher FJ 3rd. (1994) Kruppel-associated boxes are potent transcriptional repression domains. Proc Natl Acad Sci USA 91:4509–4513
47. Witzgall R, O'Leary E, Leaf A, Onaldi D, Bonventre JV (1994) The Krüppel-associated box-A (KRAB-A) domain of zinc finger proteins mediates transcriptional repression. Proc Natl Acad Sci USA 91:4514–4518
48. Birtle Z, Ponting CP (2006) Meisetz and the birth of the KRAB motif. Bioinformatics 22:2841–2845
49. Thomas JH, Emerson RO, Shendure J (2009) Extraordinary molecular evolution in the PRDM9 fertility gene. PLoS One 4:e8505
50. Baudat F, Buard J, Grey C, de Massy B (2010) [Prdm9, a key control of mammalian recombination hotspots]. Med Sci (Paris) 26:468–470
51. Vaquerizas JM, Kummerfeld SK, Teichmann SA, Luscombe NM (2009) A census of human transcription factors: function, expression and evolution. Nat Rev Genet 10:252–263
52. Krebs CJ, Larkins LK, Khan SM, Robins DM (2005) Expansion and diversification of KRAB zinc-finger genes within a cluster including Regulator of sex-limitation 1 and 2. Genomics 85:752–761
53. Nowick K, Hamilton AT, Zhang H, Stubbs L (2010) Rapid sequence and expression divergence suggests selection for novel function in primate-specific KRAB-ZNF genes. Mol Biol Evol 27:2606–2617
54. Cooper GM, Nickerson DA, Eichler EE (2007) Mutational and selective effects on copy-number variants in the human genome. Nat Genet 39:S22–29
55. Stankiewicz P, Lupski JR (2010) Structural variation in the human genome and its role in disease. Annu Rev Med 61:437–455
56. Hamilton AT, Huntley S, Kim J, Branscomb E, Stubbs L (2003) Lineage-specific expansion of KRAB zinc-finger transcription factor genes: implications for the evolution of vertebrate regulatory networks. Cold Spring Harb Symp Quant Biol 68:131–140

57. Nowick K, Stubbs L (2010) Lineage-specific transcription factors and the evolution of gene regulatory networks. Brief Funct Genomics 9:65–78
58. Shannon M, Ashworth LK, Mucenski ML, Lamerdin JE, Branscomb E, Stubbs L (1996) Comparative analysis of a conserved zinc finger gene cluster on human chromosome 19q and mouse chromosome 7. Genomics 33:112–120
59. Hamilton AT, Huntley S, Tran-Gyamfi M, Baggott DM, Gordon L, Stubbs L (2006) Evolutionary expansion and divergence in the ZNF91 subfamily of primate-specific zinc finger genes. Genome Res 16:584–594
60. Shannon M, Hamilton AT, Gordon L, Branscomb E, Stubbs L (2003) Differential expansion of zinc-finger transcription factor loci in homologous human and mouse gene clusters. Genome Res 13:1097–1110
61. Hardwick RJ, Tretyakov MV, Dubrova YE (2009) Age-related accumulation of mutations supports a replication-dependent mechanism of spontaneous mutation at tandem repeat DNA Loci in mice. Mol Biol Evol 26:2647–2654
62. Nietfeld W, Conrad S, van Wijk I, Giltay R, Bouwmeester T, Knochel W, Pieler T (1993) Evidence for a clustered genomic organization of FAX-zinc finger protein encoding transcription units in *Xenopus laevis*. J Mol Biol 230:400–412
63. Kim SS, Chen YM, O'Leary E, Witzgall R, Vidal M, Bonventre JV (1996) A novel member of the RING finger family, KRIP-1, associates with the KRAB-A transcriptional repressor domain of zinc finger proteins. Proc Natl Acad Sci USA 93:15299–15304
64. Moosmann P, Georgiev O, Le Douarin B, Bourquin JP, Schaffner W (1996) Transcriptional repression by RING finger protein TIF1 beta that interacts with the KRAB repressor domain of KOX1. Nucleic Acids Res 24:4859–4867
65. Duan J, Xia Q, Cheng D, Zha X, Zhao P, Xiang Z (2008) Species-specific expansion of C2H2 zinc-finger genes and their expression profiles in silkworm, Bombyx mori. Insect Biochem Mol Biol 38:1121–1129
66. Badis G, Berger MF, Philippakis AA, et al. (2009) Diversity and complexity in DNA recognition by transcription factors. Science 324:1720–1723
67. Ellington AD, Szostak JW (1990) In vitro selection of RNA molecules that bind specific ligands. Nature 346:818–822
68. Oliphant AR, Brandl CJ, Struhl K (1989) Defining the sequence specificity of DNA-binding proteins by selecting binding sites from random-sequence oligonucleotides: analysis of yeast GCN4 protein. Mol Cell Biol 9:2944–2949
69. Tuerk C, Gold L (1990) Systematic evolution of ligands by exponential enrichment: RNA ligands to bacteriophage T4 DNA polymerase. Science 249:505–510
70. Noyes MB, Meng X, Wakabayashi A, Sinha S, Brodsky MH, Wolfe SA (2008) A systematic characterization of factors that regulate Drosophila segmentation via a bacterial one-hybrid system. Nucleic Acids Res 36:2547–2560
71. Dekelver RC, Choi VM, Moehle EA, et al. (2010) Functional genomics, proteomics, and regulatory DNA analysis in isogenic settings using zinc finger nuclease-driven transgenesis into a safe harbor locus in the human genome. Genome Res 20:1133–1142
72. Kim HJ, Lee HJ, Kim H, Cho SW, Kim JS (2009) Targeted genome editing in human cells with zinc finger nucleases constructed via modular assembly. Genome Res 19:1279–1288
73. Barski A, Cuddapah S, Cui K, Roh TY, Schones DE, Wang Z, Wei G, Chepelev I, Zhao K (2007) High-resolution profiling of histone methylations in the human genome. Cell 129: 823–837
74. Chen X, Xu H, Yuan P, Fang F, Huss M, Vega VB, Wong E, Orlov YL, Zhang W, Jiang J, Loh YH, Yeo HC, Yeo ZX, Narang V, Govindarajan KR, Leong B, Shahab A, Ruan Y, Bourque G, Sung WK, Clarke ND, Wei CL, Ng HH (2008) Integration of external signaling pathways with the core transcriptional network in embryonic stem cells. Cell 133:1106–1117
75. Delwel R, Funabiki T, Kreider BL, Morishita K, Ihle JN (1993) Four of the seven zinc fingers of the *Evi-1* myeloid-transforming gene are required for sequence-specific binding to GA(C/T)AAGA(T/C)AAGATAA. Mol Cell Biol 13:4291–4300

76. Finn RD, Mistry J, Tate J, Coggill P, Heger A, Pollington JE, Gavin OL, Gunesekaran P, Ceric G, Forslund K, Holm L, Sonnhammer EL, Eddy SR, Bateman A (2010) The PFAM protein families database. Nucl Acids Res 38 (Database Issue):D211–D222
77. Materna SC, Howard-Ashby M, Gray RF, Davidson EH (2006) The C2H2 zinc finger genes of Strongylocentrotus purpuratus and their expression in embryonic development. Devel Biol 300:108–120
78. Portales-Casamar E, Kirov S, Lim J, Lithwick S, Swanson MI, Ticoll A, Snoddy J, Wasserman WW (2007) PAZAR: a framework for collection and dissemination of cis-regulatory sequence annotation. Genome Biol 8:R207
79. Riechmann JL, Heard J, Martin G, Reuber L, Jiang CZ, Keddie J, Adam L, Pineda O, Ratcliffe OJ, Samaha RR, Creelman R, Pilgrim M, Broun P, Zhang JZ, Ghandehari D, Sherman BK, Yu GL (2000) *Arabidopsis* transcription factors: genome-wide comparative analysis among eukaryotes. Science 290:2105–2110
80. Schaub M, Myslinski E, Schuster C, Krol A, Carbon P (1997) Staf, a promiscuous activator for enhanced transcription by RNA polymerases II and III. EMBO J 16:173–181
81. Shrivastava A, Calame K (1994) An analysis of genes regulated by the multi-functional transcriptional regulator Yin Yang-1. Nucl Acids Res 22:5151–5155
82. Thiagalingam A, De Bustros A, Borges M, Jasti R, Compton D, Diamond L, Mabry M, Ball DW, Baylin SB, Nelkin BD (1996) RREB-1, a novel zinc finger protein, is involved in the differentiation response to Ras in human medullary thyroid carcinomas. Mol Cell Biol 16:5335–5345
83. Tsai RY, Reed RR (1998) Identification of DNA recognition sequences and protein interaction domains of the multiple-Zn-finger protein Roaz. Mol Cell Biol 18:6447–6456

Chapter 5
Homeodomain Subtypes and Functional Diversity

Thomas R. Bürglin

Abstract The homeodomain is a protein domain of about 60 amino acids that is encoded by homeobox genes. The homeodomain is a DNA binding domain, and hence homeodomain proteins are essentially transcription factors (TFs). They have been shown to play major roles in many developmental processes of animals, as well as fungi and plants. A primary function of homeodomain proteins is to regulate the expression of other genes in development and differentiation. Thousands of homeobox genes have been identified, and they can be grouped into many different classes. Often other conserved protein domains are found linked to a homeodomain. Several particular types of homeobox genes are organized into chromosomal clusters. The best-known cluster, the HOX cluster, is found in all bilaterian animals. Tetrapods contain four HOX clusters that arose through duplication in early vertebrate evolution. The genes in these clusters are called Hox genes. Lower chordates, insects and nematodes tend to have only one HOX cluster. Of particular interest is that many of the HOX cluster genes function in the process of pattern formation along the anterior-posterior body axis. Many other types of homeodomain proteins play roles in the determination of cell fates and cell differentiation. Homeobox genes thus perform key roles for all aspects of the development of an organism.

5.1 The Homeodomain

5.1.1 The Homeodomain Sequence

Since their discovery in 1984, homeobox genes and the homeodomain proteins they encode have turned out to play important roles in the developmental processes of all multicellular eukaryotes. While certainly not the only developmental control genes, they have been shown to play crucial roles from the earliest steps in

T.R. Bürglin (✉)
Department of Biosciences and Nutrition, and Center for Biosciences, Karolinska Institutet, Hälsovägen 7, Novum, SE 141 83 Huddinge, Sweden
e-mail: Thomas.burglin@ki.se

T.R. Hughes (ed.), *A Handbook of Transcription Factors*, Subcellular Biochemistry 52, DOI 10.1007/978-90-481-9069-0_5, © Springer Science+Business Media B.V. 2011 embryogenesis – such as setting up an anterior-posterior gradient in the egg of the fruit fly *Drosophila melanogaster* – to the very latest steps in cell differentiation – such as the differentiation of neurons in the nematode *Caenorhabditis elegans*.

The homeobox was originally described as a conserved DNA motif of about 180 base pairs and encodes the about 60 amino acid long homeodomain. The first genes found to encode homeodomain proteins were homeotic genes from *Drosophila*, from which the name "homeo" box was derived. Soon thereafter, homologues from vertebrates were discovered and the similarity to yeast mating type factors was described [1–7]. It should be noted that not all homeobox genes are homeotic genes, and not all homeotic genes are homeobox genes.

As more and more homeobox-containing genes were isolated the range of diversity increased: additional motifs upstream and/or downstream of the homeodomain were discovered, and variants of the homeodomain were found that had insertions in the homeodomain. Now, with full genome sequences available, we know the full diversity of homeobox genes in particular organisms and have insights into the variations that can be found in a homeodomain. Nevertheless, the structural features that define a homeodomain are conserved. Figure 5.1a shows a consensus sequence that is based on a compilation of homeodomain sequences, and Fig. 5.1b shows the variability that is observed at the different positions. Two positions are almost totally invariant, the tryptophan (W) residue at position 48 and the asparagine (N) residue at position 51. Occasionally, the tryptophan (W) can be substituted by a phenylalanine (F). Three other positions predominantly contain one amino acid, but occasionally permit substitutions: position 16 is normally a leucine (L) or another small hydrophobic residue; positions 20 and 49 are normally phenylalanine (F) residues. Position 53 is almost invariably an arginine (R), though lysine (K) is possible. The 60 positions, especially the more conserved ones, define the typical homeodomain. Residues in the DNA-binding region of helix 3 are especially conserved and constitute a "trademark" for the homeodomain. The sequence compilation shown here is biased towards Antennapedia-like sequences. Nevertheless, even plant homeodomain sequences show a similar pattern of conserved residues, although amino acid frequencies at particular positions may vary from the one presented here.

5.1.2 Structure of the Homeodomain

The underlying reason for the conservation of particular amino acids is the structure of the homeodomain. Several structures have been determined using either NMR or X-Ray analysis, for example, *Drosophila* Antennapedia (Antp) [8, 9], fushi tarazu (ftz) [10], engrailed (en) [11], or mammalian PBX1 [12], Oct-1 [13, 14], HNF1alpha [15, 16], or yeast MATalpha2 [17]. The core of the homeodomain consists of three alpha-helices (Fig. 5.2). The NMR studies of the Antennapedia homeodomain identified a kink in the third helix, at position 52/53, such that this helix is considered to consist of two separate helices, helix 3 and helix 4. The three alpha-helices are represented as shaded cylinders above the consensus sequence in Fig. 5.1a.

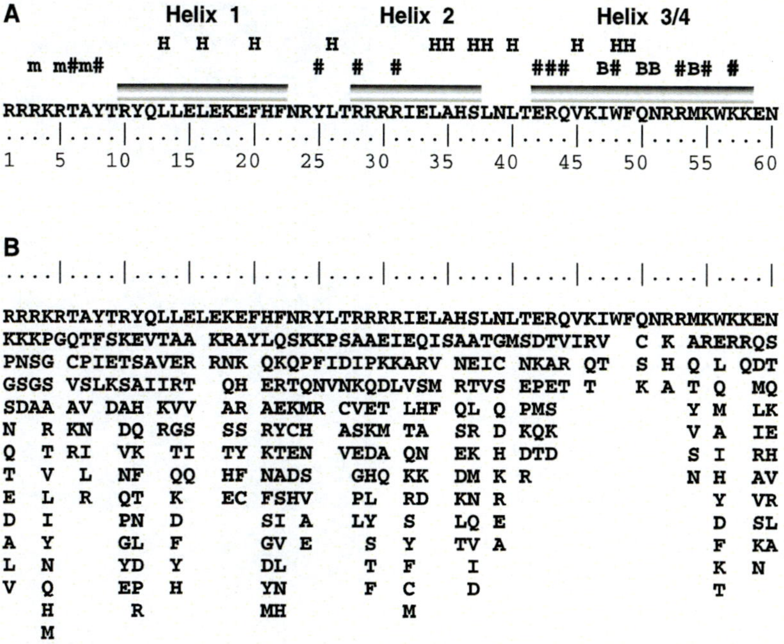

Fig. 5.1 (**a**) A homeodomain consensus sequence based on 346 homeodomain sequences. The three alpha-helices (a composite derived from the structures of the Antp, engrailed and MATalpha2 homeodomains) are shown schematically as cylinders above the consensus. Special symbols mark amino acids positions that play a role for intramolecular or protein-DNA contacts of the homeodomain; they are shown above the consensus and have the following meanings: amino acids designated "H" contribute to the hydrophobic core that is responsible for the tertiary structure of the homeodomain; residues designated "B" contact bases in the major groove and are responsible for sequence specific DNA contacts; residues designated "m" make contacts in the minor groove; the pound sign (#) indicates residues that contact the sugar-phosphate backbone of the DNA. (**b**) Amino acids encountered at a given position in the homeodomain. For each position the amino acid most frequently encountered is listed at the *top*, while other amino acids are listed beneath in decreasing order of their frequency of occurrence. Amino acids occurring fewer than five times are not shown. Note: the 346 sequences selected for this compilation are biased towards animal, in particular Antennapedia-like homeodomain sequences

The various symbols (H, B, #, m) above the helices in Fig. 5.1a give a summary of various contacts made by the amino acids of the homeodomain as deduced from structural and genetic data. Variations occur depending on the type of homeodomain. A hydrophobic core holds the three helices together. The residues marked by "H" take part in the formation of this hydrophobic core. Helix 2 and helix 3 are connected by a tight turn called a helix-turn-helix motif that is also found in prokaryotic gene regulatory proteins [2]. Helix 1 is connected to helix 2 by a loop. Helix 1 lies approximately parallel to helix 2, and crosses also over helix 3. At the crossover area, the highly conserved residues in helix 1 interact with the highly conserved residues tryptophan and phenylalanine of helix 3.

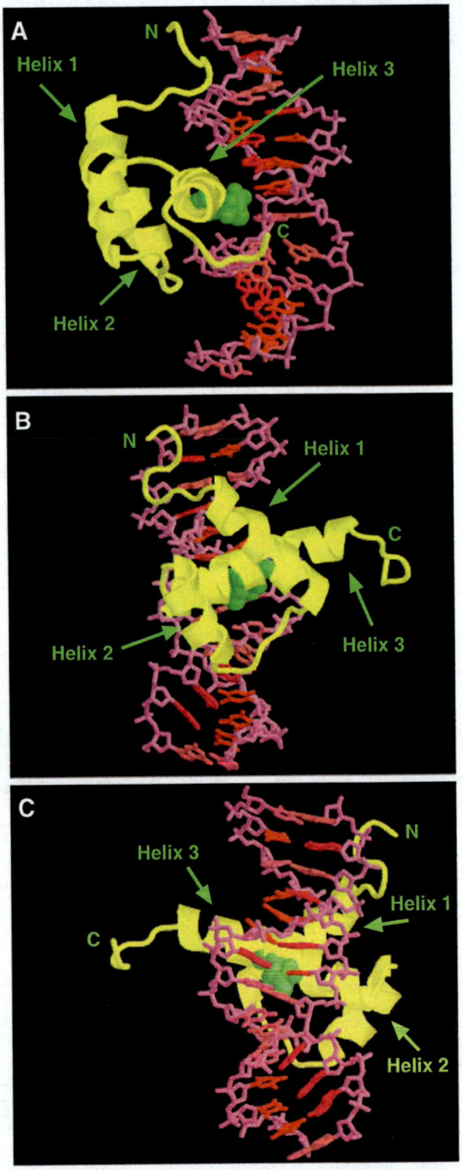

Fig. 5.2 Three views of the NMR structure of the Antennapedia homeodomain [9], PDB accession number: 1AHD. A, B, and C display three different views of the homeodomain – DNA structure that are rotated around the helical axis of the DNA. A. View along helix 3 in the major grove.
B. Homeodomain in front of the DNA.
C. Homeodomain behind the DNA. The DNA backbone is shown in *purple*, with the bases in four different shades of *red*. The homeodomain is shown in *yellow* as a cartoon, the three spirals representing the alpha helices. The alpha helices and the N- and C-termini are indicated. The glutamine residue at position 50 of the homeodomain is shown as space-filling model in *green*

Quite a number of homeobox genes have now been found whose homedomain is different in length from the typical 60 amino acids, and often they are referred to as *atypical*. For example, in TALE homeobox genes (see below) three extra amino acids are found in the loop between helix 1 and helix 2. Extra amino acids have also been found in several proteins between helix 2 and helix 3, for example, in the liver TF LFB1(HNF1alpha), in the prospero protein, in fly defective proventriculus (dve), or in plant HD-ZIP III proteins (see below). "Atypical" by itself is not a useful descriptor for the classification of homeobox genes. In fact, insertions and deletions in the homeodomain have occurred many times independently in evolution. Even within well-defined typical classes or families, such divergence is possible; for example, *C. elegans ceh-36* belongs to the Otx family, yet has three extra residues between helix 2 and 3.

5.1.3 DNA-Binding Properties of the Homeodomain

NMR and X-ray analysis of DNA-protein complexes for several homeodomain proteins have allowed the identification of the residues critical for sequence specific contacts and contacts to the DNA backbone. These studies have been complemented by in vivo and in vitro studies on the DNA-binding properties of homeodomains, and confirm the importance of some of the DNA-amino acid contacts that have been deduced by the structural studies. Helix 3, also termed the recognition helix, lies in the major groove of the DNA and plays the most important role in making sequence specific contacts (Fig. 5.2). The residues which make sequence-specific contacts are indicated with a "B" in Fig. 5.1a above the consensus. Other residues make non-sequence-specific contacts to the DNA backbone (indicated by "#"). The latter residues are not only found in helix 3, but also in helix 2, and the amino terminus. Further, the amino terminus reaches into the minor groove of the DNA, where the residues marked "m" can contact the DNA in the minor groove. A very critical residue that determines sequence specificity is found at position 9 of helix 3 (i.e., position 50). Exchanging this amino acid causes drastic changes in the type of DNA sequence that is recognized by a particular homeodomain [18]. Several sequences that are bound by homeodomains have been identified. A binding site for Antp is AGCCATTAGA, with the core being ATTA (TAAT on the other strand) [21]. This core binding site is too small for providing sufficient specificity to activate only the intended target genes. Different types of homeodomain proteins employ different strategies to solve this problem: some combine several DNA-binding domains in a single protein, some form homodimers, and many form heteromeric complexes with homeodomain or other types of TFs.

5.1.4 Classification of Homeodomains

The purpose of comparing and classifying the sequences of homeobox-containing genes is to determine evolutionary relationships between the different genes so as to identify bona fide orthologs and paralogs, and place them in a more comprehensive framework that reveals structural and functional relationships. With the advent of many completed genomes, this is now a much easier task. A number of different classification schemes have been used over the years. We arrange the homeodomain sequences into logical groups using the terms "superclass", "class" and "family", as we did in previous compilations [19–22], which are the basis for this chapter. The use of these terms reflects the hierarchy of relationships, such that a superclass encompasses several classes, and a class can be subdivided further into families. Slightly different classification schemes are used in other publications, and we adopt some of these naming conventions and their lists [23]. In general the conclusions and divisions are similar, but as new information and sequences appear, revisions are sometimes necessary.

Figure 5.3 shows a comparative tree that has been derived from a much larger tree. It presents selected genes of different classes and families. Phylogenetic trees provide an easy means of grouping homeodomain sequences for classification. All superclasses and classes, as well as most families presented here, are well conserved in evolution, and members can be found in deuterostomes as well as protostomes. In fact, a very useful criterion for the identification of families is that they should be conserved over a long time period. The ideal definition for a family would be that its members were derived from a single gene in the ancestor of protostomes and deuterostomes; classes would comprise several families with common features. Generally, one can say that classes of homeobox genes have less than 50–55% sequence identity within their homeodomains to other classes. Within a particular family, the homeodomain can be 80–90% identical from flies to vertebrates.

Many of the homeodomain classes have additional conserved protein domains or motifs outside of the homeodomain (see Chapter 12 for a discussion of TF effector and auxiliary domains). Such sequence motifs, which sometimes can be even more conserved than the homeodomain itself, provide extra criteria to differentiate the various classes of homeobox genes. Figure 5.4 shows a schematic diagram of homeobox gene families and classes in animals that encode special conserved sequence motifs outside of the homeodomain. For example, homeodomain proteins of several families of the Antennapedia superclass contain a short "Hexapeptide" motif upstream of the homeodomain. The sections that follow introduce the various classes of homeobox genes.

Fig. 5.3 (continued) homeobox genes for selected families and classes are given. The vertebrate Hox cluster genes can be grouped into 13 different paralogue groups that are indicated in the figure (PG). Parentheses on the right side mark families, classes, and superclasses. The Antennapedia group of genes can be split into two groups, those which are similar to the Hox cluster genes (HOXL) and those which are similar to the NK cluster genes (NKL). Not all genes here are assigned to their corresponding classes or families due to their divergent nature (e.g., Ce LIN-39, Ce MAB-5, d zen), or are more divergent so that a clear assignment becomes difficult (e.g., Ce CEH-7). The column at the right indicates those families that have a Hexapeptide upstream of the homeodomain. Species: d: *Drosophila*; m: mouse; h: human; r: rat; Ce: *C. elegans*; Cv: hydra; Xl: *Xenopus laevis*; Sm: *Schistosoma mansoni*; Sc: *Saccharomyces cerevisiae*; Um: *Ustilago maydis*; Cc: *Coprinus cinereus*; pAt: *Arabidopsis thaliana*

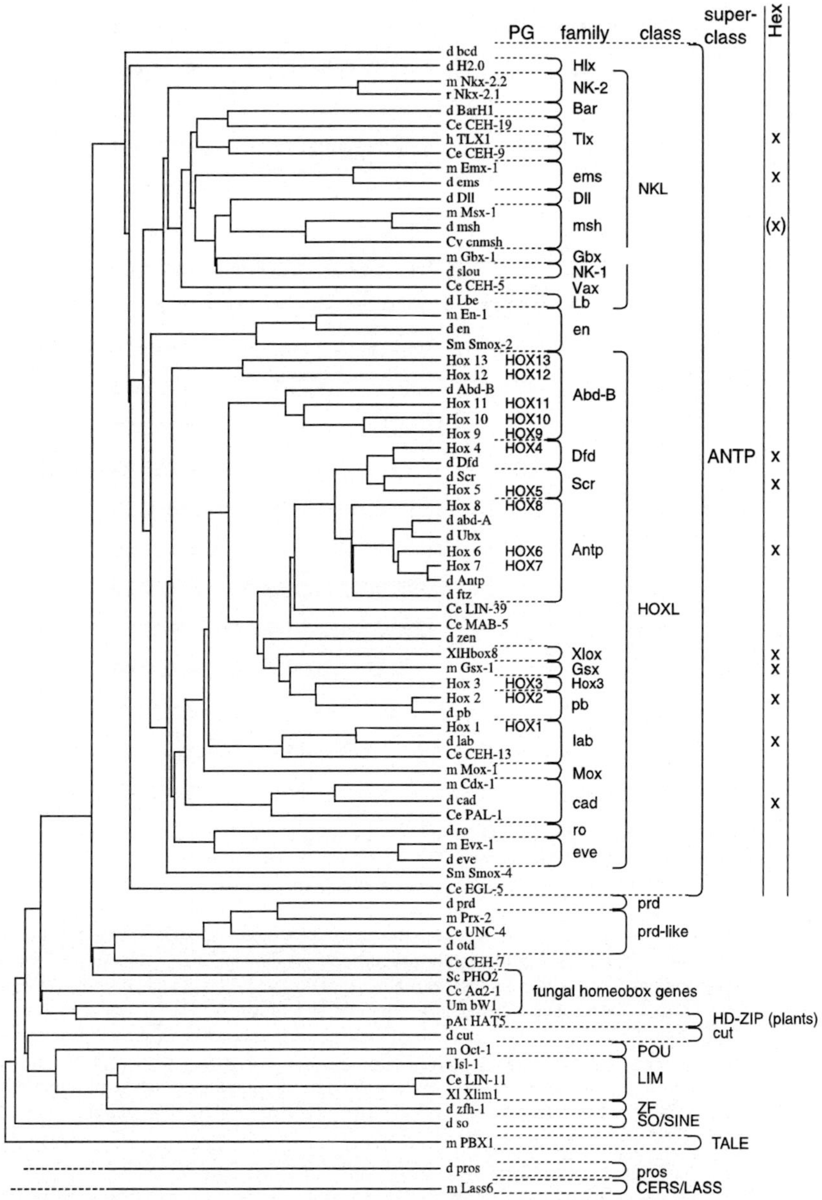

Fig. 5.3 Comparative tree showing different families and classes of homeobox genes. This simple dendrogram which is derived from a much larger tree that was generated from aligned homeodomain sequences. The more similar two homeodomain sequences are to each other, the shorter the horizontal distance is from the branch point to the endpoint. At the leaves examples of particular

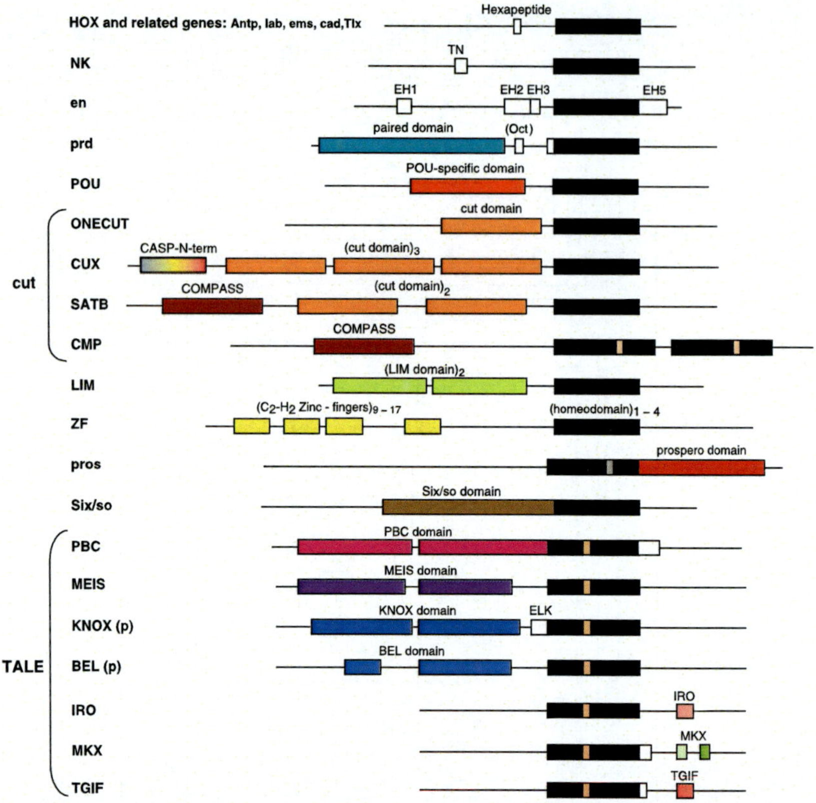

Fig. 5.4 Schematic representation of families and classes of animal homeobox genes encoding conserved motifs outside of the homeodomain. On the *left* are the names of selected different families and classes. The *black box* represents the homeodomain, and insertions within the homeodomain, either between helix 1 and 2 or helix 2 and 3, are indicated in different colors. The other colored boxes represent conserved sequence motifs specific to individual classes. The length of the boxes is approximately proportional to the size of the domains. The connecting linker regions (*black lines*) are not to scale. The Hexapeptide is found in several different families within the Antennapedia superclass, mainly in Hox genes. The TN motif and the Octamer (Oct) motif do not occur in all members of their respective families. In several instances the number of domains can vary. For example, in the ZF class, the number of zinc-fingers, as well as the number of homeodomains can vary substantially, and the zinc-fingers can be interspersed with the homeodomains. This figure is not comprehensive; some motifs have been omitted here

5.2 The Antennapedia Superclass

This is a large group of genes that are related to one of the first homeobox genes discovered, Antennapedia. A number of these genes are organized into gene clusters, i.e. the HOX cluster, the ParaHox cluster, which is a "sister" of the HOX cluster, and the NK cluster. However, many homeobox families within the Antennapedia superclass do not reside in a cluster and are referred to as "dispersed". Common features of these genes are that they do not encode large conserved domains outside of the homeodomain, but only small motifs. The Hexapeptide motif is found in genes of the HOX and ParaHox cluster, as well as in some NK cluster genes and in dispersed genes, such as the ems family (Figs. 5.3 and 5.4). The Hexapeptide is usually separated by an intron from the homeodomain. Other small motifs are found in NKL homeodomain proteins, i.e. the TN motif, or in en homeodomain proteins (Fig. 5.4). Within the Antennapedia superclass, two major groups can be distinguished. One group is comprised of the genes that are most similar to the Hox cluster genes (HOXL), such as the Hox, Mox, Evx, and ParaHox genes. The second group is comprised of the NKL class genes, comprising the NK cluster genes and gene families such as NK1(slou), NK3(bap), Dll, ems, BarH, msh, Hlx, or Tlx (Fig. 5.3).

5.2.1 The HOX Cluster

The first, and also best-known homeobox genes are the homeotic selector genes of *Drosophila melanogaster* that are organized into two complexes, the Antennapedia complex (ANT-C) and the Bithorax complex (BX-C), summarily referred to as the homeotic gene complex (HOM-C). In *Drosophila* only one such cluster is found, while tetrapods have four paralogous clusters, each on a different chromosome (Fig. 5.5) [24]. These clusters have been termed HOX clusters in vertebrates, and the usage of naming these clusters "HOX clusters" has now been expanded to include other animal phyla, including *Drosophila.* It should be noted that the word Hox refers to the particular types of homeobox genes found in the HOX cluster, especially those in vertebrates, and it is not a general term for homeobox gene. The four vertebrate HOX clusters originally arose through duplications of a single cluster at some point in early vertebrate evolution. Indeed, in the cephalochordate *Amphioxus* only a single cluster is present [25]. On the other hand, one has to note that in teleost fish extra HOX clusters (as well as extra homeobox genes and extra copies of many other genes) are found due to an extra genome duplication event [26–28].

Figure 5.5a shows the *C. elegans*, *Drosophila* and vertebrate HOX clusters. The genes in the four vertebrate clusters (termed HOXA, HOXB, HOXC, HOXD) can be aligned such that 13 groups are formed that are highly related based on their sequence. These groups are referred to as paralogue groups 1–13 (HOX1 to HOX13), and individual genes in mice are named *HoxA-1*, *HoxA-2* and so on. Each cluster is missing genes for some of the paralogue groups, indicative that during or after the duplication events some genes were lost. Overall, there are 39 Hox genes in mouse and human. In vertebrates, but not flies, the *Evx* genes are also closely linked to the HOX cluster, indicating that these genes are part of the ancestral Hox cluster. Likewise, the Mox genes are also associated with the HOX cluster, albeit at the opposite end [29].

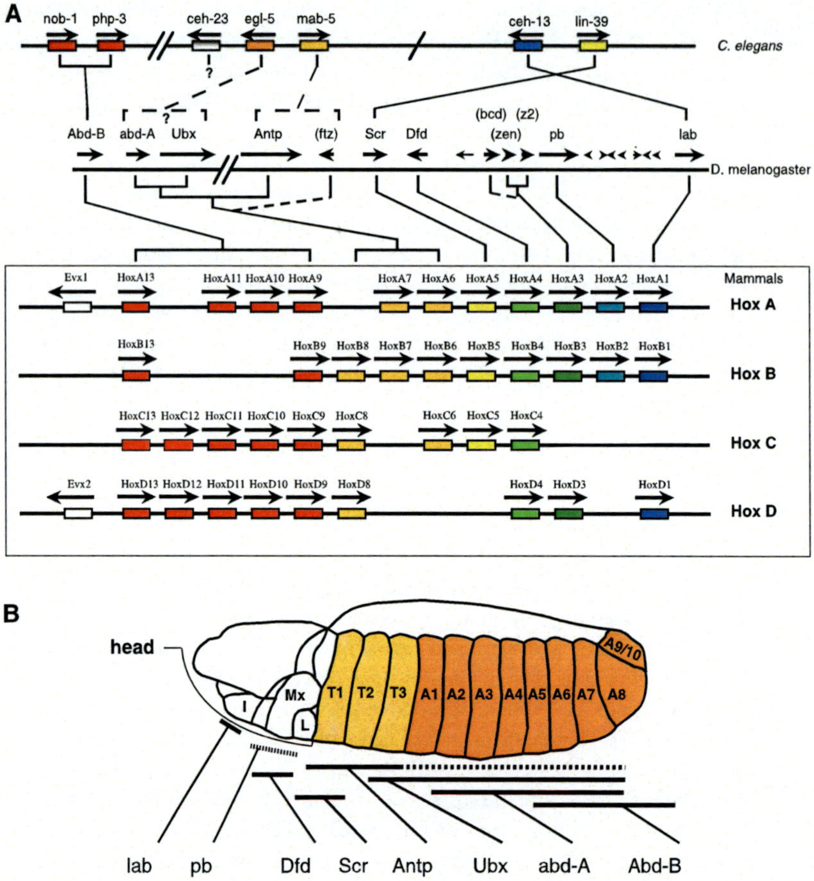

Fig. 5.5 (**a**) Organization of the *C. elegans*, *Drosophila melanogaster* and mouse/human HOX clusters. At the *top* is a schematic representation of the *C. elegans* HOX cluster. The AbdB genes (*php-3*, *nob-1*) have split far from the rest of the cluster. Of the remaining five genes in the cluster only four genes can be assigned to fly/vertebrate homologues, in the case of *mab-5* and *egl-5* only tentatively. The relationship of the homeobox gene *ceh-23* is unclear. In the *center* is a representation of the *Drosophila melanogaster* HOX cluster, composed of the Antennapedia complex (*right*) and the Bithorax complex (*left*); *large arrows* indicate the individual transcription units of homeobox genes, while *small arrows* represent non-homeobox genes, and gene names of non-homeotic homeobox genes are given in *parentheses*. At the *bottom* of the panel the organization of the four mammalian – based on mouse and human – HOX clusters (HOXA, HOXB, HOXC, HOXD) is shown. The *arrows* indicate the orientation of transcription. *Black lines* and *brackets* between the *C. elegans*, *Drosophila melanogaster* and mammalian clusters mark the homologous genes. *Dashed lines* indicate less certain sequence relationships or derived sequences. (**b**) Schematized expression domains for the *Drosophila* HOX genes as mapped onto a 10-h embryo. *Bars* indicate the approximate areas of expression for individual genes. Labels: I, Mx, L: intercalary, maxillary and labial segments of the presumptive head; T1 – T3: thoracic segments; A1 – A10: abdominal segments. After [112]

The lines and brackets in Fig. 5.5a indicate the relationships between the vertebrate and fly *Hox* genes. Most genes have a simple one-to-one correspondence from flies to vertebrates. For example, the *Drosophila* gene *labial* is the orthologue of the HOX1 genes *HoxA-1*, *HoxB-1*, and *HoxD-1* in mice. However, the fly gene *Abd-B* has five co-orthologous gene groups in vertebrates, HOX9 through HOX13, due to a separate expansion of the Abd-B family genes in the deuterostome lineage. Further, in the center of the cluster, i.e., HOX6 to HOX8 in vertebrates, and *ftz* to *AbdA* in flies, a single ancestral gene may have been present that gave rise to several genes independently in each phylum. A degenerate HOX cluster containing seven homeobox genes has been found in the nematode *C. elegans*. Two Abd-B family genes have split from the cluster, while the remaining five homeobox genes have also been separated into two subclusters interrupted by a series of unrelated genes (Fig. 5.5a). *C. elegans* apparently has lost several homeobox genes, since in several other nematode species additional homeobox genes are present in the HOX cluster [30, 31].

The genes of the HOX cluster share another intriguing feature: the physical clustering correlates with the way the Hox genes are expressed along the anterior-posterior body axis (Fig. 5.5b). Thus genes at one end of the cluster are expressed and function in the anterior body region, while, as one progresses along the chromosome, the genes are expressed and function further and further toward the posterior of the animal. This colinearity of genes and expression pattern is even more striking when one considers that the Hox genes in tetrapods are all transcribed in the same direction [32].

The Hexapeptide, also known as Pentapetide, was first described in HOX cluster genes. All genes in the HOX cluster, apart from the Abd-B family of genes encode this short conserved peptide motif upstream of the homeodomain. The Hexapeptide has a conserved core sequence of six amino acids [19].

5.2.1.1 Labial (Lab/Hox1) Family

The labial family of genes is located at the 3′ end of the cluster. The *Drosophila lab* homeodomain is 80–85% identical to the HOX1 group genes, and the homeodomains of the lab family are 55–67% identical to other HOX cluster genes. Genes of this family encode a distinct Hexapeptide sequences upstream of the homeodomain.

5.2.1.2 Proboscipedia (Pb/Hox2) Family

The vertebrate paralogue group HOX2 is closely related to the *Drosophila* gene *pb*, but is missing in *C. elegans*.

5.2.1.3 Zerknüllt (Zen/Hox3) Family

The zen genes in *Drosophila* are quite divergent from the vertebrate Hox3 genes. However, analysis of the corresponding HOX cluster genes in other arthropods suggests that the *Drosophila* genes *zen* and *zen2* are the homologues of the HOX3 genes [33]. They have taken on novel roles during development, i.e. dorsal ventral patterning, rather than anterior-posterior patterning. Another *Drosophila* gene not involved in anterior-posterior patterning is *bicoid* (*bcd*). *bcd* is the most divergent gene in the *Drosophila* HOX cluster (Fig. 5.5a). Its functional role is as a maternal morphogen in the early *Drosophila* embryo, where the protein forms a gradient in the egg. Recent evidence indicates that *bcd* has arisen by a duplication from a HOX3 gene during insect evolution, and has undergone rapid sequence divergence as an adaptation to its novel function [34–35]).

5.2.1.4 Deformed (Dfd/Hox4) Family

This family comprises the *Drosophila Deformed* gene and the vertebrate HOX4 group genes, and is quite well conserved from flies to vertebrates.

5.2.1.5 Sex-Combs-Reduced (Scr/Hox5) Family

The *Drosophila Scr* gene and the HOX5 group gene are very similar to the Antp family genes, but probably are bona fide orthologues. The homeodomain of the *C. elegans* HOX cluster gene *lin-39* is about equally similar to that of Dfd family and *Scr* family homeodomains, but may be orthologous to Scr.

5.2.1.6 Antennapedia (Antp) Family

This family comprises the genes in the center of the cluster i.e. the fly genes *Antennapedia (Antp)*, *Ultrabithorax* (*Ubx*), *abdominal-A* (*abd-A*), *fushi tarazu* (*ftz*), and the vertebrate genes HOX6, HOX7, and HOX8. The homeodomains of this family are strikingly conserved between flies and vertebrates (up to 98%). Most likely these genes derive from a single ancestral gene. The *Drosophila* gene *ftz* is not involved in anterior-posterior patterning, but in other insects and in arthropods the orthologous gene is better conserved and in some species it still seems to play a role in pattern formation [36–38].

5.2.1.7 Abdominal-B (Abd-B/Hox9-Hox13) Family

The Abd-B family of genes is located at the 5′ end of the cluster. They are more divergent than other cluster genes and they do not encode a Hexapeptide. While *Drosophila* has only one gene, *Abd-B*, there are five paralogue groups, HOX9 to HOX13, in vertebrates. The Abd-B homeodomain is 52–75% identical to the various human and mouse Abd-B family homeodomains, thus these genes can be very divergent. The further removed from the center of the HOX cluster the paralogue groups are, the more divergent their homeodomain sequences become. *HoxD-9* to *HoxD-13* play important roles during pattern formation of limbs in vertebrates [39]. *C. elegans* has two Abd-B genes, *php-3* and *nob-1*, although the latter is rather divergent. A third gene, *egl-5*, is variously grouped either as an Abd-B gene or as a divergent Antp family gene (Fig. 5.5a).

5.2.1.8 Even-Skipped (Eve/Evx) Family

The vertebrate *Evx* genes are located at the 5′ end of the HOX cluster (Fig. 5.5a), suggesting that this gene family is part of the original cluster, although in flies and *C. elegans* these genes are not linked to the HOX cluster. The eve genes do not contain a Hexapeptide upstream of the homeodomain and their homeodomain sequences are different from the Antp and Abd-B family homeodomains (Fig. 5.3).

5.2.1.9 Mox (Meox) Family

The Mox genes were found to be linked to the HOX cluster in vertebrates, and may therefore also have been part of the ancestral HOX cluster. However, these genes are more similar to Abd-B and cad genes and may be derived by duplication from the posterior genes. The ortholog in flies is *buttonless*.

5.2.2 The ParaHOX Cluster

The ParaHox gene cluster is comprised of three gene families, Gsx, Xlox (Pdx), and caudal (cad/Cdx) (Fig. 5.6a). Their sequences are similar to the Hox cluster genes and they contain a Hexapeptide. The cluster organization is not as highly conserved as the Hox cluster, but has been conserved, for example, in *Amphioxous* [40–42]. The ParaHox cluster probably arose through a duplication event from

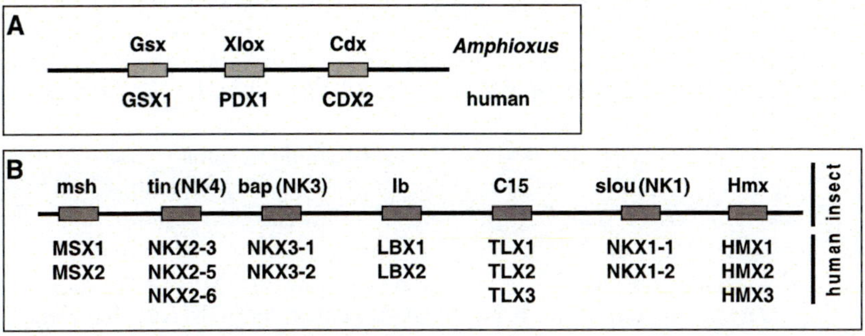

Fig. 5.6 (**a**) Schematized gene arrangement for the ParaHox gene cluster as found originally in *Amphioxus*. It is comprised of three gene families Gsx, Xlox (Pdx) and Cdx. Of the human paralogs, only one set remains in a cluster, indicated in the lower half. (**b**) Putative organization of the ancestral NK gene cluster (Tin-C) as inferred from the NK gene organization of the cluster in *Drosophila, Anopheles gambiae, Tribolium castaneum,* and *Apis mellifera* (upper half). Note that *Drosophila* has two paralogous *ladybird* (*lb*) genes: *ladybird early* (*lbe*) and *ladybird late* (*lbl*). *msh*: muscle-specific homeobox gene; *tin*: *tinman* (aka NK4); *bap*: *bagpipe* (aka NK3); *slou*: *slouch* (aka NK1). In mammals this cluster has fragmented substantially, only a few linkages remain. In the lower half the human orthologues of the insect gene cluster are indicated

an ancestral HOX cluster. The ParaHox genes also function in anterior-posterior patterning, with the Gsx genes being the most anterior, and the cad genes being the most posterior [43].

5.2.3 NKL Class Genes

5.2.3.1 The NK Cluster

The NK cluster of homeobox genes, also called Tinman complex [44], was discovered in *Drosophila*, where a series of homeobox genes are clustered in the chromosomal region 93DE (Fig. 5.6b) [45]. Corresponding clusters are present in other insects, such as mosquito, honeybee, and Tribolium, although some rearrangements have occurred [44]. In vertebrates this cluster has only been conserved in a more fragmentary nature, and many of the genes have dispersed in the genome [41]. A main function of the NK cluster genes is in pattern formation and development of the mesoderm [45]. Based on current evidence the ancestral cluster contained the following NK gene families: NK1 (fly *slou*); TLX (fly C15), which encodes a Hexapeptide upstream of the homeodomain; LBX (fly *lbe* and *lbl*); NK3 (fly *bap*); NK4 (fly *tin*); Msx (fly *msh*); HMX.

5.2.3.2 Dispersed NKL Genes

Many NK type genes are not found in clusters. In humans the following families have been defined: BarHl, Barx, Bsx, Dbx, Dlx, Dlx (in fly *Distal-less*, *Dll*), Emx (fly *empty spiracles*; *ems*), possibly En, Hhex, Hlx, Nanog, Nkx2.1, Nkx2.2, Nk6, Vax, and Ventx. A feature often shared between these families and also with those in the cluster is the presence of a TN motif upstream of the homeodomain [46].

5.2.4 Other Types of Antennapedia Superclass Like Genes

Some of the Antennapedia superclass gene families cannot confidently be assigned to either the HOXL or NKL class. One such family is defined by the fly gene *engrailed* (*en*, vertebrate En). Engrailed genes encode a series of small conserved motifs outside of the homeodomain one of which is similar to the TN motif of NKL homeobox genes (Fig. 5.4). The Gbx, Noto and Mnx families are also rather divergent and cannot easily be assigned to either the HOXL or NKL class.

5.3 Paired (PRD) and PRD-Like Classes

5.3.1 PRD Class

The homeodomain of PRD class genes is characterized by a serine residue at position 50, which impacts DNA-binding specificity. In addition, upstream of the homeodomain is a highly conserved domain of about 130 amino acids, the PRD (or Pax) domain. The PRD domain itself has been shown to be DNA-binding domain and its structure shows it to consist of two globular domains with three alpha-helices each [47]. A number of genes have been found that encode only a PRD domain, but no homeodomain. In vertebrates, the genes containing a PRD domain are called Pax genes, irrespective of whether they have a homeodomain or not [48, 49]. The PRD domain bears resemblance to transposases, and it is thought that a PRD-like homeobox gene captured a transposase early in metazoan evolution, which then evolved into the PRD domain [50]. The PRD class has nine members in mammals, and has been grouped into different families based on the PRD domain. The *Pax4/6 family* is well conserved in evolution, encompassing such genes as *Pax-6* in vertebrates and *eyeless* (*ey*) in flies. The *Pax-6* genes play a role in eye development across bilateria, despite the large differences in eye structure and development between different phyla [51–53]. The *Pax3/7 family* includes the founding member, *Drosophila paired*, as well as *gsb-d*, *gsb-p* and the vertebrate genes *Pax-3* and *Pax-7*. A small conserved region (the Octapeptide) is present between the homeodomain and the PRD domain (Fig. 5.4). The *Pax2/5/8 family* is unusual in that these genes encode only the first third of a homeodomain. *Pax1/9* family lacks a homeodomain completely. This is thought to be a secondary loss.

5.3.2 PRD-Like Class

A substantial number of homeobox genes encode a homeodomain that is similar to the PRD class homeodomains, but they do not encode a PRD domain, nor do they have a serine residue at position 50 of the homeodomain. The PRD-like homeodomains display large sequence diversity, and the PRD class itself is actually only one group within that diverse set. A large number of families have been defined in mammals [23, 56]. Some of these encode the classical glutamine at position 50 of the homeodomain, while a number of them encode a lysine (e.g., Gsc, Mix). In humans, the following families have been defined: Alx, Argfx, Arx, Dmbx, Dprx, Drgx, Dux, Esx, Gsc, Hesx, Hopx, Isx, Leutx, Mix, Nobox, Otp, Otx, Phox, Pitx, Prop, Prrx, Rax, Rhox, Sebox, Shox, Tprx, Uncx, and Vsx. The Dux family has two homeodomains, hence the name (Double homeobox). Many of these are conserved to invertebrates; however, Argfx, Dprx, Tprx, Leutx are not found in invertebrates. Nobox and HopX, which contains an atypical homeodomain, are only tentatively assigned to the PRD-like genes, since they are so divergent. Some families have quite extensive sequence conservation outside of the homeodomain. For example, the Vsx family (aka CHX10) has a motif of about 60 amino acids immediately downstream of the homeodomain that is conserved in bilateria [54–55].

Within the PRD-like class we can also find evidence of gene duplication and diversification. For example, the *Odysseus* (*OdsH*) gene, which is a recent duplication from an unc4 (Uncx) family homeobox gene, is evolving rapidly in *Drosophila* species and is involved in hybrid male sterility [57]. Another example are the *Rhox* homeobox genes which dramatically expanded in mouse to 33 genes and are clustered in a region on the X chromosome. They are expressed during embryogenesis and in adult reproductive tissues [58, 59].

5.4 POU Class

The POU class was originally defined based on the four genes *Pit-1*, *Oct-1*, *Oct-2* and *unc-86* (POU) [60]. The POU-specific domain is an approximately 80 amino acid long conserved domain upstream of the homeodomain with a variable linker in between. The POU-specific domain is required for cooperative, high affinity DNA-binding and has so far always been found in association with a POU homeodomain. The POU homeodomain is characterized by a cysteine residue in position 50. The structure of the POU domain (that is the POU-specific domain and the POU homeodomain) bound to DNA has been determined. The POU-specific domain consists of four alpha-helices. Helix 2 and 3 fold like a helix-turn-helix motif, although the loop is larger (e.g. [61, 62]). The genes have been grouped into six families, POU-I to POU-VI, and many play important roles in nervous system development. Members of the POU-II family are the well-known mammalian TFs *Oct-1* and *Oct-2*. A highly divergent homeobox gene in vertebrates, HDX, may be derived from POU genes.

5.5 HNF Class

The HNF class was originally defined by the mammalian TF LFB1 (HNF1alpha) [63]. The homeodomain contains extra residues between helix 2 and helix 3. While this class is not found in *Drosophila*, a conserved domain upstream of the homeodomain is present. Structural analysis of this domain showed that it is similar to the POU domain [16], suggesting that the HNF class is probably a highly divergent derivate of a POU class gene.

5.6 CUT Superclass

The first gene of this superclass discovered was the *Drosophila* homeobox gene *cut*, which has three copies of a conserved domain of about 80 amino acids, the cut domain, upstream of the homeodomain [64]. Other homeobox genes with cut domains were subsequently discovered and four distinct classes can be defined. The structure of the cut domain has been determined. It consists of essentially five alpha helices which form a globular domain. The third helix lies in the major groove of the DNA and provides sequence specific contacts [65–67]. The evolution of this group of genes is complicated, since a number of domain shuffling events have occurred [68, 69].

5.6.1 CUX Class

Unlike other homeodomain proteins, the CUX proteins have a histidine residue at position 50 of the homeodomain. Biochemical analysis of the human member CDP (CCAAT displacement protein) provided the first evidence that the cut domain is a DNA-binding domain [70]. The CUX class genes have a most unusual structural organization, because, apart from the cut domain, their amino terminus is actually shared with another gene, CASP, which is a Golgi membrane protein [71]. The amino-terminal half of CASP can either splice to the cut-homeodomain part of the Cux genes, or it can splice to the carboxy-terminal part of the CASP protein, giving rise to a fully functional CASP protein. At some point in evolution an ancestral CUX gene has been functionally intertwined with the CASP gene through alternative splicing. This organization is found in *C. elegans* and vertebrates, but in *Drosophila*, the CASP gene has been lost [68].

5.6.2 ONECUT Class

Genes in this class have only a single cut domain. This likely represents the most ancestral condition.

5.6.3 SATB Class

The genes of the SATB class encode two highly divergent cut domains and a highly divergent homeodomain. In addition, they have a COMPASS (CMP) domain at their amino terminus. SATB class genes have presently only been found in vertebrates. In contrast to other homeobox genes that act as regular TFs, SATB1 has been shown to be a special global gene regulator that is involved in chromatin remodeling (e.g. [72]).

5.6.4 COMPASS (CMP) Class

The CMP class of homeobox genes is an unusual group of homeobox genes that encode a CMP domain upstream of two homeodomains. The two homeodomains arose through duplication from a common ancestor. These homeodomains are distinct from those of other classes, because of the extra residues in the loop region between helix 2 and helix 3. Even though the CMP genes do not encode cut domains, the CMP domain is shared with SATB genes. Members of the CMP class have been found in invertebrates and *Amphioxus*, but not in vertebrates [68, 69]. Perhaps the CMP genes gave rise to the SATB genes through some domain shuffling in early vertebrates.

5.7 ZF Class

Genes in this class contain classical zinc-finger domains of the C_2H_2 (two cysteine residues – two histidine residues) type, which are DNA-binding domains (see Chapter 4). In humans, five families have been defined: Afhx, Azfh, Zeb, Tshz, and Zhx/Homez. Some of these genes encode many copies of both the ZF and homeodomains. For example, human *ATBF1* encodes 17 zinc-fingers and four homeodomains [73]. The homeodomains of these proteins tend to be very divergent; presumably the evolutionary constraints are relaxed due to the large number of DNA-binding domains present in a single protein.

5.8 LIM Class

LIM homeobox genes encode two LIM domains upstream of the homeodomain. This class was first defined in two *C. elegans* genes, *lin-11* and *mec-3*, and in the rat TF *Isl-1* (*lin-11, Isl-1, mec-3* = LIM) [74]. The LIM domain is about 60 amino acids long and contains conserved cysteine and histidine residues, and has been shown to be a distinct type of zinc-finger, different from the one found in ZF class homeobox genes. Six families have been described: Lhx2/9 (*apterous*), Lhx1/5, Lhx3/4, Lhx6/8 (*arrowhead*), Lmx, and Isl (Islet) [75].

The LIM domain is also present in genes that do not encode a homeodomain. The rhombotin genes (LMO, i.e. LIM-only) encode two LIM domains that are similar to the two LIM domains of LIM homeobox genes. This gene is ancient and has been found in a tandem gene cluster with several LIM homeobox genes in *Trichoplax adhaerens* [75]. On the other hand, a large number of other proteins have been found that contain only more divergent LIM domains. Examples include mammalian CRIP, a cysteine-rich intestinal protein; ESP1, an estradiol-stimulated protein in brain; hCRP, a human cysteine-rich protein; zyxin, a cytoskeletal protein; and MLP, a regulator of myogenesis. CRIP and ESP1 have only one LIM domain, while zyxin contains three LIM domains (see also [19]). The LIM domain is a protein–protein interaction domain [76]. This domain can interact with the LIM-binding protein Ldb, which is an important co-factor for the LIM homeodomain proteins [77].

5.9 SIX/SO (SINE) Class

This class was originally defined by the *Drosophila* gene *sine oculis* and the mouse genes *Six1* and *Six2* [78]. These homeobox genes encode a distinct typical homeodomain, which have a lysine at position 50 of the homeodomain. Upstream of the homeodomain is the highly conserved 120 amino acid long Six/so domain.

The Six/so class is divided into three families, Six1/2, Six3/6, and Six4/5, which are all conserved between flies, worms, and vertebrates [79].

5.10 PROS (Prospero) Class

The PROS class has been named after the *Drosophila* gene *prospero* [80, 81]. The homeodomain is highly divergent and has three extra amino acids between helix 2 and helix 3. Downstream of the homeodomain is a conserved sequence motif of about 100 amino acids, the Prospero domain, that reaches to the carboxy-terminus [82, 83]. The structure of the Prospero homeodomain and the Prospero domain has been determined. The two domains form a single structural unit that is required for sequence specific DNA binding [84].

5.11 TALE Superclass

The TALE (three amino acid loop extension) homeobox genes are characterized by having three extra residues in the loop between helix 1 and helix 2 of the homeodomain [22, 85, 86]. This group is very ancient, being present in wide range of eukaryotic kingdoms [87]. Five classes are found in animals and two in plants. Many of the TALE superclass homeobox genes encode an isoleucine at position 50 of the homeobox, though alanine (IRO) and glycine (PBC) are also found.

5.11.1 PBC (PBX) Class

This class with a characteristic 180 amino acid motif upstream of the homeodomain, the PBC domain, has been found in mammalian PBX genes, *C. elegans ceh-20*, and *Drosophila extradenticle* [88, 89]. The extended loop between helix 1 and helix 2 of the homeodomain of PBC proteins can interact with the Hexapeptide of Hox genes, for example those of the Hox1/lab family [90].

5.11.2 MEIS Class

The MEIS class can be divided into two families, *MEIS* and *PREP* (aka Pknox). Both families share a MEIS (aka HM) domain upstream of the homeodomain, which is about 130 amino acids long. The MEIS proteins Homothorax (*Drosophila* Hth), as well as vertebrate MEIS proteins, have been shown to interact with PBC class homeobox genes through their MEIS domain. Thus MEIS and PBC proteins can from heterodimers. Further, it has been shown that the interaction of Hth with Exd is responsible for translocating Exd from the cytoplasm to the nucleus [91, 92]. Hence, dimerization controls the activity of the protein complex. Further, MEIS class proteins can also interact with Hox cluster proteins that are expressed in the posterior, and triple complexes of PBC, MEIS and Hox proteins have also been reported (e.g. [93–95]).

5.11.3 IRO (IRX) Class

The IRO class was named after three *Drosophila* genes, *araucan*, *mirror*, and *caupolican*, which are located in the iroquois complex [96]. In mammals there are two paralogous Irx gene clusters with three genes each, but this is likely to be an independent evolutionary diversification [97]. A 15 amino acid motif, the IRO box, is found downstream of the homeodomain.

5.11.4 MKX Class

The homeodomain of the MKX class proteins is most similar to that of the IRO class. However, it the MKX proteins have three different small motifs downstream of the homeodomain, which are unique to the MKX class [22].

5.11.5 TGIF Class

The TGIF class of TALE homeodomains has been first defined by the vertebrate TG-interacting factor (TGIF) [98]. Fly and vertebrate TGIF genes share an additional 20 amino acids immediately downstream of the homeodomain, and a 12 amino acid motif is found further C-terminally.

5.12 Other Types of Animal Homeobox Genes

As has emerged from the gene descriptions above, some homeobox genes within particular phylogenetic branches are highly derived and are often difficult to classify. For example, the vertebrate Hdx genes may be derived from POU homeobox genes, and the vertebrate Nobox and HopX genes are also difficult to place due to their divergent nature. In other phylogenetic branches, e.g. *C. elegans*, highly divergent homeobox genes are also present. The homeobox gene *ceh-7* seems to be a divergent member of the PRD-like class, but lacks orthologs in other phyla [99].

An unusual case are the CerS (ceramide synthase) genes, also known as the LASS (longevity assurance) genes. Several of these genes, but not all, contain a divergent homeodomain in flies and vertebrates [100, 101], while yeast homologs lack a homeodomain. What is striking is that the CerS proteins contain transmembrane regions, hence the homeodomain seem unlikely to act as a DNA-binding domain in these proteins. Likely, the homeodomain was incorporated into a CerS gene during a domain shuffling event, and now has a different function.

5.13 Fungal Homeobox Genes

A small number of homeobox genes are known in the yeast *Saccharomyces cerevisiae*. Two of these genes, *MATa1* and *MATalpha2* are part of the mating type locus (MAT) [102]. *MATa1* encodes a typical homeodomain, and *MATalpha2* a TALE homeodomain. Such a dyad of a typical and TALE homeobox gene is also found other fungi, such as *Ustilago maydis*, *Schizophyllum commune* and *Coprinus cinereus* [103]. In microsporidia there is also a closely linked pair of a TALE and a normal homeobox gene [104]. In yeast, MATa1 and MATalpha2 form heterodimers, and MATalpha2 also forms homodimers, allowing regulation of different sets of target genes. None of the fungal genes have any of the additional domains found in animals. But, it can be estimated that there were not more than two TALE homeobox genes and two to three typical homeobox genes present in the first fungal ancestors [104].

5.14 Plant Homeobox Genes

In plants, fourteen distinct classes of homeobox genes are found that have been conserved between moss, monocots and dicots [105] (Fig. 5.7). One large group, HD-ZIP, consists of homeobox genes that have a leucine-zipper, a protein interaction motif, downstream of the homeodomain. This group can be further subdivided into four classes, HD-ZIP I, II, III, and IV. Genes of the HD-ZIP III and IV classes encode a START domain, a lipid-binding domain, downstream of the homeodomain. The HD-ZIP III class is further distinguished by having a MEKHLA domain at the very C-terminus that was derived from a PAS domain of bacterial origin, most likely from the chloroplast [105]. The other classes are PLINC, WOX, DDT, PHD, NDX, LD, PINTOX, SAWADEE, KNOX and BEL. The homeodomains of several classes, i.e. HD-ZIP III, WOX, NDX, and SAWADEE are atypical. All these classes encode additional domains and motifs that are distinct for each class (Fig. 5.7). Many of these domains are found in other proteins without a homeodomain, and in a number of instances they are associated with other domains not found in homeodomain proteins. Several of them are conserved between the plant/animal divide, i.e. PHD, DDT, WUS, and START. Further, a number of the domains have conserved cysteine and/or histidine residues, indicating that they are different types of zinc fingers, i.e. PHD, D-TOX ZF, PLINC, and SAWADEE. Plants have two ancient classes of TALE homeobox genes, the KNOX class and the BEL class. Both encode large bipartite domains upstream of the homeodomain, termed KNOX and BEL domain, respectively. The KNOX domain can be aligned with the MEIS domain, showing that these

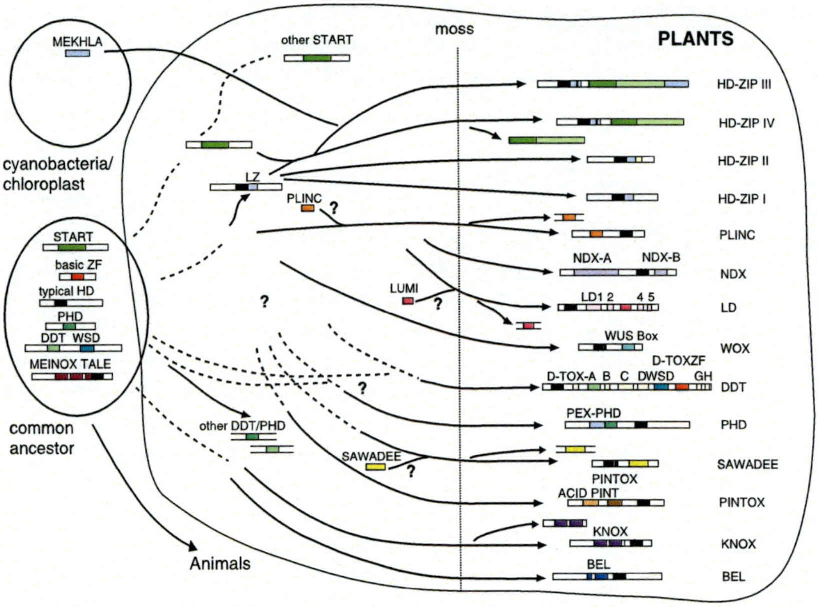

Fig. 5.7 Plant homeodomain diversification into 14 classes. Putative derivations are indicated by *arrows* and the different domains are colored. Each class has distinctive domains and motifs associated with the homeodomain. A number of different domains and at least two types of homeodomains must have been present in the last common ancestor between plants and animals

two gene classes in plants and animals have arisen from a common ancestral TALE homeobox gene that also encoded – what we termed the MEINOX domain [86, 106]. BEL and KNOX proteins have been shown to interact, and this interaction is mediated through their BEL and KNOX domains, respectively [107–110].

5.15 Origin and Diversification of Homeobox Genes

The different classes of homeobox genes found in plants and animals have arisen mainly independently of each other. Only the TALE homeobox genes, in particular the MEIS and KNOX class of homeobox genes in animals and plants, respectively, can be traced to a common ancestor in primitive eukaryotes [106]. Homeobox genes have been found now also in many protists [87]. Even in these organisms one can find typical homeodomains as well as TALE homeodomain containing genes. Thus, the primitive, early eukaryotes must have had already at least two distinct types of homeodomains. A single "Urhomeobox" gene must have given rise to the typical and TALE homeobox genes, but at present we do not know whence it came from. The homeodomain is structurally related to bacterial helix-turn-helix proteins, so in some protozoa or early eukaryote a helix-turn-helix protein seems to have become the first homeodomain protein.

In animals, the diversification of the homeobox genes has led to a large proliferation of different classes and families. It has been quite surprising to find that the sea anemone *Nematostella vectensis* has many of the classes of homeobox genes found in bilateria [111]. The last common ancestor of protostomes and deuterostomes seems to have had at least 70 different homeobox genes. Many homeobox genes have been quite well preserved in evolution, but many cases are now known in which some phyla have lost particular families. Conversely, new and rapidly evolving homeobox genes have and will be discovered that are specific to particular groups of animals. Present day organisms such as *Drosophila* and *C. elegans* have around 100 homeobox genes [113], while the simple chordate *Amphioxus* has about 130 [114] and, in part due to the large scale duplications in early vertebrate evolution, humans have about 235 [23]. In plants the numbers of homeobox genes within a species is similar to that in animals; for example, *Arabidopsis thaliana* has 110 homeobox genes [105], and, as in animals, they play important roles in development. While there are many other types of TFs encoded in a genome, the homeobox genes are certainly playing a pivotal role in shaping the evolution of multicellular organisms.

References

1. Carrasco AE, McGinnis W, Gehring WJ, De Robertis EM (1984) Cloning of an *X. laevis* gene expressed during early embryogenesis coding for a peptide region homologous to *Drosophila* homeotic genes. Cell 37:409–414
2. Laughon A, Scott MP (1984) Sequence of a *Drosophila* segmentation gene: protein structure homology with DNA-binding proteins. Nature 310:25–31
3. McGinnis W, Garber RL, Wirz J, Kuroiwa A, Gehring WJ (1984a) A homologous protein-coding sequence in *Drosophila* homeotic genes and its conservation in other metazoans. Cell 37:403–408
4. McGinnis W, Hart CP, Gehring WJ, Ruddle FH (1984b) Molecular cloning and chromosome mapping of a mouse DNA sequence homologous to homeotic genes of Drosophila. Cell 38:675–680
5. McGinnis W, Levine MS, Hafen E, Kuroiwa A, Gehring WJ (1984c) A conserved DNA sequence in homoeotic genes of the *Drosophila* Antennapedia and bithorax complexes. Nature 308:428–433
6. Scott MP, Weiner AJ (1984) Structural relationships among genes that control development: Sequence homology between the *Antennapedia*, *Ultrabithorax*, and *fushi tarazu* loci in *Drosophila*. Proc Natl Acad Sci USA 81:4115–4119
7. Shepherd JCW, McGinnis W, Carrasco AE, De Robertis EM, Gehring WJ (1984) Fly and frog homoeo domains show homologies with yeast mating type regulatory proteins. Nature 310:70–71
8. Qian YQ, Billeter M, Otting G, Müller M, Gehring WJ, Wüthrich K (1989) The structure of the Antennapedia homeodomain determined by NMR spectroscopy in solution: comparison with prokaryotic repressors. Cell 59:573–580
9. Billeter M, Qian YQ, Otting G, Müller M, Gehring W, Wüthrich K (1993) Determination of the nuclear magnetic resonance solution structure of an Antennapedia homeodomain-DNA complex. J Mol Biol 234:1084–1093

10. Qian YQ, Furukubo-Tokunaga K, Resendez-Perez D, Müller M, Gehring WJ, Wüthrich K (1994) Nuclear magnetic resonance solution structure of the *fushi tarazu* homeodomain from *Drosophila* and comparison with the *Antennapedia* homeodomain. J Mol Biol 238:333–345
11. Kissinger CR, Liu B, Martin-Blanco E, Kornberg TB, Pabo CO (1990) Crystal structure of an engrailed homeodomain-DNA complex at 2.8 Å resolution: a framework for understanding homeodomain–DNA interactions. Cell 63:579–590
12. Piper DE, Batchelor AH, Chang CP, Cleary ML, Wolberger C (1999) Structure of a HoxB1-Pbx1 heterodimer bound to DNA: role of the hexapeptide and a fourth homeodomain helix in complex formation. Cell 96:587–597
13. Assa-Munt N, Mortishire-Smith RJ, Aurora R, Herr W, Wright PE (1993) The solution structure of the Oct-1 POU-specific domain reveals a striking similarity to the bacteriophage λ repressor DNA-binding domain. Cell 73:193–205
14. Dekker N, Cox M, Boelens R, Verrijzer CP, van der Vliet PC, Kaptein R (1993) Solution structure of the POU-specific DNA-binding domain of Oct-1. Nature 362:852–855
15. Ceska TA, Lamers M, Monaci P, Nicosia A, Cortese R, Suck D (1993) The X-ray structure of an atypical homeodomain present in the rat liver transcription factor LFB1/HNF1 and implications for DNA binding. EMBO J 12:1805–1810
16. Chi YI, Frantz JD, Oh BC, Hansen L, Dhe-Paganon S, Shoelson SE (2002) Diabetes mutations delineate an atypical POU domain in HNF-1alpha. Mol Cell 10:1129–1137
17. Wolberger C, Vershon AK, Liu B, Johnson AD, Pabo CO (1991) Crystal structure of MATα2 Homeodomain-operator complex suggests a general model for homeodomain-DNA interactions. Cell 67:517–528
18. Hanes SD, Brent R (1989) DNA specificity of the bicoid activator protein is determined by homeodomain recognition helix residue 9. Cell 57:1275–1283
19. Bürglin TR (1994) A comprehensive classification of homeobox genes. In Guidebook to the Homeobox Genes, Duboule D, ed. (Oxford, Oxford University Press), pp. 25–71
20. Bürglin TR (2005) Homeodomain proteins. In Encyclopedia of Molecular Cell Biology and Molecular Medicine, Meyers RA, ed. (Weinheim, Wiley-VCH Verlag GmbH & Co.), pp. 179–222
21. Gehring WJ, Affolter M, Bürglin TR (1994) Homeodomain proteins. Annu Rev Biochem 63:487–526
22. Mukherjee K, Bürglin TR (2007) Comprehensive analysis of animal TALE homeobox genes: new conserved motifs and cases of accelerated evolution. J Mol Evol 65:137–153
23. Holland PW, Booth HA, Bruford EA (2007) Classification and nomenclature of all human homeobox genes. BMC Biol 5:47
24. Duboule D, Morata G (1994) Colinearity and functional hierarchy among genes of the homeotic complexes. Trends Genet 10:358–364
25. Garcia-Fernàndez J, Holland PWH (1994) Archetypal organization of the amphioxus *Hox* gene cluster. Nature 370:563–566
26. Amores A, Force A, Yan YL, Joly L, Amemiya C, Fritz A, Ho RK, Langeland J, Prince V, Wang YL, et al (1998) Zebrafish hox clusters and vertebrate genome evolution. Science 282:1711–1714
27. Crow KD, Stadler PF, Lynch VJ, Amemiya C, Wagner GP (2006) The "fish-specific" Hox cluster duplication is coincident with the origin of teleosts. Mol Biol Evol 23:121–136
28. Prohaska SJ, Stadler PF (2004) The duplication of the Hox gene clusters in teleost fishes. Theory Biosci 123:89–110
29. Pollard SL, Holland PWH (2000) Evidence for 14 homeobox gene clusters in human genome ancestry. Curr Biol 10:1059–1062
30. Aboobaker A, Blaxter M (2003a) Hox gene evolution in nematodes: novelty conserved. Curr Opin Genet Dev 13:593–598
31. Aboobaker AA, Blaxter ML (2003b) Hox gene loss during dynamic evolution of the nematode cluster. Curr Biol 13:37–40
32. Kmita M, Duboule D (2003) Organizing axes in time and space; 25 years of colinear tinkering. Science 301:331–333

33. Damen WGM, Tautz D (1998) A Hox class 3 orthologue from the spider *Cupiennius salei* is expressed in a Hox-gene-like fashion. Dev Genes Evol 208:586–590
34. Stauber M, Jäckle H, Schmidt-Ott U (1999) The anterior determinant *bicoid* of *Drosophila* is a derived *Hox* class 3 gene. Proc Natl Acad Sci USA 96:3786–3789
35. Stauber M, Prell A, Schmidt-Ott U (2002) A single *Hox3* gene with composite *bicoid* and *zerknüllt* expression characteristics in non-Cyclorrhaphan flies. Proc Natl Acad Sci USA 99:274–279
36. Brown SJ, Hilgenfeld RB, Denell RE (1994) The beetle *Tribolium castaneum* has a fushi tarazu homolog expressed in stripes during segmentation. Proc Natl Acad Sci USA 91: 12922–12926
37. Damen WG (2002) *fushi tarazu*: a Hox gene changes its role. Bioessays 24:992–995
38. Dawes R, Dawson I, Falciani F, Tear G, Akam M (1994) *Dax*, a locust Hox gene related to *fushi-tarazu* but showing no pair-rule expression. Development 120:1561–1572
39. Zakany J, Duboule D (2007) The role of *Hox* genes during vertebrate limb development. Curr Opin Genet Dev 17:359–366
40. Brooke NM, Garcia-Fernàndez J, Holland PWH (1998) The ParaHox gene cluster is an evolutionary sister of the Hox gene cluster. Nature 392:920–922
41. Luke GN, Castro LF, McLay K, Bird C, Coulson A, Holland PW (2003) Dispersal of NK homeobox gene clusters in amphioxus and humans. Proc Natl Acad Sci U S A 100: 5292–5295
42. Mulley JF, Chiu CH, Holland PW (2006) Breakup of a homeobox cluster after genome duplication in teleosts. Proc Natl Acad Sci USA 103:10369–10372
43. Garcia-Fernàndez J (2005) The genesis and evolution of homeobox gene clusters. Nat Rev Genet 6:881–892
44. Cande JD, Chopra VS, Levine M (2009) Evolving enhancer-promoter interactions within the tinman complex of the flour beetle, *Tribolium castaneum*. Development 136:3153–3160
45. Jagla K, Bellard M, Frasch M (2001) A cluster of *Drosophila* homeobox genes involved in mesoderm differentiation programs. Bioessays 23:125–133
46. Smith ST, Jaynes JB (1996) A conserved region of engrailed, shared among all en-, gsc-, Nk1-, Nk2- and msh-class homeoproteins, mediates active transcriptional repression in vivo. Development 122:3141–3150
47. Xu W, Rould MA, Jun S, Desplan C, Pabo CO (1995) Crystal structure of a paired domain-DNA complex at 2.5 Å resolution reveals structural basis for Pax developmental mutations. Cell 80:639–650
48. Gruss P, Walther C (1992) Pax in development. Cell 69:719–722
49. Walther C, Guenet J-L, Simon D, Deutsch U, Jostes B, Goulding MD, Plachov D, Balling R, Gruss P (1991) Pax: a murine multigene family of paired box-containing genes. Genomics 11:424–434
50. Breitling R, Gerber JK (2000) Origin of the paired domain. Dev Genes Evol 210:644–650
51. Callaerts P, Halder G, Gehring WJ (1997) *Pax-6* in development and evolution. Annu Rev Neurosci 20:483–532
52. Halder G, Callaerts P, Gehring WJ (1995) Induction of ectopic eyes by targeted expression of the *eyeless* gene in *Drosophila*. Science 267:1788–1792
53. Quiring R, Walldorf U, Kloter U, Gehring WJ (1994) Homology of the *eyeless* gene of *Drosophila* to the *Small eye* gene in mice and *Aniridia* in humans. Science 265:785–789
54. Chow RL, Snow B, Novak J, Looser J, Freund C, Vidgen D, Ploder L, McInnes RR (2001) *Vsx1*, a rapidly evolving *paired*-like homeobox gene expressed in cone bipolar cells. Mech Dev 109:315–322
55. Liu ISC, Chen J, Ploder L, Vidgen D, van der Kooy D, Kalnins VI, McInnes RR (1994) Developmental expression of a novel murine homeobox gene (*Chx10*): evidence for roles in determination of the neuroretina and inner nuclear layer. Neuron 13:377–393
56. Galliot B, de Vargas C, Miller D (1999) Evolution of homeobox genes: Q50 Paired-like genes founded the Paired class. Dev Genes Evol 209:186–197

57. Sun S, Ting CT, Wu CI (2004) The normal function of a speciation gene, *Odysseus*, and its hybrid sterility effect. Science 305:81–83
58. MacLean JA 2nd, Wilkinson MF (2010) The *Rhox* genes. Reproduction 140:195–213
59. Maclean JA 2nd, Chen MA, Wayne CM, Bruce SR, Rao M, Meistrich ML, Macleod C, Wilkinson MF (2005) *Rhox*: a new homeobox gene cluster. Cell 120:369–382
60. Herr W, Sturm RA, Clerc RG, Corcoran LM, Baltimore D, Sharp PA, Ingraham HA, Rosenfeld MG, Finney M, Ruvkun G, et al. (1988) The POU domain: a large conserved region in the mammalian *pit-1, oct-1, oct-2,* and *Caenorhabditis elegans unc-86* gene products. Genes Dev 2:1513–1516
61. Jacobson EM, Li P, Leon-del-Rio A, Rosenfeld MG, Aggarwal AK (1997) Structure of Pit-1 POU domain bound to DNA as a dimer: unexpected arrangement and flexibility. Genes Dev 11:198–212
62. Klemm JD, Rould MA, Aurora R, Herr W, Pabo CO (1994) Crystal structure of the Oct-1 POU domain bound to an Octamer site: DNA recognition with tethered DNA-binding modules. Cell 77:21–32
63. Frain M, Swart G, Monaci P, Nicosia A, Stämpfli S, Frank R, Cortese R (1989) The liver-specific transcription factor LF-B1 contains a highly diverged homeobox DNA binding domain. Cell 59:145–157
64. Blochlinger K, Bodmer R, Jack J, Jan LY, Jan YN (1988) Primary structure and expression of a product from *cut*, a locus involved in specifying sensory organ identity in *Drosophila*. Nature 333:629–635
65. Iyaguchi D, Yao M, Watanabe N, Nishihira J, Tanaka I (2007) DNA recognition mechanism of the ONECUT homeodomain of transcription factor HNF-6. Structure 15:75–83
66. Yamaguchi H, Tateno M, Yamasaki K (2006) Solution structure and DNA-binding mode of the matrix attachment region-binding domain of the transcription factor SATB1 that regulates the T-cell maturation. J Biol Chem 281:5319–5327
67. Yamasaki K, Akiba T, Yamasaki T, Harata K (2007) Structural basis for recognition of the matrix attachment region of DNA by transcription factor SATB1. Nucleic Acids Res 35:5073–5084
68. Bürglin TR, Cassata G (2002) Loss and gain of domains during evolution of cut superclass homeobox genes. Int J Dev Biol 46:115–123
69. Takatori N, Saiga H (2008) Evolution of CUT class homeobox genes: insights from the genome of the amphioxus, *Branchiostoma floridae*. Int J Dev Biol *52*:969–977
70. Neufeld EJ, Skalnik DG, Lievens PM-J, Orkin SH (1992) Human CCAAT displacement protein is homologous to the *Drosophila* homeoprotein *cut*. Nature Genet 1:50–55
71. Gillingham AK, Pfeifer AC, Munro S (2002) CASP, the alternatively spliced product of the gene encoding the CCAAT-displacement protein transcription factor, is a Golgi membrane protein related to giantin. Mol Biol Cell 13:3761–3774
72. Dickinson LA, Dickinson CD, Kohwi-Shigematsu T (1997) An atypical homeodomain in SATB1 promotes specific recognition of the key structural element in a matrix attachment region. J Biol Chem 272:11463–11470
73. Morinaga T, Yasuda H, Hashimoto T, Higashio K, Tamaoki T (1991) A human alpha-fetoprotein enhancer-binding protein, ATBF1, contains four homeodomains and seventeen zinc fingers. Mol Cell Biol 11:6041–6049
74. Freyd G, Kim S, Horvitz RH (1990) Novel cysteine-rich motif and homeodomain in the product of the *Caenorhabditis elegans* cell lineage gene *lin-11*. Nature 344: 876–879
75. Srivastava M, Larroux C, Lu DR, Mohanty K, Chapman J, Degnan BM, Rokhsar DS (2010) Early evolution of the LIM homeobox gene family. BMC Biol 8:4
76. Schmeichel KL, Beckerle MC (1994) The LIM domain is a modular protein-binding interface. Cell 79:211–219
77. Agulnick AD, Taira M, Breen JJ, Tanaka T, Dawid IB, Westphal H (1996) Interactions of the LIM-domain-binding factor Ldb1 with LIM homeodomain proteins. Nature 384: 270–272

78. Oliver G, Wehr R, Jenkins NA, Copeland NG, Cheyette BNR, Hartenstein V, Zipursky SL, Gruss P (1995) Homeobox genes and connective tissue patterning. Development 121: 693–705
79. Dozier C, Kagoshima H, Niklaus G, Cassata G, Bürglin TR (2001) The *Caenorhabditis elegans* Six/sine oculis class homeobox gene *ceh-32* is required for head morphogenesis. Dev Biol 236:289–303
80. Chu-Lagraff Q, Wright DM, McNeil LK, Doe CQ (1991) The *prospero* gene encodes a divergent homeodomain protein that controls neuronal identity in *Drosophila.* Development Suppl. 2:79–85
81. Vaessin H, Grell E, Wolff E, Bier E, Jan LY, Jan YN (1991) *prospero* is expressed in neuronal precursors and encodes a nuclear protein that is involved in the control of axonal outgrowth in *Drosophila.* Cell 67:941–953
82. Bürglin TR (1994) A *Caenorhabditis elegans prospero* homologue defines a novel domain. Trends Biochem Sci 19:70–71
83. Oliver G, Sosa-Pineda B, Geisendorf S, Spana EP, Doe CQ, Gruss P (1993) Prox 1, a *prospero*-related homeobox gene expressed during mouse development. Mech Dev 44:3–16
84. Ryter JM, Doe CQ, Matthews BW (2002) Structure of the DNA binding region of prospero reveals a novel homeo-prospero domain. Structure (Camb) 10:1541–1549
85. Bertolino E, Wildt S, Richards G, Clerc RG (1996) Expression of a novel murine homeobox gene in the developing cerebellar external granular layer during its proliferation. Dev Dyn 205:410–420
86. Bürglin TR (1997) Analysis of TALE superclass homeobox genes (MEIS, PBC, KNOX, Iroquois, TGIF) reveals a novel domain conserved between plants and animals. Nucl Acids Res 25:4173–4180
87. Derelle R, Lopez P, Le Guyader H, Manuel M (2007) Homeodomain proteins belong to the ancestral molecular toolkit of eukaryotes. Evol Dev 9:212–219
88. Bürglin TR, Ruvkun G (1992) New motif in PBX genes. Nature Genet 1:319–320
89. Rauskolb C, Peifer M, Wieschaus E (1993) *extradenticle*, a regulator of homeotic gene activity, is a homolog of the homeobox-containing human proto-oncogene *pbx1*. Cell 74:1101–1112
90. Mann RS, Chan S-K (1996) Extra specificity from *extradenticle*: the partnership between HOX and PBX/EXD homeodomain proteins. Trends Genet 12:258–262
91. Pai C-Y, Kuo T-S, Jaw TJ, Kurant E, Chen C-T, Bessarab DA, Salzberg A, Sun YH (1998) The Homothorax homeoprotein activates the nuclear localization of another homeoprotein, Extradenticle, and suppresses eye development in *Drosophila.* Genes Dev 12: 435–446
92. Rieckhof GE, Casares F, Ryoo HD, Abu-Shaar M, Mann RS (1997) Nuclear translocation of Extradenticle requires *homothorax*, which encodes an Extradenticle-related Homeodomain protein. Cell 91:171–183
93. Jacobs Y, Schnabel CA, Cleary ML (1999) Trimeric association of Hox and TALE homeodomain proteins mediates *Hoxb2* hindbrain enhancer activity. Mol Cell Biol 19: 5134–5142
94. Mann RS, Affolter M (1998) Hox proteins meet more partners. Curr Opin Genet Dev 8: 423–429
95. Shen WF, Rozenfeld S, Kwong A, Kom ves LG, Lawrence HJ, Largman C (1999) HOXA9 forms triple complexes with PBX2 and MEIS1 in myeloid cells. Mol Cell Biol 19:3051–3061
96. Gómez-Skarmeta J-L, Diez del Corral R, de la Calle-Mustienes E, Ferrés-Marcó D, Modolell J (1996) *araucan* and *caupolican*, two members of the novel Iroquois complex, encode homeoproteins that control proneural and vein-forming genes. Cell 85:95–105
97. Kerner P, Ikmi A, Coen D, Vervoort M (2009) Evolutionary history of the iroquois/Irx genes in metazoans. BMC Evol Biol 9:74
98. Bertolino E, Reimund B, Wildt-Perinic D, Clerc RG (1995) A novel homeobox protein which recognizes a TGT core and functionally interferes with a retinoid-responsive motif. J Biol Chem 270:31178–31188

99. Kagoshima H, Cassata G, Bürglin TR (1999) A *Caenorhabditis elegans* homeobox gene expressed in the male tail, a link between pattern formation and sexual dimophism? Dev Genes Evol 209:59–62
100. Mizutani Y, Kihara A, Igarashi Y (2005) Mammalian Lass6 and its related family members regulate synthesis of specific ceramides. Biochem J 390:263–271
101. Pewzner-Jung Y, Ben-Dor S, Futerman AH (2006) When do Lasses (longevity assurance genes) become CerS (ceramide synthases)?: Insights into the regulation of ceramide synthesis. J Biol Chem 281:25001–25005
102. Astell CR, Ahlstrom-Jonasson L, Smith M, Tatchell K, Nasmyth KA, Hall BD (1981) The sequence of the DNAs coding for the mating-type loci of *Saccharomyces cerevisiae*. Cell 27:15–23
103. Kahmann R, Bölker M (1996) Self/nonself recognition in fungi: old mysteries and simple solutions. Cell 85:145–148
104. Bürglin TR (2003) The homeobox genes of *Encephalitozoon cuniculi* (Microsporidia) reveal a putative mating-type locus. Dev Genes Evol 213:50–52
105. Mukherjee K, Brocchieri L, Bürglin TR (2009) A comprehensive classification and evolutionary analysis of plant homeobox genes. Mol Biol Evol 26:2775–2794
106. Bürglin TR (1998) The PBC domain contains a MEINOX domain: Coevolution of Hox and TALE homeobox genes? Dev Genes Evol 208:113–116
107. Bellaoui M, Pidkowich MS, Samach A, Kushalappa K, Kohalmi SE, Modrusan Z, Crosby WL, Haughn GW (2001) The *Arabidopsis* BELL1 and KNOX TALE homeodomain proteins interact through a domain conserved between plants and animals. Plant Cell 13:2455–2470
108. Chen H, Rosin FM, Prat S, Hannapel DJ (2003) Interacting transcription factors from the three-amino acid loop extension superclass regulate tuber formation. Plant Physiol 132: 1391–1404
109. Muller J, Wang Y, Franzen R, Santi L, Salamini F, Rohde W (2001) In vitro interactions between barley TALE homeodomain proteins suggest a role for protein-protein associations in the regulation of Knox gene function. Plant J 27:13–23
110. Smith HMS, Boschke I, Hake S (2002) Selective interaction of plant homeodomain proteins mediates high DNA-binding affinity. Proc Natl Acad Sci USA 99:9579–9584
111. Ryan JF, Burton PM, Mazza ME, Kwong GK, Mullikin JC, Finnerty JR (2006) The cnidarian-bilaterian ancestor possessed at least 56 homeoboxes. Evidence from the starlet sea anemone, *Nematostella vectensis*. Genome Biol 7:R64
112. McGinnis W, Krumlauf R (1992) Homeobox genes and axial patterning. Cell 68:283–302
113. Reece-Hoyes JS, Deplancke B, Shingles J, Grove CA, Hope IA, Walhout AJ (2005) A compendium of *Caenorhabditis elegans* regulatory transcription factors: a resource for mapping transcription regulatory networks. Genome Biol 6:R110
114. Takatori N, Butts T, Candiani S, Pestarino M, Ferrier DE, Saiga H, Holland PW (2008) Comprehensive Survey and classification of homeobox genes in the genome of amphioxus, *Branchiostoma floridae*. Dev Genes Evol 218:579–590

Chapter 6
Nuclear Receptors: Small Molecule Sensors that Coordinate Growth, Metabolism and Reproduction

Keith Pardee, Aleksandar S. Necakov, and Henry Krause

Abstract One of the largest groups of metazoan transcription factors (TFs), the Nuclear Receptor superfamily, regulates genes required for virtually all aspects of development, reproduction and metabolism. Together, these master regulators can be thought of as a fundamental operating system for metazoan life. Their most distinguishing feature is a structurally conserved domain that acts as a switch, powered by the presence of small diffusible ligands. This ligand-responsive regulation has allowed the Nuclear Receptors to help their hosts adapt to a wide variety of physiological niches and roles, making them one of the most evolutionarily successful TF families. Originally discovered as receptors for steroid hormones, the Nuclear Receptor field has grown to encompass much more than traditional endocrinology. For example, recent work has highlighted the role of Nuclear Receptors as major regulators of metabolism and biological clocks. By monitoring endogenous metabolites and absorbed xenobiotics, these receptors also coordinate rapid, system-wide responses to changing metabolic and environmental states. While many new Nuclear Receptor ligands have been discovered in the past couple of decades, approximately half of the 48 human receptors are still orphans, with a significantly higher percentage of orphans in other organisms. The discovery of new ligands has led to the elucidation of new regulatory mechanisms, target genes, pathways and functions. This review will highlight both the common as well as newly emerging traits and functions that characterize this particularly unique and important TF family.

6.1 Introduction

The ability of Nuclear Receptor (NR) ligands to move relatively freely within the body, tissues and cells sets the NRs apart from other signaling systems. G

K. Pardee (✉)
Department of Biomedical Engineering, Boston University, Boston, MA 02215, USA; Wyss Institute for Biologically Inspired Engineering, Harvard University, Boston, MA 02215, USA
e-mail: keith.pardee@wyss.harvard.edu

Keith Pardee and Aleksandar S. Necakov are co-first authors.

T.R. Hughes (ed.), *A Handbook of Transcription Factors*, Subcellular Biochemistry 52,
DOI 10.1007/978-90-481-9069-0_6,

protein-coupled receptors, for example, must transduce their ligand signals from the cell surface to the nucleus through an extended chain of intermediary signaling proteins. Accordingly, the coupling of ligand-binding and DNA-binding capabilities allows NRs to directly sense the concentration of their ligands within the cell and at target gene promoters [1, 2].

NRs have been shown to act as key regulators in a diverse range of developmental and homeostatic pathways. These include embryogenesis, growth, vascular tone, detoxification, circadian rhythm, glucose and lipid homeostasis, reproduction and behavior. Consistent with their role in most, if not all, fundamental biological processes, mutations in NR genes also play a role in most human disease states including obesity, inflammation, autoimmune disorders, cardiovascular disease and cancer [3, 4]. Importantly, many of these disease states can be prevented or treated by the use of natural or synthetic NR ligands. However, only a relatively small number of all NRs have a known natural ligand. As a consequence, in an effort to identify novel therapeutics, significant attempts have been made towards the identification of both endogenous and synthetic ligands for the orphan members of the NR superfamily [5, 6].

6.2 Nuclear Receptor Domain Architecture

Early work on NRs revealed a common domain architecture, traditionally referred to by the letters A-F, from the N- to C-terminal ends respectively (Fig. 6.1). However, only the C and E domains are broadly conserved. The C domain encompasses the highly conserved DNA binding domain (DBD), and the E domain the less well conserved ligand-binding domain (LBD). These are linked by a flexible hinge region (D domain), which varies in length and sequence. Like the D domain, the A/B domains are also poorly conserved and relatively unstructured [7]. This N-terminal region is also often referred to as the Activation Function-1 (AF-1) domain, due to its general role in transcriptional activation. Some receptors also contain an extended carboxyl-terminal domain, referred to as the F-domain, which appears to have a general role in transcriptional repression [8, 9]. While there is interaction and interdependence between these domains, these regions are largely semi-autonomous and modular [10].

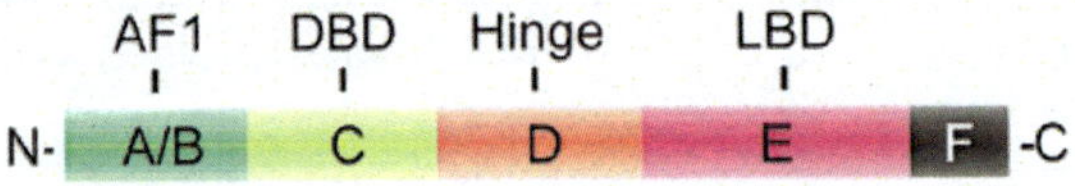

Fig. 6.1 Nuclear receptor domain structure. Beginning at the N-terminus, nuclear receptors include the N-terminal domain (A/B), DNA binding domain (C), Hinge region (D), ligand binding domain (E) and C-terminal domain (F)

6.3 NR Phylogeny and Classes

NRs have been shown to exist in most metazoan clades, including sponges, echinoderms, tunicates, arthropods and vertebrates, and are therefore believed to be present throughout the Metazoa. As such the NR superfamily represents one of the largest families of transcriptional regulators in metazoans [11–13]. Comparison of the DBD and LBD sequences has led to the classification of NRs into six subfamilies (Table 6.1) [14]. These are denoted by the prefix "NR" followed by a three-digit code. The first digit refers to the subfamily number, the second digit to the group letter and the final digit to the specific gene number [15]. As an example, the receptor Rev-erbα was given the code NR1D1, which corresponds to subfamily I, group D, and being the first gene identified in this group. This nomenclature also encompasses insect receptors. For example the *Drosophila* homolog to Rev-erbα, E75, is identified by the code NR1D3.

6.4 NR Evolution

A major question regarding the evolution of NRs is how they managed to gain the ability to bind ligands that are functionally relevant to the genes they control. For example, in most of the characterized examples, ligands are either metabolic precursors, products or targets of the gene products regulated by the receptor. Clues to this conundrum have been provided by the presence of NR genes and target genes in more primitive organisms, and the NR sequences themselves.

NRs are not found in plants, fungi or protozoa. However, if the DBD and LBD are considered separately, the yeast zinc-finger homologs containing LIM (eg: PXL1) and GATA (eg: GZF3, GATs 1-4 and DAL80) domains, and the yeast membrane protein Pex11p, have partial sequence and structural alignment with these respective NR domains. Based on these similarities, it has been proposed that NRs may have arisen from the fusion of these or related proteins in pre-metazoan eukaryotes as early as 635 million years ago [11, 24, 25]. Another protein family with interesting functional parallels to the NRs is the fungal binuclear zinc cluster TFs. Like the NRs, this family of fungal proteins can operate as monomers or dimers and is modulated by small molecules, including nutrients, metabolites and xenobiotics. The domain structure of these proteins is also remarkably similar to the NRs, beginning with an N-terminal zinc cluster DBD followed by a linker and LBD [26, 27].

Species at the base of the metazoan clade have provided further insight into the NR ancestral state [11, 28, 29]. The earliest metazoan species, the sponges, only have receptors from subfamily II (ie: HNF4/RXR), while moderately early metazoans, such as *Hydra sp.* and *Anemonia sp.,* contain a larger number of subfamily II receptors (Coup-tf, TLL, TR2/4) and a putative member of subfamily VI (GCNF ortholog) [28, 30, 31]. However, all six NR subfamilies are found within most other levels of metazoan phylogeny, suggesting that they underwent their first

Table 6.1 Summary of human nuclear receptors with their known endogenous ligands and *Drosophila* homologs

Subfamily 1: Thyroid hormone receptor-like

Group	Human receptor	Isoform	Symbol	Endogenous ligand	Fly ortholog (%ID DBD/LBD, ligand)
A	Thyroid hormone receptor	TRα	NR1A1	Thyroid hormone	
		TRβ	NR1A2	Thyroid hormone	
B	Retinoic acid receptor	RARα	NR1B1	Retinoic acid	
		RARβ	NR1B2	Retinoic acid	
		RARγ	NR1B3	Retinoic acid	
C	Peroxisome proliferator-activated receptor	PPARα	NR1C1	Fatty acids, leukotriene B4	
		PPARβ/δ	NR1C2	Fatty acids	
		PPARγ	NR1C3	Fatty acids	
D	Rev-erbα and β	Rev-erbα	NR1D1	Heme, NO, CO	E75 (80/25, Heme, NO, CO), E78 (69/23)
		Rev-erbβ	NR1D2	Heme, NO, CO	
F	RAR-related orphan receptor	RORα	NR1F1	Cholesterol, cholesterol sulphate	DHR3 (76/35, Cholesterol)
		RORβ	NR1F2	All trans retinoic acid	
		RORγ	NR1F3	Retinoic acid	
H	Liver X receptor	LXRα	NR1H3	Oxysterols	EcR (64/37, 20-hydroxyecdysone)
		LXRβ	NR1H2	Oxysterols	
	Farnesoid X receptor	FXR	NR1H4	Bile acids, lanosterol	EcR (72/28, 20-hydroxyecdysone)
I	Vitamin D receptor	VDR	NR1I1	Vitamin D_3, bile acids	DHR96 (55/20, Cholesterol)
	Pregnane X receptor	PXR	NR1I2	Xenobiotics	
	Constitutive androstane receptor	CAR	NR1I3	Androstane, xenobiotics	

Table 6.1 (continued)

Group	Human receptor	Isoform	Symbol	Endogenous ligand	Fly ortholog (%ID DBD/LBD, ligand)
Subfamily 2: Retinoid X receptor-like					
A	Hepatocyte nuclear factor-4	HNF4α	NR2A1	Fatty acids	DHNF4 (89/61)
		HNF4γ	NR2A2	Fatty acids	
B	Retinoid X receptor	RXRα	NR2B1	9-cis-Retinoic acid; Heme?	Ultraspiracle (84/43, Phosphatidyl-ethanolamine)
		RXRβ	NR2B2	9-cis-Retinoic acid	
		RXRγ	NR2B3	9-cis-Retinoic acid	
C	Testicular receptor	TR2	NR2C1	Orphan	DHR78 (67/23)
		TR4	NR2C3	Orphan	
E	Human homologue of the *Drosophila* tailless gene	TLX	NR2E1	Orphan	Tailless (80/34), Dissatisfaction (74/35)
	Photoreceptor cell-specific nuclear receptor	PNR	NR2E3	Orphan	DHR51 (70/47, Heme, NO, CO) DNR83 (60/20)
F	Chicken ovalbumin upstream promoter-transcription factor	COUP-TFI	NR2F1	Orphan	Seven up (89/92)
		COUP-TFII	NR2F2	Orphan	
	V-erbA-related gene IV-erbA-related	EAR-2	NR2F6	Orphan	
Subfamily 3: Estrogen receptor-like					
Group	Human receptor	Isoform	Symbol	Endogenous ligand	Fly ortholog (%ID DBD/LBD, ligand)
A	Estrogen receptor	ERα	NR3A1	Estradiol-17β	
		ERβ	NR3A2	Estradiol-17β	
B	Estrogen-related receptor	ERRα	NR3B1	Orphan	DERR (88/34)
		ERRβ	NR3B2	Orphan	
		ERRγ	NR3B3	Orphan	
C	Glucocorticoid receptor	GR	NR3C1	Cortisol	
	Mineralocorticoid receptor	MR	NR3C2	Aldosterone	

Table 6.1 (continued)

	Progesterone receptor	PR	NR3C3	Progesterone	
	Androgen receptor	AR	NR3C4	Testosterone	
Subfamily 4: Nerve growth factor IB-like					
Group	Human receptor	Isoform	Symbol	Endogenous ligand	Fly ortholog (%ID DBD/LBD, ligand)
A	Nerve growth FACTOR IB	NGFIB	NR4A1	Orphan	
	Nuclear receptor related 1	NURR1	NR4A2	Orphan	DHR38 (93/59)
	Neuron-derived orphan receptor 1	NOR1	NR4A3	Orphan	
Subfamily 5: Steroidogenic factor-like					
Group	Human receptor	Isoform	Symbol	Endogenous ligand	Fly ortholog (%ID DBD/LBD, ligand)
A	Steroidogenic factor 1	SF1	NR5A1	Phospholipid	Ftz-f1 (88/28), DHR39 (60/26)
	Liver receptor homolog-1	LRH-1	NR5A2	Phospholipid	Ftz-f1 (89/35), DHR39 (62/25)
Subfamily 6: Germ cell nuclear factor-like					
Group	Human receptor	Isoform	Symbol	Endogenous ligand	Fly ortholog (%ID DBD/LBD, ligand)
A	Germ cell nuclear factor	GCNF	NR6A1	Orphan	DHR4 (61/21)
Subfamily 0: Miscellaneous					
Group	Human receptor	Isoform	Symbol	Endogenous ligand	Fly ortholog (%ID DBD/LBD, ligand)
B	Dosage-sensitive sex reversal, adrenal hypoplasia critical region, on chromosome X, gene 1	DAX1	NR0B1	Orphan	
	Small heterodimer partner	SHP	NR0B2	Orphan	
C	Nuclear receptors with two DNA binding domains	2DBD-NR		Orphan	

Data for Table 6.1 compiled from [1, 16–23].

round of diversification before the metazoans experienced significant phylogenetic radiation. The second wave of vertebrate NR diversification occurred much later and followed the divergence of invertebrate and vertebrate lineages. This event generated the paralogous groups and much of the diversity observed within human receptors. It is this multiplicity that has led to isoform-specific tissue expression, function and ligand binding profiles for many of the vertebrate receptors [11, 16, 32].

The current consensus is that the primordial NR was not ligand regulated. Instead this feature appears to have developed independently multiple times during evolutionary history. Although receptors may have first evolved as apo-proteins, it has been suggested that the first "ligands" were permanent co-factors. Recalling that the common ancestral NR may have belonged to subfamily II [29], this makes a lot of sense. This subfamily contains HNF4 and USP, two receptors that appear to bind non-exchangeable structural co-factors rather than conventional ligands [23, 33–35]. If the LBD of the common ancestral receptor relied on a co-factor, this constraint may have contributed to the structural conservation we see in LBDs across the receptor group. Furthermore, such an ancestral receptor provides a plausible model for the development of reversible ligand binding. The exchange of such a co-factor with structurally similar compounds from the cellular environment may have been the origin of ligand binding [11, 36, 37]. As one possible example, the NRD1 orthologue in flies, E75, requires heme as a structural component [19], whereas the vertebrate counterparts, the Rev-erbs, can exchange heme readily with no apparent effects on stability [22, 38, 39].

6.5 Co-evolution of Receptors and Ligands

A genome-wide comparison of NR/ligand pairs has also led to the understanding that in many cases there is no correlation between receptor subfamilies and the biosynthetic origin of their ligands. The most dramatic example of this can be found in subfamily I. Although the TRs, RARs, PPARs and VDRs are all closely related by sequence, their ligands are all synthesized in highly divergent metabolic pathways and differ highly in structure (Table 6.1). The same lack of correlation is also apparent with the RXR and RAR receptors. Although these bind structurally and biosynthetically similar ligands, 9-*cis* retinoic acid and all-*trans* retinoic acid respectively, they are some of the most distantly related human receptors [11, 40].

A few key observations can be made based on the distribution of receptors and ligands. Most importantly, NRs and their ligands did not, in general, co-evolve. Second, this lack of correlation between receptor and ligand suggests that the original coupling of receptors and ligands likely resulted from beneficial fortuitous interactions [11, 40, 41]. However, there is evidence that once the receptor/ligand pair was functionally coupled, ligand binding specificity and affinity may have co-evolved through mutations to either the receptor or the enzymes of ligand biosynthesis [42]. The vertebrate steroid receptors provide an interesting example. The ancestral steroid receptor is the Estrogen Receptor (ER), which can be found in both protostome and deuterostome species. From this one primordial steroid receptor, vertebrate-specific diversification gave rise to the six variants in present day vertebrates. Genome sequence analyses suggests that these transitions were accompanied by the evolution of steroidogenic and steroid-specific catabolic enzymes, producing new potential ligands and genes for the new paralogs to bind and regulate [43].

Interestingly, *Drosophila* appears to have lost the original steroid-binding receptors. However, it has developed a parallel steroidal signaling system centered around ecdysteroids and the subfamily I receptor, Ecdysone Receptor (EcR) [40, 42]. It appears that the EcR and other sterol binding members of subfamily I likely evolved from a common ancestor, referred to as proto-FXR/LXR/EcR, which acquired the ability to bind steroids independent of the subfamily III steroid receptors [32, 40]. The insect EcR from subfamily I serves a functionally parallel role to the vertebrate sex steroid receptors of subfamily III in vertebrates. So remarkably, both vertebrates and invertebrates have evolved independent steroid-based developmental NRs [1, 44].

6.6 DBD Structure and Function

The DBD serves as a gene locator for the receptor by docking to specific hexanucleotide sequences or response elements (REs) in the promoter/enhancer regions of gene targets. Sequence conservation is highest in the DBD, which is due presumably to a need to conserve binding site specificity as well as structural stability within such a small domain [9, 37]. Making up the core of the DBD are two zinc finger motifs, each containing four cysteine residues that together coordinate a single zinc atom. These cysteine–zinc interactions stabilize the domain in place of a hydrophobic core. The N-terminal helix of the DBD interacts with the major groove of the DNA, and thus it is this sequence, called the P-box, which defines the DNA binding specificity of the receptor. The second helix lies perpendicular to helix 1, and contributes to domain stability and dimerization with the partner DBD.

NRs have been categorized into four classes based on their mode of DNA binding (Fig. 6.2) [45]. The first, class I, defines the mechanism of action for the steroid receptors. For most class I NRs, ligand binding occurs in the cytoplasm, which triggers the shedding of chaperones and translocation to the nucleus. Once there, they bind to inverted hexanucleotide repeats as homodimers in a head-to-head configuration. Class II NRs form heterodimers with RXR and bind to hexanucleotide direct repeats in a head-to-tail configuration. These NRs bind to DNA independent of ligand status. As apo-receptors, they silence gene expression, and in the presence of a ligand, transcription is activated [36, 45]. Many of the NRs in classes III and IV are orphans, and not surprisingly, remain less understood and more heterogeneous. Like class I receptors, class III receptors homodimerize but only bind promoters with hexameric direct repeats. Class IV receptors can bind as either monomers or dimers, but are unique in that they bind only single hexameric sites. As the orphans become better understood their regulatory

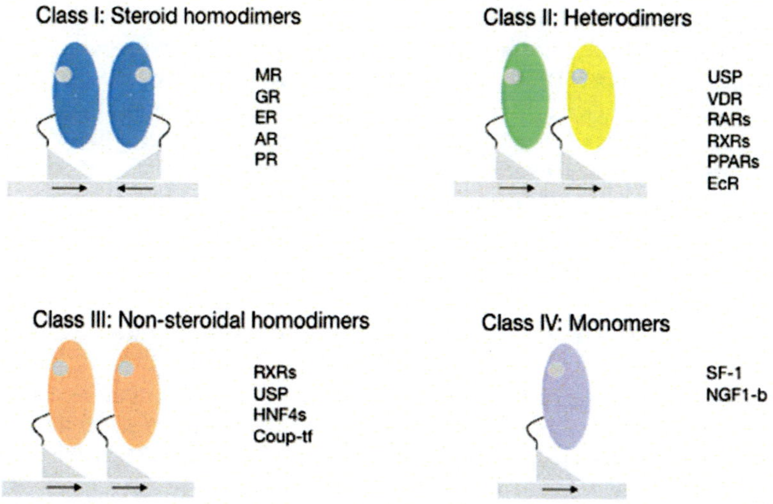

Fig. 6.2 Four Nuclear Receptor classes based on DNA response elements. Nuclear receptors can be grouped into four classes according to their ligand binding, DNA binding, and dimerization properties: steroid receptors, RXR heterodimers, homodimeric orphan receptors, and monomeric orphan receptors. Steroid receptors bind to DNA at inverted response element (RE) repeats as homodimers. RXR heterodimers bind to DNA at direct RE repeats. Homodimeric orphan receptors bind to DNA at direct RE repeats. Monomeric orphan receptors bind to single REs as individual monomers. Shown are representative receptors for each group with known ligands. Adapted from [45]

features will likely become even more enmeshed with the class I and II receptors [9, 16, 36, 37].

Given the high sequence similarity between the hexameric REs recognized by NR DBDs, a key factor in determining their target gene specificity is the orientation and spacing of REs within promoters [9, 16, 36, 46]. In the absence of dimerization, NR monomers derive further DNA specificity through interactions between a DNA-binding motif C-terminal of the DBD (called the C-terminal extension) and DNA sequence immediately 5′ of the hexameric RE. For many NR monomers, this interaction contributes significant specificity and stability to DNA binding [47–50]. Even so, further specificity cues are required in vivo to discriminate between the tens or hundreds of thousands of potential binding sites and those that are functional. Recent genome-wide binding studies suggest that the average NR is bound to a subset of approximately 5,000–10,000 binding sites within a particular cell type, approximately 10% of which contribute to changes in gene expression levels (reviewed in [51]). These studies also suggest that additional cell-specific target gene selection is provided by differences in chromatin accessibility and cofactors that help tether and stabilize NRs on active sites [52–56].

Although the classical view of NR DNA binding holds that the DBD is responsible only for site-specific recognition and binding to DNA, it should be noted that recent evidence from structural studies has indicated that binding of the DBD to specific RE sequences may be more than just a mechanism for localizing the receptor to the correct DNA sequence [57]. In fact, it has been hypothesized that the exact hexanucleotide sequence of the RE affects not only the overall affinity of the receptor for its RE site, but also influences the three dimensional (3-D) configuration of the receptor, thereby regulating NR activity through the binding of certain ancillary factors [57, 58]. Recent studies have also shown that interactions between different NR domains can influence DNA contacts and binding site specificity [59].

6.7 LBD Structure and Functional Classifications

The LBD, as the name indicates, is responsible for binding of the receptor's cognate ligand(s) and can be thought of as a molecular switch that mediates the transcription output. Although the sequence conservation of this domain between receptors can be as low at 15%, the 3-D structure is nonetheless universally conserved. The secondary structure, in most solved structures, is composed of 12 helices and three short β-strands. Described as an α-helical sandwich, the LBD comprises three antiparallel layers of helices that form the sides and central layer of the fold (Fig. 6.3) [10]. The central and generally hydrophobic core of this globular domain is absent in the lower half of the domain, and it is this non-polar cavity that forms the ligand-binding pocket. The sides of this pocket are composed of the outer layers of the α-helical sandwich, and the front and back are formed by helix 12 and two to three β-strands, respectively.

As alluded to earlier, the LBD also serves as a primary mediator for the self-assembly of receptors into homo or heterodimers. This dimerization, mediated primarily by helices 9 and 10, contributes to the specificity of DNA binding by correctly spacing and orienting the DBD subunits [9, 60]. Interestingly, while dimerization interfaces between respective LBDs and DBDs are well established, the importance of interdomain contacts was only recently shown with the solution of the first intact DBD/LBD dimer structure. In this PPARγ/RXRα structure, the DBD and LBD of the opposite heterodimer partners also form dimerization interfaces that contribute to the stability of the complex [59].

Another critical function of the LBD is to serve as a platform for the binding and assembly of transcriptional co-activator or co-repressor complexes. These proteins are recruited or shed from the LBD surface depending on the ligand-binding state of the domain. Early crystal structures showed that much of the structural basis for these transitions lies with the positioning of helix 12, which moves in the presence of ligand to close off the ligand binding pocket, and to redesign the cofactor binding grooves. In the absence of ligand, co-repressors containing LXXI/HIXXXI/L motifs (also referred to as a co-repressor nuclear-receptor [CoRNR] box) can bind, whereas in the presence of an agonist ligand, co-activators containing LXXLL motifs can bind [9, 10]. An example of such interactions is clearly demonstrated by the apo-, agonist- and antagonist-bound forms

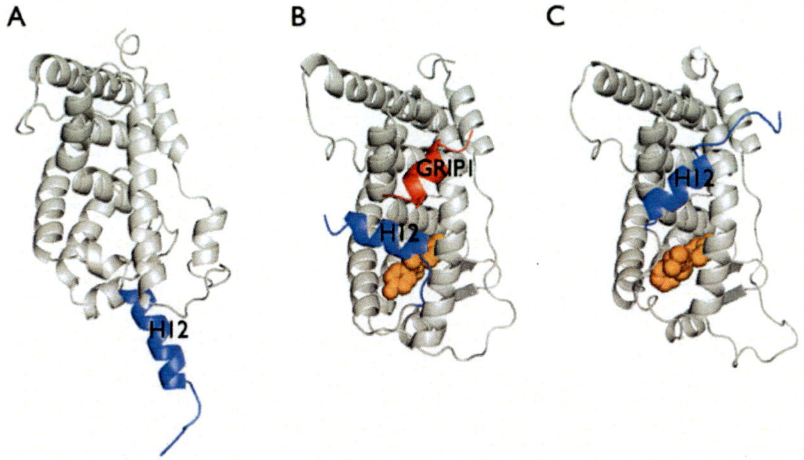

Fig. 6.3 Structural basis for ligand-response. The structures shown demonstrate the canonical apo, agonist-bound and antagonist-bound conformations of NR LBDs [63]. (**a**) The unliganded form of RXRα shows H12 extending away from the body of the LBD and H11 partially occupying the ligand binding pocket (RXRα; PDB: 1LDB). (**b**) Agonist bound ERα, in association with a co-activator GRIP1 peptide, is in a transcriptionally active conformation (ERα/diethylstilbestrol; PDB 3ERD; [64]). (**c**) Antagonist bound ERα in a transcriptionally inactive conformation. The molecular extension of 4-hydroxytamoxifen protrudes from the ligand-binding pocket to displace AF-2/helix 12, which instead occupies the hydrophobic groove and blocks co-activator binding (ERα/4-hydroxytamoxifen; PDB 3ERT; [64]). Helix 12 is in *blue*, the GRIP1 peptide is in *red*, agonist/antagonist ligands are in *orange* and the main body of the LBD is in *grey*

of the ERα and RAR LBDs (Fig. 6.3). Co-repressors tend to recruit chromatin condensing complexes, whereas co-activators recruit chromatin opening and RNA polymerase II holoenzyme recruiting complexes [61]) (Fig. 6.4). Although the majority of well-characterized cofactors are those that interact with the LBD, there are also numerous NR–specific interactions made by the variable A/B domains. Comprehensive reviews on NR cofactors have recently been published [61, 62].

As orphan receptors have become adopted, the conventionality of this LBD response has been challenged. The human receptors CARβ and RORβ, for instance, appear to be constitutively active in the absence of a ligand. In what could be described as the inverse of the NR model, their respective endogenous ligands androstane and all-trans retinoic acid, repress the high basal transcription levels of the apo receptors [65, 66]. There are also likely to be many NRs that require a ligand in order to fully repress their target genes, as appears to be the case for the NR1D receptors (E75 and Rev-erbs). These NRs lack the canonical helix 12, and appropriately, appear to function as dedicated repressors. As more ligands are discovered, further variations on the theme defined by the steroid receptors are likely to be discovered.

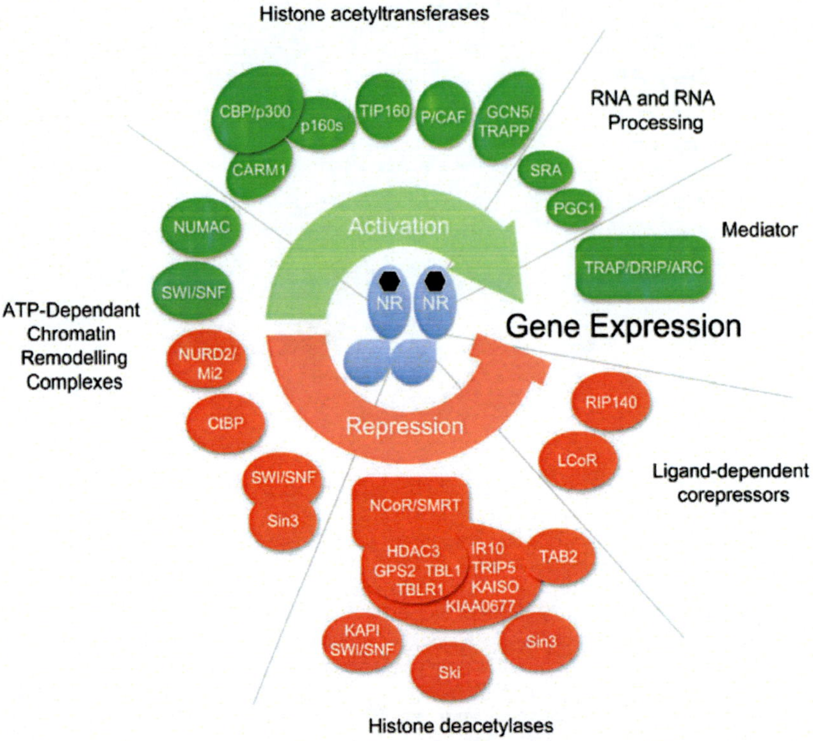

Fig. 6.4 Coactivator and corepressor complexes. Coactivator complexes (*green*) include factors that contain ATP-dependent chromatin remodeling histone arginine methyltransferase and histone acetyltransferase activities. They may also contain factors involved in RNA processing as well as components of the Mediator complex. Conversely, co-repressors (*red*) recruit histone deacetylases and other chromatin-condensing enzymes and cofactors. Adapted from [67]

6.8 NR Ligands

Ligands for NRs are small and hydrophobic, giving them the general ability to move relatively freely between tissues and cells. The binding of hydrophobic ligands also contributes to the stability of the cognate receptor, completing the LBD hydrophobic core and setting up further intra- and inter-molecular interactions [10, 68]. The ligand responsive nature of NRs has meant that, long before the advent of modern pharmacology, NRs have been probed by the chemical diversity surrounding them. These interactions include fortuitous ecological compounds, and now man-made pollutants, as well as compounds actively synthesized by plants and animals for chemical defence. An intriguing example of this is the fly EcR and the evolution of ecdysteroids outside of insect taxa. Ecdysteroids are the steroids responsible for the timing of development in insects [1, 44]. However, ecdysteroids have also been identified in many sessile species, such as soft coral and many plants, at concentrations 2–5 fold greater than what is found in insects. While these organisms do not contain an EcR, they synthesize precise chemical mimics of the insect ecdysteroid 20-hydroxyecdysone (20E), as well as many other biologically active variations of the insect hormone. This inappropriate EcR activation disrupts the insects developmental program, resulting in lethality [69–72].

As evidenced by the use of plants as the source of our first pharmaceuticals, the ability of natural products to mimic the structure of receptor ligands is not limited to interactions with the insect world. Some examples of interactions with mammalian NRs are the isoflavones (phytoestrogens) from legumes, which interact with the ER, and have been shown to reduce the risk of certain cancers and heart disease [73]. The plant sterol guggulsterone is another interesting example, in which interaction with FXR has been shown to reduce serum cholesterol in mammals [74]. Interestingly ecdysteroids also have a physiological effect on mammals, with positive effects on muscle strength, lipid metabolism and immunity being some of the most cited [75]. Such interactions hold both risk to human health, as in the case of endocrine disruptors, and potential benefit in the form of new drugs. Accordingly, NR-based drug discovery and toxicology screens are actively probing the natural environment for interacting compounds [76–78]. These interactions are discussed in greater detail below (section 6.12).

Many NR drugs do not simply agonize or antagonize the receptor, but have pharmacological selectivity that comes from the disruption of specific receptor/co-regulator interactions that are either responsible for only a subset of receptor functions or are cell-type specific [16]. One of the best-studied examples of these selective nuclear receptor modulators (SNuRMs) is the anti-cancer drug tamoxifen. While tamoxifen serves as an antagonist to combat ER positive cancer in breast tissue, it conversely serves as an ER agonist in the bone and uterus, where ER activity is still needed. The tissue-specific nature of this response is, at least in part, the result of differential co-factor distribution, with the co-activator SRC1 at higher levels in the uterus and bone [15, 16, 79, 80]. Similar drugs are being developed for other NRs, and hold promise for overcoming the side effects of current treatments. Such new selective modulators provide a model for not only the next generation of NR drugs, but may also inspire similar strategies for other therapeutic targets [5, 81].

With the discoveries of recently de-orphaned NR ligands has come the realization of new ligand types and interaction mechanisms. Some of the first endocrine NRs to be characterized, such as ER, TR and VDR, were found to bind their ligands with high affinity, but also to readily release or exchange their ligands. The more recently deorphanized metabolic NRs tend to bind physiologically abundant ligands, with a lower affinity. Several other NRs, such as HNF4α/γ and the fly NRs E75 and USP, bind molecules that appear to serve as permanent co-factors or prosthetic groups [19, 23, 33–35]. This diversity of ligand types and interactions is continuing to grow. One of the most surprising and unusual is the E75/Rev-erb ligand heme and its retention in the LBD pocket via coordinate bonds between the heme iron and amino acid side-chains. Even more unusual is the ability of E75/Rev-erb-heme to bind the diatomic gases Nitric oxide and Carbon monoxide, which displace one of the coordinate bonds [19, 22, 38, 39]. The unconventional and unexpected nature of these new ligands may explain, in part, the recalcitrance of the remaining orphans to reveal their ligand identities. There are almost certainly new surprises waiting in the wings.

6.9 The Orphan Receptors

Several solved LBD structures lack a ligand-binding pocket, which has led to the suggestion that many orphan NRs may be authentic orphan receptors with no ligand counterpart. For example, Nurr1, and its fly homolog DHR38 [82, 83], and Rev-erb β [84], when purified from bacterial expression systems, contained no ligand or pocket. Accordingly, it was suggested that these LBDs function simply as transcriptionally active platforms for constitutive cofactor binding [16, 68, 85]. This interpretation was strengthened by observations that these and other NR LBDs can recruit cofactors in the absence of ligand. One consequence of these findings has been a decrease in drug development programs directed against NRs by the major pharmaceutical companies.

Recent studies, however, have challenged this commonly held point of view. For example, the Rev-erb LBDs have since been shown to be capable of binding heme, with significant LBD structural changes made to accommodate this relatively large molecule [22]. This type of flexibility, in terms of LBD pocket size and shape, has also been found with a number of other NRs [22, 86–89]. Thus, not only may orphan LBDs exist in apo and bound forms, but some may also be capable of binding multiple and diverse ligand types, as observed with PXR [90] and EcR [86]. These ligands could serve as agonists or antagonists that elicit both quantitative and qualitative differences in activities. An exciting consideration is that some of these ligands may exist only in certain tissues, with unique outcomes on cofactor recruitment, target gene selection and ensuing levels of expression.

6.10 Other Modes of Nuclear Receptor Activity Modulation

Beyond the ligand-binding pocket, there are many established ligand-independent modes of control that influence the transcriptional activity of NRs. As with other TFs, post-translational modifications are a significant contributor to NR responses and responsiveness. Phosphorylation, sumoylation, ubiquitynation and acetylation can all influence receptor stability, localization and cofactor interactions [36, 68, 91–93]. The widespread nature of post-translational signaling and the growing recognition of ligand-independent modes of control have led to calls to broaden research efforts beyond the LBD towards the consideration of a "multivalent allosteric switch" that reacts to a wide range of inputs [16]. The details of these alternative modes of regulation, however, are likely to be receptor-specific in terms of the degree and mechanism of action, whereas LBD–ligand interactions will tend to have more universal and pervasive consequences.

6.11 Nuclear Receptor Functions

NRs have generally been classified functionally into one of two groups, endocrine or metabolic/xenobiotic. However, it is becoming increasingly clear that NRs have a large number of functions that bridge these broad roles, as well as many others. In fact, these initial classifications are relatively uninformative and misleading. Nevertheless, for historical and clarification purposes, these groups are described below.

The endocrine receptors, which include the Androgen Receptor (AR), Mineralocorticoid Receptor (MR), Glucocorticoid Receptor (GR), Progesterone Receptor (PR), ERα and ERβ are largely recognized for their roles in developmental and reproductive biology [16]. Orthologues for these NRs in *Drosophila*, *C. elegans* and *Ciona*, are largely absent. In fact, aside from the orphan dERRs, the steroid subfamily III is completely absent from these genomes [40]. When this absence was first noted, it was assumed that steroid receptors must be a product of vertebrate evolution [12, 32]. However, as mentioned earlier, ER orthologues are found in early metazoans. Thus the fly and worm clade, termed the Ecdysozoans, appears to have undergone loss of all but one subfamily III steroid receptor gene [13].

The metabolic and xenobiotic sensors, comprised of the PPARs, FXRs, LXRs, RORs, Rev-erbs, and HNF4s, allow organisms to respond to metabolic imbalances and changes in their environment. Ligands for these receptors are often nutritionally important compounds or intermediates and products of key biochemical pathways [3]. Together, these receptors have been shown to form a network that ensures energy and metabolic homeostasis [22, 38, 39, 94, 95] (Fig. 6.5). In their surveillance of metabolism, many of these NRs regulate the genes involved in the production, destruction or trafficking of their own ligands. For example, to aid in metabolite clearance, metabolic NRs upregulate catalytic enzymes, such as P450s, to transform excess compounds into less active/more soluble intermediates. These same receptors also promote pathways and transporters involved in the ultimate elimination of these compounds [96–98].

The xenobiotic receptors (PXR, CAR, ERRs) form a parallel system that monitors the chemical diversity surrounding the organism for chemical threats in the environment. In the same way the metabolic receptors respond to an oversupply of endogenous metabolites, these receptors respond to toxic threats by upregulating catabolic enzymes and transporters [3–5, 99]. Together, these two receptor systems, metabolic and xenobiotic, form a sensing and response network throughout the body, and are particularly important in the gut. As one of the largest interfaces with the outside chemical world, these NRs help the enteric tract cue genetic responses to our changing nutritional status as well as pathogenic and toxic challenges [94, 100].

It has been 20 years since the first metabolic NRs were identified. The RXR/RARs and the PPARs were found to bind retinoic acid and fatty acid metabolites, respectively [101–104]. In the time since, an entire subgroup of receptors has been identified as being regulated by endogenous metabolites. While dietary and membrane lipids, heme and metabolic waste products may not have originally

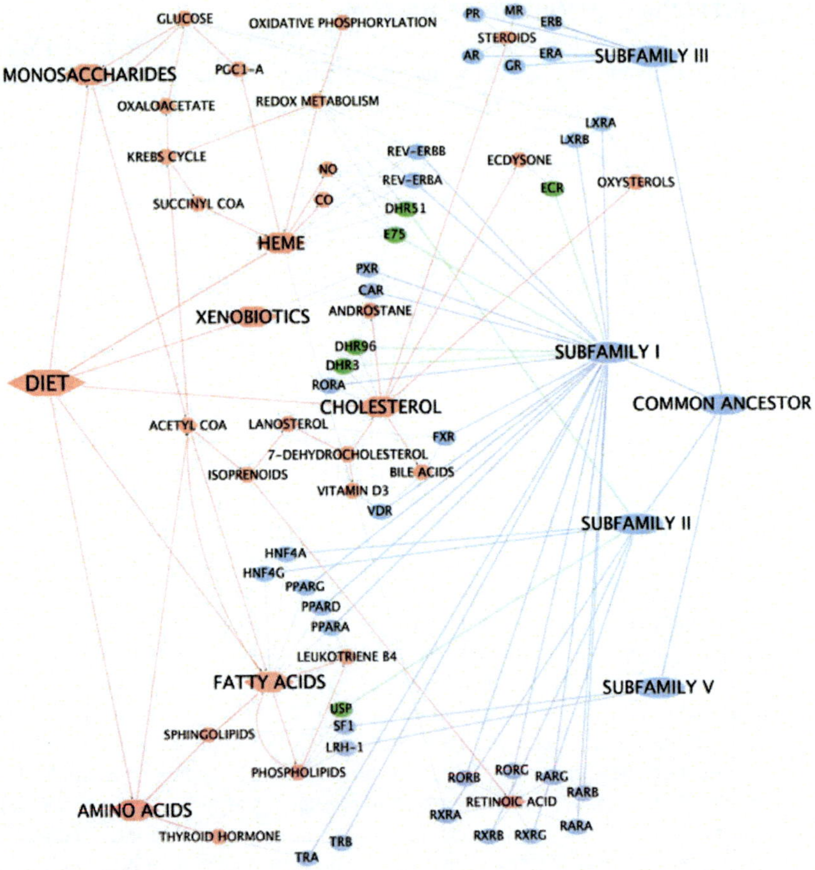

Fig. 6.5 Nuclear receptors form a network of sensors that synchronize target gene expression with diverse small molecule signaling and metabolic flux. In their surveillance of hormone signaling and disparate arms of metabolism, NRs integrate the chemical signaling environment of the cell with metabolic, developmental and reproductive gene expression. In a form of feedback regulation, many of these NRs regulate the genes involved in the regulation of their own ligands and other related metabolites. Examples of both receptor/ligand co-evolution (steroid receptors) and the independent evolution of receptor/ligand pairs (retinoids with RARs/RORs/RXRs) can be found in the superfamily. Originating with the dietary uptake, the metabolic pathways connect individual metabolites, marked by *red hexagons*. Nuclear receptors are indicated by *ovals*, which are either *blue* (human) or *green* (fly) and linked by phylogenetic relationships. Binding between ligands and receptors are indicated by *grey lines*. Orphan receptors that have yet to have cognate ligands identified are not shown

had the appeal of highly specific endocrine hormones, the importance of these ligands and their receptors has now been realized. These metabolic NRs are partly responsible for the now widely understood intercalation of metabolism with virtually all aspects of development and physiology [3, 94]. For example, recent work has highlighted the potential role of metabolic NRs in the regulation of circadian rhythm and development [44, 105, 106]. In the fly, a network of NRs that are expressed in response to developmental pulses of ecdysone have been reported to bind and be transcriptionally regulated by a wide ranging collection of metabolites ([19, 21, 107–108]; DHR3: Krause unpublished results). The addition of metabolite sensing to the NR-mediated regulation of fly development brings a substantial layer of information-rich signaling to hormonal timing. These metabolic ligands have redefined the relationship between the NRs of the ecdysone response pathway from an autonomous system set in motion by an ecdysone pulse [109], to one that is responsive to the state of the organism and its environment.

Of the metabolic receptors, perhaps the fly receptor E75 and its human homologs, Rev-erbα and β, are the most novel. Until these receptors were recently de-orphaned, known NR ligands were limited to steroid hormones, fatty acids and other dietary and non-dietary lipids [3]. E75 and the Rev-erbs bind heme as a ligand and/or prosthetic group that allows for gas (NO, CO) and redox responsive transcriptional regulation [19, 22, 38, 39]. Like lipids, heme has long been recognized as an important molecule in metabolism. It is required for oxygen and carbon dioxide transport, for cytochrome function in the mitochondria and for the neutralization of reactive oxygen species (ROS) arising as a consequence of metabolism. It is also a required component of the cytochrome P450s that produce and break down most lipids, including those that serve as the ligands of most nuclear receptors [110–117].

In much the same way that E75, DHR3 and their ligands coordinate the developmental process of the fly, heme/gas/redox and cholesterol, respectively, serve as metabolic indicators to the mammalian molecular clock through the NRs Rev-erbα/β and RORα. As a regulatory couple, the Rev-erbs and RORα entrain the expression of other clock proteins to these fundamental measures of cell metabolism [22, 38, 39, 46, 118]. These inputs contribute to the more established modes of circadian entrainment, such as photoperiod, which together comprise a system of independent measures that coordinates metabolism with sleep wake cycles. This newly recognized capacity to monitor heme/gas/redox gives the NR superfamily access to regulatory signaling at the core of mammalian physiology. Heme abundance oscillates during the circadian cycle and, importantly, also functions as a prosthetic group to other circadian proteins, including NPAS2, Period2, and Clock [112–116, 119]. Given that heme is so central to respiration and other central metabolic processes, and that its abundance oscillates over time, it appears that heme serves as a fundamental measure of the diurnal metabolic state and as such provides feedback through the Rev-erbs and other clock proteins, to entrain the molecular clock to an organism's diurnal metabolic flux [22]. The circadian clock also appears to be in control of lunar and annual functions, which also need to be linked closely to nutrient availability and temperature fluctuations [120, 121].

Like heme, the other E75/Rev-erb regulators, redox and gas, are also generated in a circadian manner [117, 122, 123]. Redox homeostasis can be affected by the generation of reactive oxygen species (ROS), a large proportion of which arise not surprisingly from mitochondrial respiration. The redox state of a cell, or organelles, is dependent on the ratio of ROS generated by metabolic activity and the abundance of antioxidants, both of which cycle diurnally (reviewed in [124–126]) and accordingly serve as an important measure of metabolic activity. Aside from the damage that ROS can cause, these molecules have also become recognized as important signaling molecules. Interestingly, ROS signaling is commonly associated with stress response [127], which like the molecular clock, is coordinated in the hypothalamus [128, 129]. Thus, as was mentioned earlier, distinct functions for NRs are often difficult to prescribe. In addition to a clear metabolic role, E75 and the Rev-erbs may also fulfill functions of stress-response, much like GR, or even xenobiotic surveillance of environmental oxidative stress.

As mentioned, the gases NO and CO also cycle with circadian periodicity. Heme is an essential component of both NO and CO producing enzymes, respectively Nitric oxide synthase and Heme oxygenase. Thus, not surprisingly, both NO and CO production have also been shown to oscillate diurnally [117, 122, 123]. The membrane permeable and transient nature of NO and CO gases conform to the ideal signaling properties of many other specialized NR ligands, and further connect the NR superfamily to a broad range of physiology.

It is interesting to note that, in retrospect, many of the processes influenced by Rev-erb proteins and their ROR counterparts, such as circadian rhythm, metabolism and inflammation, have long been known to involve NO/CO gas signaling [130]. For eample, there is an inverse relationship between heme and nitric oxide in the transactivation of NF-κB, a gene known to be regulated by Rev-erbα [131]. While heme leads to an increased activation of NF-κB [132], NO inhibits its activity [133, 134]. Likewise, NO and CO also affect the establishment and growth of cholesterol-rich plaques within arteries [135, 136], where RORs and Rev-erbs also play major roles [130, 137, 138]. This coincidence extends further to include mood/behavior disorders and obesity etiologies that have been associated with aberrant Rev-erb expression and/or circadian rhythm [139–143]. Taken together, one can imagine a scenario where the Rev-erb/heme/redox/gas signaling axis acts to coordinate overall energy management with diurnal cycles, feeding behaviour, local tissue metabolism and other related processes.

Our recent work in *Drosophila* shows that NO signaling via the Rev-erb/ROR orthologues, DHR3 and E75, also controls the timing of larval molts and metamorphosis, suggesting that transitions between growth, diapause and reproductive phases of the life cycle are also coordinated by these receptors (Caceres et al, in prep). Interestingly, disruption of these interactions results in either morbid obesity or wasting, depending on the direction of the genetic, ligand or chemical manipulation. The ability of these and other NRs to control and respond to dietary and circadian variations has likely played a major role in the ability of metazoa to adapt to so many diverse ecological niches. This assumption is consistent with the appearance of NRs during the Cambrian explosion and the ability of these new organisms to develop multicellularity and invade new ecosystems and environments [144]. Part of this diversification involved the evolution of new endocrine and metabolic organs that further enabled nutrient selection, uptake, storage and management, as well as efficient means of optimizing and balancing the growth and reproductive phases of the lifecycle. Accordingly, recent evidence has shown that both stem cell pluripotency and differentiation into various cell and tissue types is also guided by the actions of a number of receptors [145–147].

As these new realms of ligand diversity and NR functions now show, the NR regulated processes of development, growth, metabolism and reproduction are deeply intertwined and reciprocally regulated, and it is the integration of these systems throughout the body that defines the NR superfamily. Also deeply related to these functions are the associated behaviors that make food consumption, reproduction and survival in different ecosystems possible. As with xenobiotic responses, the ability to mount immune responses to environmental pathogens is also under the control of NRs [94, 148, 149], and is modulated in a clock-dependent fashion [150].

In summary, it is clear that NR functions can no longer be categorized simply into hormonal or metabolic roles. A new subdivision into distinct functional categories is going to be challenging. This challenge will likely only grow as the roles of the less studied receptors, particularly the orphans, become better understood.

6.12 Medical Impact

Since Elwood Jensen's landmark discovery of the receptor for estrogen (ER), the degree to which NRs feature in the cause and prevention of diseases has become increasingly clear. ER on its own has been implicated as a major player in a broad range of disease states. As with the other early identified endocrine receptors, these include sexual, developmental and growth disorders, as well as a variety of cancers (reviewed in [151–157]). More recently, prominent roles for ER in obesity, behavioral disorders and aging have also been uncovered. Likewise, a survey of the literature reveals roles for most of the other NRs in virtually all aspects of human disease [4, 5, 158]. These diseases can be instigated by a variety of means including genetic mutations, endocrine tissue disruption, drugs and toxins, inappropriate diet, autoimmune disorders, lack of sunlight or the complexities of aging. For many of the same reasons that NRs and their ligands can cause disease, they can also play positive roles in disease prevention or cure. For example, Vitamin D and omega-3 fatty acids have recently been shown to have major beneficial effects on cancer prevention, immunity, metabolism, mood and memory.

Considering that many NRs control the expression of genes that promote cellular growth or differentiation, it is not surprising that NRs play a major role in cancer onset and progression, as well as prevention and therapy. A classic example is the role of ERs in breast cancer, and the use of antagonists such as tamoxifen or raloxifene to treat it. Similarly, many other NRs have since been linked to the onset, progression and treatment of many other cancers (reviewed in [159]). Recent examples of NRs and ligands being used to treat or prevent cancer include VDR, RAR, RXR and their cognate ligands to reestablish programmed cell death in various tumor types [160–162].

NRs play a major role in diseases stemming from defects in immunity. These include a large variety of autoimmune diseases, asthma, acnes, and numerous other inflammatory reactions. Since its discovery in 1948, the GR ligand cortisone has been used to treat many of these diseases and reactions [159, 163–165]. More recently, selective GR agonists (SEGRAs) have garnered considerable attention as potential therapeutics in the treatment of autoimmunity [166]. In addition to GR, several other NRs have been shown to be involved in the regulation of immune responses (reviewed in [165]). For example, ER has been implicated as a potential target in regulating autoimmune responses that underlie multiple sclerosis [167], and FXR, PXR and VDR, originally characterized for their roles as bile acid and xenobiotic sensors, have emerged as potent modulators of immune and inflammatory reactions in entero-hepatic tissues (reviewed in [168]). PPARγ, LXRα and β, VDR, NURR1, and RAR have also now been shown to have important regulatory functions in immune cells (reviewed in [165, 169, 170]), with PPARγ recently shown to also play a prominent role in multiple sclerosis (reviewed in [171]).

Considering their roles as core components of metabolic homeostasis, it is not surprising that NRs contribute significantly to metabolic diseases. The fundamental importance of these regulatory networks is becoming increasingly clear in light of the rapidly rising, near-epidemic levels of metabolic disorders that comprise "metabolic syndrome". These include obesity, diabetes, cardiovascular disease, hyperlipidemia, atherosclerosis and hypertension [3, 4]. The following projections from the World Health Organization (WHO) provide some insight into the frequency of metabolic disorders worldwide [172].

- Globally in 2005 approximately 1.6 billion adults (age 15+) were overweight and at least 400 million adults were obese.
- In 2005, an estimated 1.1 million people died from diabetes.
- More than 220 million people currently have diabetes.
- By 2015, approximately 2.3 billion adults are expected to be overweight and more than 700 million obese.

Although the most sensible long-term solution to this problem lies in prevention, molecular medicine has shown tremendous promise in offering a means of treatment in the late stages of these diseases, and in extreme cases where dietary modification on its own is insufficient to restore health.

Although it is clear that the types and volumes of food currently consumed in modern societies are a major contributor to the current metabolic disease pandemic, it appears that a number of industry generated (synthetic?) compounds may also be to blame. It has been known for some time that industrial compounds such as Bisphenol A (BPA), and contraceptive contamination of wastewater runoffs, affect ER and ERR activities in animals and humans. However, it is becoming increasingly clear that these and other endocrine disrupting compounds also affect a number of other NRs, including GR, TR, PPARs and RXRs, with striking effects on adipocyte proliferation, differentiation and function (reviewed in [173–175]). Several new screening strategies capable of identifying these NR-targeted "obesogens" have recently been described [176–178]. These approaches hold great promise towards the identification of obesogenic compounds in industrial, agricultural and municipal effluents and byproducts.

Another emerging area in which NRs have been shown to play a critical role is circadian rhythm and associated sleep-based disorders. As mentioned previously, the role of the Rev-erbs and the RORs in the mammalian circadian clock has become increasingly evident. The identification of several other NRs including ER, RAR, PPARα and γ, and EAR2 as regulators of the circadian clock has helped to further demonstrate that NR signaling and metabolism form an integral part of the circadian timing system (reviewed in [179–183]). Considering the importance of circadian rhythm in the regulation of metabolism, obesity and depression, it will be important to fully explore the medical implications of these circuits and mechanisms.

Circadian and metabolic clocks also play a key role in controlling lifespan. Without exception, excessive dietary intake leads to life threatening diseases, while reduced caloric intake has been shown to prolong life span [184–189]. Cholesterol, lipid metabolism and NRs have also been linked to a variety of other age-associated neuronal diseases such as Alzheimer's, Parkinson's, Niemann Pick, Fragile X and Huntington disease (reviewed in [21, 190–194]). Correspondingly, several NRs have been identified as playing particularly important roles in related neuronal processes. Examples include ER, which has been shown to regulate cognition and synaptic plasticity [195, 196], LXR, which functions as a major regulator of genes involved in cholesterol homeostasis and which has been implicated in Alzheimer's [197], the PPARs, which function as regulators of aging through their roles in lipid homeostasis [198, 199], and VDR which elicits neuroprotective functions and plays a beneficial role in both the developing brain and in adult cognition [200, 201]).

Considering the well-established interplay between circadian rhythm, metabolism, cancer, and immunity it will be important to further understand how NRs regulate and integrate these superficially distinct processes. Further NR ligand identification should help provide new insights into the pathways and processes that give rise to these diseases, as well as new means to prevent and treat them.

6.13 Conclusions

It seems increasingly clear from research in model organisms, and with the latest round of ligand discoveries, that the general role of NRs is to match rates of growth, development, and reproduction to the available dietary and physical offerings provided by unique environmental niches. The co-diversification of NR proteins and ligands has allowed metazoa to adapt and differentiate their lifecycles, diets, metabolism and behaviors to meet the challenges of diverse and hostile environments.

While the importance of NRs in mammalian physiology and disease has helped spur considerable research and progress, we still know relatively little about the majority of NRs outside this metazoan class. There is still much to learn about all of the existing orphans, and about the roles of NRs in numerous tissues and developmental stages. One particularly challenging frontier will be the brain, where both

NRs and their ligands are particularly abundant. Behaviors linked to metabolism, development, sexual diversification and reproduction will likely be controlled by these NRs and their metabolites.

Although a growing body of research is finding that NRs are subject to many forms of ligand-independent regulation, which play important roles in controlling and fine-tuning NR activities and functions, it will likely continue to be the discovery of new ligands that drives this field forward at a maximal pace. These ligands will switch on the lights that illuminate new and unexpected roles and pathways.

References

1. King-Jones K, Thummel CS (2005) Nuclear receptors – a perspective from *Drosophila*. Nat Rev Genet 6:311–323
2. Willson TM, Moore JT (2002) Genomics versus orphan nuclear receptors – a half-time report. Mol Endocrinol 16:1135–1144
3. Chawla A, Repa JJ, Evans RM, Mangelsdorf DJ (2001) Nuclear receptors and lipid physiology: opening the X-files. Science 294:1866–1870
4. Francis GA, Fayard E, Picard F, Auwerx J (2003) Nuclear receptors and the control of metabolism. Annu Rev Physiol 65:261–311
5. Chen T (2008) Nuclear receptor drug discovery. Curr Opin Chem Biol 12:418–426
6. Huang P, Chandra V, Rastinejad F (2010) Structural overview of the nuclear receptor superfamily: insights into physiology and therapeutics. Annu Rev Physiol 72:247–272
7. Warnmark A, Treuter E, Wright AP, Gustafsson JA (2003) Activation functions 1 and 2 of nuclear receptors: molecular strategies for transcriptional activation. Mol Endocrinol 17:1901–1909
8. Olefsky JM (2001) Nuclear receptor minireview series. J Biol Chem 276:36863–36864
9. Bain DL, Heneghan AF, Connaghan-Jones KD, Miura MT (2007) Nuclear receptor structure: implications for function. Annu Rev Physiol 69:201–220
10. Weatherman RV, Fletterick RJ, Scanlan TS (1999) Nuclear-receptor ligands and ligand-binding domains. Annu Rev Biochem 68:559–581
11. Escriva H, Bertrand S, Laudet V (2004) The evolution of the nuclear receptor superfamily. Essays Biochem 40:11–26
12. Laudet V (1997) Evolution of the nuclear receptor superfamily: early diversification from an ancestral orphan receptor. J Mol Endocrinol 19:207–226
13. Thornton JW, Need E, Crews D (2003) Resurrecting the ancestral steroid receptor: ancient origin of estrogen signaling. Science 301:1714–1717
14. Escriva Garcia H, Laudet V, Robinson-Rechavi M (2003) Nuclear receptors are markers of animal genome evolution. J Struct Funct Genomics 3:177–184
15. Germain P, Staels B, Dacquet C, Spedding M, Laudet V (2006) Overview of nomenclature of nuclear receptors. Pharmacol Rev 58:685–704
16. Gronemeyer H, Gustafsson JA, Laudet V (2004) Principles for modulation of the nuclear receptor superfamily. Nat Rev Drug Discov 3:950–964
17. Noy N (2007) Ligand specificity of nuclear hormone receptors: sifting through promiscuity. Biochemistry 46:13461–13467
18. Robinson-Rechavi M, Carpentier AS, Duffraisse M, Laudet V (2001) How many nuclear hormone receptors are there in the human genome? Trends Genet 17:554–556
19. Reinking J, Lam MM, Pardee K, Sampson HM, Liu S, et al. (2005) The *Drosophila* nuclear receptor e75 contains heme and is gas responsive. Cell 122:195–207
20. de Rosny E, de Groot A, Jullian-Binard C, Borel F, Suarez C, et al. (2008) DHR51, the *Drosophila melanogaster* Homologue of the Human Photoreceptor Cell-Specific Nuclear Receptor, Is a Thiolate Heme-Binding Protein. Biochemistry 47:13252–13260

21. Horner MA, Pardee K, Liu S, King-Jones K, Lajoie G, et al. (2009) The *Drosophila* DHR96 nuclear receptor binds cholesterol and regulates cholesterol homeostasis. Genes Dev 23:2711–2716
22. Pardee KI, Xu X, Reinking J, Schuetz A, Dong A, et al. (2009) The structural basis of gas-responsive transcription by the human nuclear hormone receptor REV-ERBbeta. PLoS Biol 7:e43
23. Potier N, Billas IM, Steinmetz A, Schaeffer C, van Dorsselaer A, et al. (2003) Using nondenaturing mass spectrometry to detect fortuitous ligands in orphan nuclear receptors. Protein Sci 12:725–733
24. Barnett P, Tabak HF, Hettema EH (2000) Nuclear receptors arose from pre-existing protein modules during evolution. Trends Biochem Sci 25:227–228
25. Clarke ND, Berg JM (1998) Zinc fingers in *Caenorhabditis elegans*: finding families and probing pathways. Science 282:2018–2022
26. Naar AM, Thakur JK (2009) Nuclear receptor-like transcription factors in fungi. Genes Dev 23:419–432
27. Woodson JD, Chory J (2008) Coordination of gene expression between organellar and nuclear genomes. Nat Rev Genet 9:383–395
28. Grasso LC, Hayward DC, Trueman JW, Hardie KM, Janssens PA, et al. (2001) The evolution of nuclear receptors: evidence from the coral Acropora. Mol Phylogenet Evol 21:93–102
29. Wiens M, Batel R, Korzhev M, Muller WE (2003) Retinoid X receptor and retinoic acid response in the marine sponge Suberites domuncula. J Exp Biol 206:3261–3271
30. Gauchat D, Escriva H, Miljkovic-Licina M, Chera S, Langlois MC, et al. (2004) The orphan COUP-TF nuclear receptors are markers for neurogenesis from cnidarians to vertebrates. Dev Biol 275:104–123
31. Reitzel AM, Tarrant AM (2009) Nuclear receptor complement of the cnidarian Nematostella vectensis: phylogenetic relationships and developmental expression patterns. BMC Evol Biol 9:230
32. Escriva H, Delaunay F, Laudet V (2000) Ligand binding and nuclear receptor evolution. Bioessays 22:717–727
33. Dhe-Paganon S, Duda K, Iwamoto M, Chi YI, Shoelson SE (2002) Crystal structure of the HNF4 alpha ligand binding domain in complex with endogenous fatty acid ligand. J Biol Chem 277:37973–37976
34. Tocchini-Valentini GD, Rochel N, Escriva H, Germain P, Peluso-Iltis C, et al. (2009) Structural and functional insights into the ligand-binding domain of a nonduplicated retinoid X nuclear receptor from the invertebrate chordate amphioxus. J Biol Chem 284:1938–1948
35. Wisely GB, Miller AB, Davis RG, Thornquest AD Jr., Johnson R, et al. (2002) Hepatocyte nuclear factor 4 is a transcription factor that constitutively binds fatty acids. Structure 10:1225–1234
36. McEwan IJ (2004) Sex, drugs and gene expression: signalling by members of the nuclear receptor superfamily. Essays Biochem 40:1–10
37. Schwabe JW, Teichmann SA (2004) Nuclear receptors: the evolution of diversity. Sci STKE 2004:pe4
38. Raghuram S, Stayrook KR, Huang P, Rogers PM, Nosie AK, et al. (2007) Identification of heme as the ligand for the orphan nuclear receptors REV-ERBalpha and REV-ERBbeta. Nat Struct Mol Biol 14:1207–1213
39. Yin L, Wu N, Curtin JC, Qatanani M, Szwergold NR, et al. (2007) Rev-erbalpha, a heme sensor that coordinates metabolic and circadian pathways. Science 318:1786–1789
40. Bertrand S, Brunet FG, Escriva H, Parmentier G, Laudet V, et al. (2004) Evolutionary genomics of nuclear receptors: from twenty-five ancestral genes to derived endocrine systems. Mol Biol Evol 21:1923–1937
41. Escriva H, Safi R, Hanni C, Langlois MC, Saumitou-Laprade P, et al. (1997) Ligand binding was acquired during evolution of nuclear receptors. Proc Natl Acad Sci USA 94: 6803–6808

42. Markov GV, Paris M, Bertrand S, Laudet V (2008) The evolution of the ligand/receptor couple: A long road from comparative endocrinology to comparative genomics. Mol Cell Endocrinol 293:5–16
43. Baker ME (2004) Co-evolution of steroidogenic and steroid-inactivating enzymes and adrenal and sex steroid receptors. Mol Cell Endocrinol 215:55–62
44. Thummel CS (2001) Molecular mechanisms of developmental timing in *C. elegans* and *Drosophila*. Dev Cell 1:453–465
45. Mangelsdorf DJ, Thummel C, Beato M, Herrlich P, Schütz G, et al. (1995) The nuclear receptor superfamily: the second decade. Cell 83:835–839
46. Guillaumond F, Dardente H, Giguere V, Cermakian N (2005) Differential control of Bmal1 circadian transcription by REV-ERB and ROR nuclear receptors. J Biol Rhythms 20: 391–403
47. Ueda H, Sun GC, Murata T, Hirose S (1992) A novel DNA-binding motif abuts the zinc finger domain of insect nuclear hormone receptor FTZ-F1 and mouse embryonal long terminal repeat-binding protein. Mol Cell Biol 12:5667–5672
48. Wilson TE, Fahrner TJ, Milbrandt J (1993) The orphan receptors NGFI-B and steroidogenic factor 1 establish monomer binding as a third paradigm of nuclear receptor–DNA interaction. Mol Cell Biol 13:5794–5804
49. Schrader M, Becker-Andre M, Carlberg C (1994) Thyroid hormone receptor functions as monomeric ligand-induced transcription factor on octameric half-sites. Consequences also for dimerization. J Biol Chem 269:6444–6449
50. Gearhart MD, Holmbeck SM, Evans RM, Dyson HJ, Wright PE (2003) Monomeric complex of human orphan estrogen related receptor-2 with DNA: a pseudo-dimer interface mediates extended half-site recognition. J Mol Biol 327:819–832
51. Carlberg C, Seuter S (2010) Dynamics of nuclear receptor target gene regulation. Chromosoma 119:479–484
52. Ramagopalan SV, Heger A, Berlanga AJ, Maugeri NJ, Lincoln MR, et al. (2010) A ChIP-seq defined genome-wide map of vitamin D receptor binding: Associations with disease and evolution. Genome Res 20:1352–1360
53. Stender JD, Kim K, Charn TH, Komm B, Chang KC, et al. (2010) Genome-wide analysis of estrogen receptor alpha DNA binding and tethering mechanisms identifies Runx1 as a novel tethering factor in receptor-mediated transcriptional activation. Mol Cell Biol 30: 3943–3955
54. Zhao C, Gao H, Liu Y, Papoutsi Z, Jaffrey S, et al. (2010) Genome-wide mapping of estrogen receptor-beta-binding regions reveals extensive cross-talk with transcription factor activator protein-1. Cancer Res 70:5174–5183
55. Lin Z, Reierstad S, Huang CC, Bulun SE (2007) Novel estrogen receptor-alpha binding sites and estradiol target genes identified by chromatin immunoprecipitation cloning in breast cancer. Cancer Res 67:5017–5024
56. Chong HK, Infante AM, Seo YK, Jeon TI, Zhang Y, et al. (2010) Genome-wide interrogation of hepatic FXR reveals an asymmetric IR-1 motif and synergy with LRH-1. Nucleic Acids Res 38:6007–6017
57. Thompson EB, Kumar R (2003) DNA binding of nuclear hormone receptors influences their structure and function. Biochem Biophys Res Commun 306:1–4
58. Azoitei A, Spindler-Barth M (2009) DNA affects ligand binding of the ecdysone receptor of *Drosophila melanogaster*. Mol Cell Endocrinol 303:91–99
59. Chandra V, Huang P, Hamuro Y, Raghuram S, Wang Y, et al. (2008) Structure of the intact PPAR-gamma-RXR- nuclear receptor complex on DNA. Nature 456:350–356
60. Gampe RT Jr., Montana VG, Lambert MH, Miller AB, Bledsoe RK, et al. (2000) Asymmetry in the PPARgamma/RXRalpha crystal structure reveals the molecular basis of heterodimerization among nuclear receptors. Mol Cell 5:545–555
61. Perissi V, Jepsen K, Glass CK, Rosenfeld MG (2010) Deconstructing repression: evolving models of co-repressor action. Nat Rev Genet 11:109–123

62. Baek SH, Rosenfeld MG (2004) Nuclear receptor coregulators: their modification codes and regulatory mechanism by translocation. Biochem Biophys Res Commun 319:707–714
63. Bourguet W, Ruff M, Chambon P, Gronemeyer H, Moras D (1995) Crystal structure of the ligand-binding domain of the human nuclear receptor RXR-alpha. Nature 375: 377–382
64. Shiau AK, Barstad D, Loria PM, Cheng L, Kushner PJ, et al. (1998) The structural basis of estrogen receptor/coactivator recognition and the antagonism of this interaction by tamoxifen. Cell 95:927–937
65. Forman BM, Tzameli I, Choi HS, Chen J, Simha D, et al. (1998) Androstane metabolites bind to and deactivate the nuclear receptor CAR-beta. Nature 395:612–615
66. Stehlin-Gaon C, Willmann D, Zeyer D, Sanglier S, Van Dorsselaer A, et al. (2003) All-trans retinoic acid is a ligand for the orphan nuclear receptor ROR beta. Nat Struct Biol 10:820–825
67. Perissi V, Rosenfeld MG (2005) Controlling nuclear receptors: the circular logic of cofactor cycles. Nat Rev Mol Cell Biol 6:542–554
68. Nagy L, Schwabe JW (2004) Mechanism of the nuclear receptor molecular switch. Trends Biochem Sci 29:317–324
69. Adler JH, Grebenok RJ (1999) Occurrence, biosynthesis, and putative role of ecdysteroids in plants. Crit Rev Biochem Mol Biol 34:253–264
70. Baker KD, Warren JT, Thummel CS, Gilbert LI, Mangelsdorf DJ (2000) Transcriptional activation of the *Drosophila* ecdysone receptor by insect and plant ecdysteroids. Insect Biochem Mol Biol 30:1037–1043
71. Thummel CS, Chory J (2002) Steroid signaling in plants and insects – common themes, different pathways. Genes Dev 16:3113–3129
72. Bakrim A, Maria A, Sayah F, Lafont R, Takvorian N (2008) Ecdysteroids in spinach (Spinacia oleracea L.): biosynthesis, transport and regulation of levels. Plant Physiol Biochem 46:844–854
73. Ososki AL, Kennelly EJ (2003) Phytoestrogens: a review of the present state of research. Phytother Res 17:845–869
74. Urizar NL, Liverman AB, Dodds DT, Silva FV, Ordentlich P, et al. (2002) A natural product that lowers cholesterol as an antagonist ligand for FXR. Science 296:1703–1706
75. Dinan L, Lafont R (2006) Effects and applications of arthropod steroid hormones (ecdysteroids) in mammals. J Endocrinol 191:1–8
76. Lascombe I, Beffa D, Ruegg U, Tarradellas J, Wahli W (2000) Estrogenic activity assessment of environmental chemicals using in vitro assays: identification of two new estrogenic compounds. Environ Health Perspect 108:621–629
77. Doukhanina EV, Apuya NR, Yoo HD, Wu CY, Davidow P, et al. (2007) Expression of human nuclear receptors in plants for the discovery of plant-derived ligands. J Biomol Screen 12:385–395
78. Grun F, Blumberg B (2007) Perturbed nuclear receptor signaling by environmental obesogens as emerging factors in the obesity crisis. Rev Endocr Metab Disord 8:161–171
79. Shang Y, Brown M (2002) Molecular determinants for the tissue specificity of SERMs. Science 295:2465–2468
80. Jensen EV, Jordan VC (2003) The estrogen receptor: a model for molecular medicine. Clin Cancer Res 9:1980–1989
81. Li Y, Wang Z, Furukawa N, Escaron P, Weiszmann J, et al. (2008) T2384, a novel antidiabetic agent with unique peroxisome proliferator-activated receptor gamma binding properties. J Biol Chem 283:9168–9176
82. Baker KD, Shewchuk LM, Kozlova T, Makishima M, Hassell A, et al. (2003) The *Drosophila* orphan nuclear receptor DHR38 mediates an atypical ecdysteroid signaling pathway. Cell 113:731–742
83. Wang Z, Benoit G, Liu J, Prasad S, Aarnisalo P, et al. (2003) Structure and function of Nurr1 identifies a class of ligand-independent nuclear receptors. Nature 423:555–560

84. Woo EJ, Jeong DG, Lim MY, Jun Kim S, Kim KJ, et al. (2007) Structural insight into the constitutive repression function of the nuclear receptor Rev-erbbeta. J Mol Biol 373: 735–744
85. Teague SJ (2003) Implications of protein flexibility for drug discovery. Nat Rev Drug Discov 2:527–541
86. Billas IM, Iwema T, Garnier JM, Mitschler A, Rochel N, et al. (2003) Structural adaptability in the ligand-binding pocket of the ecdysone hormone receptor. Nature 426:91–96
87. Chrencik JE, Orans J, Moore LB, Xue Y, Peng L, et al. (2005) Structural disorder in the complex of human pregnane X receptor and the macrolide antibiotic rifampicin. Mol Endocrinol 19:1125–1134
88. Yamagishi K, Yamamoto K, Mochizuki Y, Nakano T, Yamada S, et al. (2010) Flexible ligand recognition of peroxisome proliferator-activated receptor-gamma (PPARgamma). Bioorg Med Chem Lett 20:3344–3347
89. Nettles KW, Bruning JB, Gil G, O'Neil EO, Nowak J, et al. (2007) Structural plasticity in the oestrogen receptor ligand-binding domain. EMBO Reports 8:563–568
90. Orans J, Teotico DG, Redinbo MR (2005) The nuclear xenobiotic receptor pregnane X receptor: recent insights and new challenges. Mol Endocrinol 19:2891–2900
91. Hammer GD, Krylova I, Zhang Y, Darimont BD, Simpson K, et al. (1999) Phosphorylation of the nuclear receptor SF-1 modulates cofactor recruitment: integration of hormone signaling in reproduction and stress. Mol Cell 3:521–526
92. Lee YK, Choi YH, Chua S, Park YJ, Moore DD (2006) Phosphorylation of the hinge domain of the nuclear hormone receptor LRH-1 stimulates transactivation. J Biol Chem 281: 7850–7855
93. Hwang EJ, Lee JM, Jeong J, Park JH, Yang Y, et al. (2009) SUMOylation of RORalpha potentiates transcriptional activation function. Biochem Biophys Res Commun 378: 513–517
94. Schmidt DR, Mangelsdorf DJ (2008) Nuclear receptors of the enteric tract: guarding the frontier. Nutr Rev 66:S88–97
95. Burris TP (2008) Nuclear hormone receptors for heme: REV-ERBalpha and REV-ERBbeta are ligand-regulated components of the mammalian clock. Mol Endocrinol 22:1509–1520
96. Lefebvre P, Cariou B, Lien F, Kuipers F, Staels B (2009) Role of bile acids and bile acid receptors in metabolic regulation. Physiol Rev 89:147–191
97. Storch J, Thumser AE (2000) The fatty acid transport function of fatty acid-binding proteins. Biochim Biophys Acta 1486:28–44
98. Sun J, Hoshino H, Takaku K, Nakajima O, Muto A, et al. (2002) Hemoprotein Bach1 regulates enhancer availability of heme oxygenase-1 gene. Embo J 21:5216–5224
99. Tremblay GB, Kunath T, Bergeron D, Lapointe L, Champigny C, et al. (2001) Diethylstilbestrol regulates trophoblast stem cell differentiation as a ligand of orphan nuclear receptor ERR beta. Genes Dev 15:833–838
100. Wahli W (2008) A gut feeling of the PXR, PPAR and NF-kappaB connection. J Intern Med 263:613–619
101. Giguere V, Ong ES, Segui P, Evans RM (1987) Identification of a receptor for the morphogen retinoic acid. Nature 330:624–629
102. Petkovich M, Brand NJ, Krust A, Chambon P (1987) A human retinoic acid receptor which belongs to the family of nuclear receptors. Nature 330:444–450
103. Mangelsdorf DJ, Umesono K, Kliewer SA, Borgmeyer U, Ong ES, et al. (1991) A direct repeat in the cellular retinol-binding protein type II gene confers differential regulation by RXR and RAR. Cell 66:555–561
104. Gottlicher M, Widmark E, Li Q, Gustafsson JA (1992) Fatty acids activate a chimera of the clofibric acid-activated receptor and the glucocorticoid receptor. Proc Natl Acad Sci U S A 89:4653–4657
105. Truman JW (1971) Hour-glass behavior of the circadian clock controlling eclosion of the silkmoth Antheraea pernyi. Proc Natl Acad Sci U S A 68:595–599

106. Truman JW, Riddiford LM, Safranek L (1974) Temporal patterns of response to ecdysone and juvenile hormone in the epidermis of the tobacco hornworm, Manduca sexta. Dev Biol 39:247–262
107. Billas IM, Moulinier L, Rochel N, Moras D (2001) Crystal structure of the ligand-binding domain of the ultraspiracle protein USP, the ortholog of retinoid X receptors in insects. J Biol Chem 276:7465–7474
108. Clayton GM, Peak-Chew SY, Evans RM, Schwabe JW (2001) The structure of the ultraspiracle ligand-binding domain reveals a nuclear receptor locked in an inactive conformation. Proc Natl Acad Sci U S A 98:1549–1554
109. Riddiford LM (1993) Hormone receptors and the regulation of insect metamorphosis. Receptor 3:203–209
110. Paoli M, Marles-Wright J, Smith A (2002) Structure-function relationships in heme-proteins. DNA Cell Biol 21:271–280
111. Schneider S, Marles-Wright J, Sharp KH, Paoli M (2007) Diversity and conservation of interactions for binding heme in b-type heme proteins. Nat Prod Rep 24:621–630
112. Dioum EM, Rutter J, Tuckerman JR, Gonzalez G, Gilles-Gonzalez MA, et al. (2002) NPAS2: a gas-responsive transcription factor. Science 298:2385–2387
113. Kaasik K, Lee CC (2004) Reciprocal regulation of haem biosynthesis and the circadian clock in mammals. Nature 430:467–471
114. Pardee K, Reinking J, Krause H (2004) Nuclear hormone receptors, metabolism, and aging: what goes around comes around. Transcription factors link lipid metabolism and aging-related processes. Sci Aging Knowledge Environ 2004:re8
115. Rogers PM, Ying L, Burris TP (2008) Relationship between circadian oscillations of Rev-erbalpha expression and intracellular levels of its ligand, heme. Biochem Biophys Res Commun 368:955–958
116. Yang J, Kim KD, Lucas A, Drahos KE, Santos CS, et al. (2008) A novel heme-regulatory motif mediates heme-dependent degradation of the circadian factor period 2. Mol Cell Biol 28:4697–4711
117. Tsiftsoglou AS, Tsamadou AI, Papadopoulou LC (2006) Heme as key regulator of major mammalian cellular functions: molecular, cellular, and pharmacological aspects. Pharmacol Ther 111:327–345
118. Kallen JA, Schlaeppi JM, Bitsch F, Geisse S, Geiser M, et al. (2002) X-ray structure of the hRORalpha LBD at 1.63 A: structural and functional data that cholesterol or a cholesterol derivative is the natural ligand of RORalpha. Structure 10:1697–1707
119. Lukat-Rodgers GS, Correia C, Botuyan MV, Mer G, Rodgers KR (2010) Heme-based sensing by the mammalian circadian protein CLOCK. Inorg Chem 49:6349–6365
120. Schibler U, Naef F (2005) Cellular oscillators: rhythmic gene expression and metabolism. Curr Opin Cell Biol 17:223–229
121. Hamblen M, Zehring WA, Kyriacou CP, Reddy P, Yu Q, et al. (1986) Germ-line transformation involving DNA from the period locus in *Drosophila melanogaster*: overlapping genomic fragments that restore circadian and ultradian rhythmicity to per0 and per- mutants. J Neurogenet 3:249–291
122. Rubio MF, Agostino PV, Ferreyra GA, Golombek DA (2003) Circadian heme oxygenase activity in the hamster suprachiasmatic nuclei. Neurosci Lett 353:9–12
123. Ferreyra GA, Cammarota MP, Golombek DA (1998) Photic control of nitric oxide synthase activity in the hamster suprachiasmatic nuclei. Brain Res 797:190–196
124. Hardeland R, Coto-Montes A, Poeggeler B (2003) Circadian rhythms, oxidative stress, and antioxidative defense mechanisms. Chronobiol Int 20:921–962
125. Tu BP, McKnight SL (2006) Metabolic cycles as an underlying basis of biological oscillations. Nat Rev Mol Cell Biol 7:696–701
126. Winterbourn CC (2008) Reconciling the chemistry and biology of reactive oxygen species. Nat Chem Biol 4:278–286

127. Winterbourn CC, Hampton MB (2008) Thiol chemistry and specificity in redox signaling. Free Radic Biol Med 45:549–561
128. Walker E, Mittal V, Tessner K (2008) Stress and the hypothalamic pituitary adrenal axis in the developmental course of schizophrenia. Annu Rev Clin Psychol 4:189–216
129. Bao AM, Meynen G, Swaab DF (2008) The stress system in depression and neurodegeneration: focus on the human hypothalamus. Brain Res Rev 57:531–553
130. Ramakrishnan SN, Muscat GEO (2006) The orphan Rev-erb nuclear receptors: a link between metabolism, circadian rhythm and inflammation? Nuclear Receptor Signaling 4:e009
131. Migita H, Morser J, Kawai K (2004) Rev-erbalpha upregulates NF-kappaB-responsive genes in vascular smooth muscle cells. FEBS Letters 561:69–74
132. Hasan RN, Schafer AI (2008) Hemin upregulates Egr-1 expression in vascular smooth muscle cells via reactive oxygen species ERK-1/2-Elk-1 and NF-kappaB. Circ Res 102: 42–50
133. Peng HB, Libby P, Liao JK (1995) Induction and stabilization of I kappa B alpha by nitric oxide mediates inhibition of NF-kappa B. J Biol Chem 270:14214–14219
134. Katsuyama K, Shichiri M, Marumo F, Hirata Y (1998) NO inhibits cytokine-induced iNOS expression and NF-kappaB activation by interfering with phosphorylation and degradation of IkappaB-alpha. Arterioscler Thromb Vasc Biol 18:1796–1802
135. Ignarro LJ (2000) Nitric oxide: biology and pathobiology, Ignarro LJ, ed. Academic Press, San Diego, CA
136. Ryter SW, Otterbein LE (2004) Carbon monoxide in biology and medicine. Bioessays 26:270–280
137. Coste H, Rodriguez JC (2002) Orphan nuclear hormone receptor Rev-erbalpha regulates the human apolipoprotein CIII promoter. J Biol Chem 277:27120–27129
138. Raspe E, Duez H, Mansen A, Fontaine C, Fievet C, et al. (2002) Identification of Rev-erbalpha as a physiological repressor of apoC-III gene transcription. J Lipid Res 43: 2172–2179
139. McClung CA (2007) Circadian genes, rhythms and the biology of mood disorders. Pharmacol Ther 114:222–232
140. Lumeng JC, Somashekar D, Appugliese D, Kaciroti N, Corwyn RF, et al. (2007) Shorter sleep duration is associated with increased risk for being overweight at ages 9 to 12 years. Pediatrics 120:1020–1029
141. Froy O (2007) The relationship between nutrition and circadian rhythms in mammals. Front Neuroendocrinol 28:61–71
142. Yin L, Wang J, Klein PS, Lazar MA (2006) Nuclear receptor Rev-erbalpha is a critical lithium-sensitive component of the circadian clock. Science 311:1002–1005
143. Chen G, Huang LD, Jiang YM, Manji HK (1999) The mood-stabilizing agent valproate inhibits the activity of glycogen synthase kinase-3. J Neurochem 72:1327–1330
144. Sladek FM (2011) What are nuclear receptor ligands? Mol Cell Endocrinol 334:3–13
145. Kumar N, Solt LA, Wang Y, Rogers PM, Bhattacharyya G, et al. (2010) Regulation of adipogenesis by natural and synthetic REV-ERB ligands. Endocrinology 151:3015–3025
146. Burke L, Downes M, Carozzi A, Giguere V, Muscat GE (1996) Transcriptional repression by the orphan steroid receptor RVR/Rev-erb beta is dependent on the signature motif and helix 5 in the E region: functional evidence for a biological role of RVR in myogenesis. Nucleic Acids Res 24:3481–3489
147. Jeong Y, Mangelsdorf DJ (2009) Nuclear receptor regulation of stemness and stem cell differentiation. Exp Mol Med 41:525–537
148. Chow EK, Razani B, Cheng G (2007) Innate immune system regulation of nuclear hormone receptors in metabolic diseases. J Leukoc Biol 82:187–195
149. Arulampalam V, Greicius G, Pettersson S (2006) The long and winding road to gut homeostasis. Curr Opin Gastroenterol 22:349–353

150. Bechtold DA, Gibbs JE, Loudon AS (2010) Circadian dysfunction in disease. Trends Pharmacol Sci 31:191–198
151. Foryst-Ludwig A, Kintscher U (2010) Metabolic impact of estrogen signalling through ERalpha and ERbeta. J Steroid Biochem Mol Biol 122:74–81
152. Riggins RB, Mazzotta MM, Maniya OZ, Clarke R (2010) Orphan nuclear receptors in breast cancer pathogenesis and therapeutic response. Endocr Relat Cancer 17:R213–231
153. Dominguez LJ, Scalisi R, Barbagallo M (2010) Therapeutic options in osteoporosis. Acta Biomed 81(Suppl 1):55–65
154. Warner M, Gustafsson JA (2010) The role of estrogen receptor beta (ERbeta) in malignant diseases – a new potential target for antiproliferative drugs in prevention and treatment of cancer. Biochem Biophys Res Commun 396:63–66
155. Rice LW (2010) Hormone prevention strategies for breast, endometrial and ovarian cancers. Gynecol Oncol 118:202–207
156. Katzenellenbogen BS, Choi I, Delage-Mourroux R, Ediger TR, Martini PG, et al. (2000) Molecular mechanisms of estrogen action: selective ligands and receptor pharmacology. J Steroid Biochem Mol Biol 74:279–285
157. Swedenborg E, Power KA, Cai W, Pongratz I, Ruegg J (2009) Regulation of estrogen receptor beta activity and implications in health and disease. Cell Mol Life Sci 66:3873–3894
158. Giguere V (1999) Orphan nuclear receptors: from gene to function. Endocrine Reviews 20:689–725
159. Conzen SD (2008) Minireview: nuclear receptors and breast cancer. Mol Endocrinol 22:2215–2228
160. Kizildag S, Ates H, Kizildag S (2009) Treatment of K562 cells with 1,25-dihydroxyvitamin D(3) induces distinct alterations in the expression of apoptosis-related genes BCL2, BAX, BCL(XL), and p21. Ann Hematol 89:1–7
161. Toma S, Isnardi L, Riccardi L, Bollag W (1998) Induction of apoptosis in MCF-7 breast carcinoma cell line by RAR and RXR selective retinoids. Anticancer Res 18:935–942
162. Altucci L, Gronemeyer H (2001) Nuclear receptors in cell life and death. Trends Endocrinol Metab 12:460–468
163. Baschant U, Tuckermann J (2010) The role of the glucocorticoid receptor in inflammation and immunity. J Steroid Biochem Mol Biol 120:69–75
164. Kleiman A, Tuckermann JP (2007) Glucocorticoid receptor action in beneficial and side effects of steroid therapy: lessons from conditional knockout mice. Mol Cell Endocrinol 275:98–108
165. Glass CK, Saijo K (2010) Nuclear receptor transrepression pathways that regulate inflammation in macrophages and T cells. Nat Rev Immunol 10:365–376
166. Schacke H, Berger M, Rehwinkel H, Asadullah K (2007) Selective glucocorticoid receptor agonists (SEGRAs): novel ligands with an improved therapeutic index. Mol Cell Endocrinol 275:109–117
167. Niino M, Hirotani M, Fukazawa T, Kikuchi S, Sasaki H (2009) Estrogens as potential therapeutic agents in multiple sclerosis. Cent Nerv Syst Agents Med Chem 9:87–94
168. Fiorucci S, Cipriani S, Mencarelli A, Renga B, Distrutti E, et al. (2010) Counter-regulatory role of bile acid activated receptors in immunity and inflammation. Curr Mol Med 10: 579–595
169. Kamen DL, Tangpricha V (2010) Vitamin D and molecular actions on the immune system: modulation of innate and autoimmunity. J Mol Med 88:441–450
170. Joseph SB, Castrillo A, Laffitte BA, Mangelsdorf DJ, Tontonoz P (2003) Reciprocal regulation of inflammation and lipid metabolism by liver X receptors. Nat Med 9:213–219
171. Heneka MT, Landreth GE (2007) PPARs in the brain. Biochim Biophys Acta 1771: 1031–1045
172. WHO (2006) World health organization fact sheet no 311. http://www.who.int/mediacentre/factsheets/fs311/en/.

173. Grun F (2010) Obesogens. Curr Opin Endocrinol Diabetes Obes 17:453–459
174. Grun F, Blumberg B (2006) Environmental obesogens: organotins and endocrine disruption via nuclear receptor signaling. Endocrinology 147:S50–55
175. Chen JQ, Brown TR, Russo J (2009) Regulation of energy metabolism pathways by estrogens and estrogenic chemicals and potential implications in obesity associated with increased exposure to endocrine disruptors. Biochim Biophys Acta 1793:1128–1143
176. Wang YF, Chao HR, Wu CH, Tseng CH, Kuo YT, et al. (2010) A recombinant peroxisome proliferator response element-driven luciferase assay for evaluation of potential environmental obesogens. Biotechnol Lett 32(12):1789–1796
177. Tiefenbach J, Moll PR, Nelson MR, Hu C, Baev L, et al. (2010) A live zebrafish-based screening system for human nuclear receptor ligand and cofactor discovery. PLoS One 5:e9797
178. Palanker L, Necakov AS, Sampson HM, Ni R, Hu C, et al. (2006) Dynamic regulation of *Drosophila* nuclear receptor activity in vivo. Development 133:3549–3562
179. Teboul M, Grechez-Cassiau A, Guillaumond F, Delaunay F (2009) How nuclear receptors tell time. J Appl Physiol 107:1965–1971
180. Duez H, Staels B (2009) Rev-erb-alpha: an integrator of circadian rhythms and metabolism. J Appl Physiol 107:1972–1980
181. Jetten AM (2009) Retinoid-related orphan receptors (RORs): critical roles in development, immunity, circadian rhythm, and cellular metabolism. Nucl Recept Signal 7:e003
182. Yang X, Lamia KA, Evans RM (2007) Nuclear receptors, metabolism, and the circadian clock. Cold Spring Harb Symp Quant Biol 72:387–394
183. Yin L, Wu N, Lazar MA (2010) Nuclear receptor Rev-erbalpha: a heme receptor that coordinates circadian rhythm and metabolism. Nucl Recept Signal 8:e001
184. Longo VD, Finch CE (2003) Evolutionary medicine: from dwarf model systems to healthy centenarians? Science 299:1342–1346
185. Partridge L, Gems D (2002) Mechanisms of ageing: public or private? Nat Rev Genet 3: 165–175
186. Jiang JC, Jaruga E, Repnevskaya MV, Jazwinski SM (2000) An intervention resembling caloric restriction prolongs life span and retards aging in yeast. Faseb J 14:2135–2137
187. Braeckman BP, Houthoofd K, Vanfleteren JR (2001) Insulin-like signaling, metabolism, stress resistance and aging in *Caenorhabditis elegans*. Mech Ageing Dev 122:673–693
188. Chapman T, Partridge L (1996) Female fitness in *Drosophila melanogaster*: an interaction between the effect of nutrition and of encounter rate with males. Proc Biol Sci 263:755–759
189. Masoro EJ (2000) Caloric restriction and aging: an update. Exp Gerontol 35:299–305
190. Bjorkhem I, Leoni V, Meaney S (2010) Genetic connections between neurological disorders and cholesterol metabolism. J Lipid Res 51:2489–2503
191. Nicolakakis N, Hamel E (2010) The Nuclear Receptor PPARgamma as a Therapeutic Target for Cerebrovascular and Brain Dysfunction in Alzheimer's Disease. Front Aging Neurosci 2:pii: 21
192. Alessandri JM, Guesnet P, Vancassel S, Astorg P, Denis I, et al. (2004) Polyunsaturated fatty acids in the central nervous system: evolution of concepts and nutritional implications throughout life. Reprod Nutr Dev 44:509–538
193. Chaturvedi RK, Beal MF (2008) PPAR: a therapeutic target in Parkinson's disease. J Neurochem 106:506–518
194. Timmermann S, Lehrmann H, Polesskaya A, Harel-Bellan A (2001) Histone acetylation and disease. Cell Mol Life Sci 58:728–736
195. Kim HJ, Casadesus G (2010) Estrogen-mediated effects on cognition and synaptic plasticity: What do estrogen receptor knockout models tell us? Biochim Biophys Acta 1800:1090–1093
196. Agrawal K, Onami S, Mortimer JE, Pal SK (2010) Cognitive changes associated with endocrine therapy for breast cancer. Maturitas 67:209–214
197. Vaya J, Schipper HM (2007) Oxysterols, cholesterol homeostasis, and Alzheimer disease. J Neurochem 102:1727–1737

198. Erol A (2007) The Functions of PPARs in Aging and Longevity. PPAR Res 2007:39654
199. Chung JH, Seo AY, Chung SW, Kim MK, Leeuwenburgh C, et al. (2008) Molecular mechanism of PPAR in the regulation of age-related inflammation. Ageing Res Rev 7:126–136
200. Buell JS, Dawson-Hughes B (2008) Vitamin D and neurocognitive dysfunction: preventing "D"ecline? Mol Aspects Med 29:415–422
201. Haussler MR, Haussler CA, Whitfield GK, Hsieh JC, Thompson PD, et al. (2010) The nuclear vitamin D receptor controls the expression of genes encoding factors which feed the "Fountain of Youth" to mediate healthful aging. J Steroid Biochem Mol Biol 121:88–97

Chapter 7
Methods for Analysis of Transcription Factor DNA-Binding Specificity In Vitro

Arttu Jolma and Jussi Taipale

Abstract Transcription of genes during development and in response to environmental stimuli is determined by genomic DNA sequence. The DNA sequences regulating transcription are read by sequence-specific transcription factors (TFs) that recognize relatively short sequences, generally between four and twenty base pairs in length. Transcriptional regulation generally requires binding of multiple TFs in close proximity to each other. Mechanistic understanding of transcription in an organism thus requires detailed knowledge of binding affinities of all its TFs to all possible DNA sequences, and the co–operative interactions between the TFs. However, very little is known about such co-operative binding interactions, and even the simple TF-DNA binding information exists only for a very small proportion of all TFs – for example, mammals have approximately 1,300–2,000 TFs [1, 2], yet the largest public databases for TF binding specificity, Jaspar and Uniprobe [3, 4] currently list only approximately 500 moderate to high resolution profiles for human or mouse. This lack of knowledge is in part due to the fact that analysis of TF DNA binding has been laborious and expensive. In this chapter, we review methods that can be used to determine binding specificity of TFs to DNA, mainly focusing on recently developed assays that allow high-resolution analysis of TF binding specificity in relatively high throughput.

7.1 Introduction

Sequences that specifically bind to a TF are known for a relatively small proportion of all TFs, and in most cases only few bound sequences have been identified. For many TFs, only a "consensus sequence", the DNA sequence that binds to the TF with highest affinity is known. However, many biologically relevant binding sites

---

A. Jolma (✉)
Department of Biosciences and Nutrition, SE-171 77, Stockholm, Sweden
e-mail: arttu.jolma@ki.se

T.R. Hughes (ed.), *A Handbook of Transcription Factors*, Subcellular Biochemistry 52,
DOI 10.1007/978-90-481-9069-0_7,

are of lower than maximal affinity, and affinity of sites also in some cases appears to be affected by biologically relevant sequence polymorphisms and/or to be under selective pressure [5, 6].

DNA binding proteins bind with relatively high affinity to their target sequences. The strong binding is a result of two different types of interactions, sequence–specific interactions with DNA bases, and non–sequence specific interactions with the DNA backbone [7–9].

The low affinity ($\sim 10^{-3}$–10^{-5} M) non-sequence specific binding to DNA allows TFs to slide along the DNA and find their target sites, whereas the high affinity ($\sim 10^{-8}$–10^{-12} M) sequence–specific interactions allow the immobilization of the TF to its target sites for sufficient time to allow regulation of transcription [7–9]. The methods reviewed here are focused on determining this sequence-specific DNA-binding of TFs (Table 7.1). We first discuss some classical methods that can find high affinity TF binding sites and/or measure TF affinity to a relatively small number of target sequences. Subsequently, we describe the more recent advances that have made it feasible to determine the "binding affinity landscape" of a TF – the relative affinity of a TF to all possible target sites.

7.2 Classical Methods for DNA-Binding Specificity Analysis

Sequence-specific TF–DNA interactions have been studied for decades inside living cells (in vivo) and in test tubes using purified and/or synthetic components (in vitro). During these years, a large number of different in vitro methods have been developed. Most of the protocols are aimed at finding target sequences for one TF at a time, and generally utilize protein in purified form. We discuss briefly here two classical methods that are still in wide use, footprinting and electrophoretic mobility shift assay (EMSA). In addition, we briefly cover methods than can be used to identify factors that bind to a single known DNA sequence. For a more in-depth review, see book by Moss and Leblanc [10].

7.2.1 Footprinting

Among the earliest effective methods for identification of TF binding sites were DNA protection assays, which are often referred to as footprinting. In these methods, DNA is labelled at one end, and incubated with proteins that bind to it. Subsequently the DNA is degraded using agents such as DNase1 [11] or hydroxyl radicals [12], resulting in a pool of DNA fragments of different size. When analyzed using gel electrophoresis, these fragments generate a typical ladder pattern ("footprint"). A DNA-bound protein will protect the area close to its binding site from the degrading agent, and the protected region can be identified by analyzing the difference between sample and control footprints. Sequences corresponding to

Table 7.1 Methods to determine binding affinity (or relative preference) of TFs to DNA sequences

Method	Sequence space	Throughput	Sources of error	Resolution	References
HT-SELEX	10^{15} (~25 bp)	High	DNA contamination, saturation of binding, non-specific DNA carryover, multimeric binding, data analysis	++++	[35–37]
ChIP-seq or DIP	10^{8} (~12 bp)	Low	Bias due to non-randomness of genome, artefacts due to endogenous proteins (ChIP-seq only), data analysis	++	[31, 32, 68]
Bacterial one-hybrid	10^{8} (~12 bp)	Medium	Saturation of selection marker expression, artefacts due to endogenous proteins, inability to control environment, data analysis	++	[33, 34, 44]
Protein binding microarrays (PBMs)	10^{6} (~10 bp)	Medium	Array artefacts, position effects, partial detection of long binding sites, data analysis	+++	[49, 50]
DNAse footprinting	10^{3}	Low	TF concentration determines stringency, measurement error	++	[11–14, 69]
Microfluidic direct binding assay (MITOMI)	10^{2}	Low	Synthetic DNA quality, quantification errors due to small sample size	++++	[59]
Surface plasmon resonance array	10^{2}	Low	Synthetic DNA quality	++++	[58]
Competition binding assay (microwell based)	10^{2}	High	Synthetic DNA quality, well-to-well variation	++++	[28, 60, 61]
Competition EMSA	10^{1}	Low	Synthetic DNA quality, measurement error, inadequately understood biochemistry	++	[16, 18, 19]

The methods are ordered on the basis of the sequence space that they can search to identify binding specificities. The length of all possible subsequences that can be analyzed in a single reaction is also given in parenthesis for the methods that can characterize the binding affinity landscape of TFs to all specifically bound sequences (top four). Throughput refers to the number of samples that can be analyzed in parallel: High, hundreds of samples per experiment using liquid handling robots; Medium, tens of samples in parallel; Low, few samples in parallel. The main sources of error and an estimate of the practically achievable resolution using each method is also indicated.

the protected areas can then be aligned to identify binding sites for the TF analyzed. This method is very sensitive to the dose of the proteins used, and increasing the concentration of the TF will result in protection of weaker binding sites. For thorough overview of footprinting analysis and conditions see Refs [13, 14]. Footprinting is also used to analyze DNA–protein interactions in vivo (see Chapter 8).

7.2.2 EMSA

EMSA is based on the observation that electrophoretic mobility of a DNA strand on native (i.e. non-denaturing) gels is affected (generally slowed) by proteins that bind to it. Briefly, TF or DBD and labeled (traditionally, radiolabeled) target sequences are mixed in a low ionic strength buffer, and incubated to allow complex formation. The reactions are then run in an electrophoretic gel to separate the protein-DNA complexes from free DNA. Typically, in a successful EMSA experiment, migration of only a very small fraction of the labeled DNA is retarded. Excess unlabeled sequences are often added to the initial reaction to compete for binding to the protein, and to demonstrate specificity for the labeled sequence. Using a single labeled high affinity sequence, relative binding activity of many different sequences can be assessed in such competitive EMSA reactions. EMSA can also be utilized to determine association constants by using a small amount of labeled DNA and a titration of protein [15].

EMSA has been used for both purified proteins and highly heterogeneous nuclear protein extracts [16]. It is also often used to determine the amount of a given TF in a sample. The proteins that bind to the DNA fragment can be identified by preincubating the reaction with antibodies, generating a larger complex and thus a "supershift" if the antibodies bind to the DNA-bound protein [17].

Although widely used, and considered by some to be a gold standard, it is also widely acknowledged that EMSA is not particularly stable or quantitative, and the physical chemistry of the reactions involved are very complex and therefore not especially well-defined. Electrophoresis in the gel-matrix stabilizes the DNA-protein complexes to some extent, reducing dissociation of the complex compared to what is observed for the same complex in free solution. This explains why, even though the gel is typically run for 2 h or longer, the retarded bands often look very sharp, rather than a smear one would expect if the complexes would dissociate during the electrophoresis. Multiple different effects are likely to contribute to the altered stability of the TF-DNA complex in the highly non-native condition that exists inside the gel matrix during electrophoresis. These include, for example, high local concentration of reactants, restriction of diffusion by the gel matrix, migration of the reactants in the applied electric field, and altered concentration of ions. However, some dissociation of the complexes will always occur, and thus competitive EMSA cannot be thought of as a pure equilibrium reaction, but rather a mixture of equilibrium reaction followed by indeterminate amount of dissociation that happens under changing and poorly understood conditions. It is also common to observe reactants retained in the well, and/or additional bands that are rationalized as non-specific binding to contaminating proteins. For many TFs, EMSA is very sensitive to reaction conditions and electrophoretic setup, and depending on the conditions, both non-specific and specific sequences can generate a detectable mobility shift [16, 18–20]. Thus, even though EMSA remains popular, its use should be limited to qualitative analysis of protein complexes (supershift), and to semiquantitative analysis of DNA-binding activity in extracts. For other applications, much simpler and more quantitative and reliable methods have been developed (see below).

7.2.3 Methods to Identify Proteins that Bind to Specific DNA Sequences

Several methods can be used to identify TFs that bind to sequence elements of interest that are, for example, overrepresented in regulatory elements known to have similar properties, or have been associated with an activity using genetic analyses. These methods include protein microarrays [21], one–hybrid interaction analysis [22] (see Section 7.3.3 below) and methods based on biochemical purification of proteins using DNA affinity matrices [23] followed by protein sequencing or mass spectrometry (MS). The MS-based approach requires a relatively large supply of relevant tissue or cellular material. In addition, due to the non-specific affinity of all TFs to DNA, and the general difficulty of purifying proteins using a single affinity step, it is hard to identify the specifically bound TF using direct MS analysis. This difficulty can be partially overcome by directly and quantitatively comparing the proteins bound to the specific sequence to the proteins bound to a control non-specific sequence using method such as SILAC (Stable Isotope Labeling by Amino acids on Cell culture). In SILAC, one sample is derived from cells labelled with heavy isotopes, and the other from unlabeled cells [24, 25]. These samples are purified in parallel, using the sequence of interest and non-specific DNA. After purification, the samples are mixed together and analyzed using liquid chromatography tandem MS instrument. Analysis of the ratio of isotope-labeled to non-labeled peptides can then be used to identify the proteins that specifically bound to the sequence of interest. Although TFs bind to DNA non-specifically, affinity of non-specific DNA is commonly between 10^3- and 10^7-fold lower than that of a high-affinity sequence. The resulting difference in TF binding can be detected using the SILAC approach. SILAC has been utilized for example to find the sequence specific TF ZBED6 that binds to a site containing a regulatory polymorphism that affects muscle growth in pigs [24, 26]. Recently, an MS-based approach has also been used to identify proteins that bind to a defined, albeit highly repetitive sequence in vivo; here DNA is cross-linked to proteins and defined sequences subsequently purified using hybridization, followed by MS [27]. Further development of MS technology may make such approaches widely applicable even in cases where sample amount is limiting and the genomic copy-number of the sequence of interest is low.

7.3 Methods that Can Determine the Binding Affinity Landscape of a TF

Traditionally, individual investigators have analyzed different TFs, often with different methods and using different conditions. The resulting loss of precision has made comparisons between data from different publications difficult, and in many cases impossible [28]. Recently several methods have been developed that allow analysis of large numbers of TFs in high resolution, using similar conditions for all factors analyzed (Table 7.1). Most of these methods also allow identification of TF binding specificities without any prior knowledge about bound sites, and in theory could also be applied for determination of the binding affinity landscape of the TF by measuring its affinity to all possible specifically bound sequences. The novel methods are in general related to different classical methods used to study TFs (e.g. filter binding, chromatin immunoprecipitation, one-hybrid analyses, SELEX; see Ref. [10] for review), but represent dramatic expansion of their capacity by utilizing microarrays (protein binding microarray [29], ChIP on chip [30], DIP-chip [31]) or massively parallel sequencing (ChIP-seq [32], high throughput bacterial one-hybrid [33, 34] and HT-SELEX [35–37]). The length of sites analyzed using these methods ranges from 10 bp for protein binding microarrays to 25 bp for HT-SELEX.

7.3.1 HT-SELEX

In SELEX (systematic evolution of ligands by exponential enrichment, also referred to as CASTing) TF binding specificity is solved by allowing a protein to select its target sites from pool of DNA strands containing randomized sequences [38, 39]. Typically proteins are bound to excess of DNA ligands until complex formation reaches equilibrium, after which the protein-DNA complexes are separated from free DNA using affinity capture, EMSA or using materials that bind to proteins but not free dsDNA.

Subsequently, the bound DNA fragments are amplified by PCR and sequenced. The resulting library that is enriched in specifically bound sequences will then be used as starting material for another round of selection. Typically three to seven such rounds of selection are performed (Fig. 7.1a). The optimal number of cycles for analysis of data needs to be experimentally determined; in the very first rounds the signal (specifically bound target sequences) to background (non-specifically carried over or bound DNA) ratio can remain very low, whereas after too many cycles the highest-affinity sites will become heavily overrepresented (Fig. 7.1b). Although modern DNA sequencers are efficient enough to allow the identification of correct consensus sequences even after first selection cycle [35–37], the building of high quality binding models often requires analysis of multiple cycles [35].

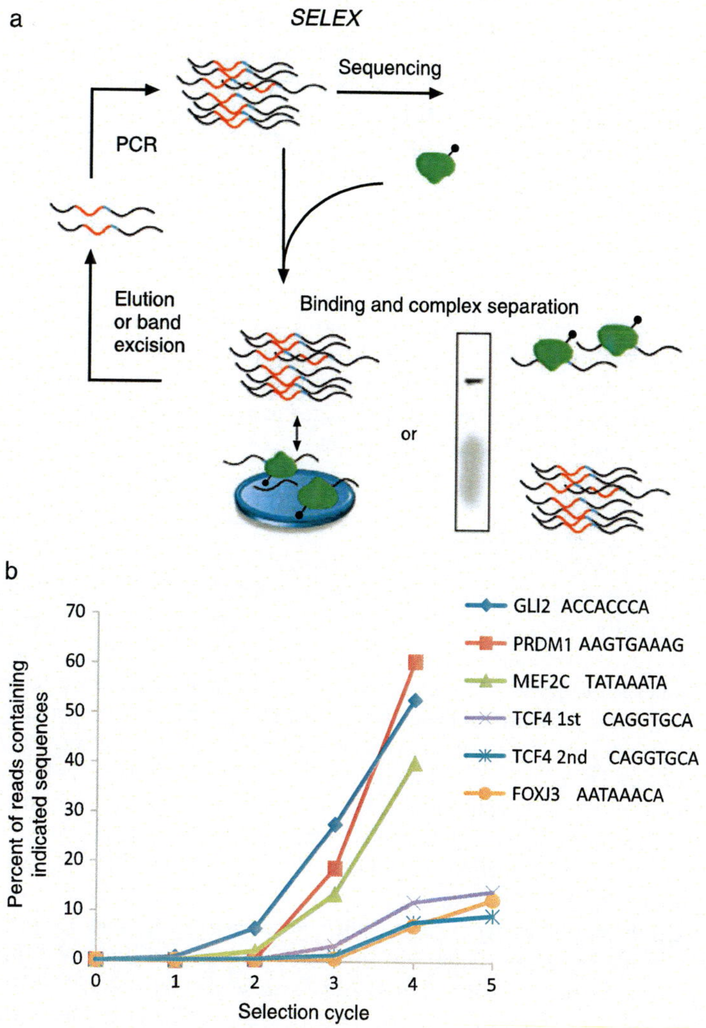

Fig. 7.1 High-Throughput SELEX. (**a**) Schematic representation of the SELEX process. The ligands for the SELEX (selection ligands) are DNA strands composed of a randomized region (*red*) flanked by constant regions (*black*). Constant regions can be made compatible with direct sequencing on massively parallel sequencers such as the Illumina Genome Analyzer. Throughput can be increased by incorporating different tag-sequences ("barcode", *blue*) into ligands used in different experiments, allowing multiplexed analysis of hundreds of selected ligands in a single sequencing reaction. Selection ligands are incubated with purified TF (*green*) until equilibrium is reached, after which the DNA complexed with the TF is separated from free DNA by affinity purification (*bottom center*) or electrophoresis (*bottom right*). Bound DNA is then amplified with PCR and sequenced and/or used in a subsequent cycle of selection. Binding specificity of the TF is then determined by analyzing the enrichment of sequences during the selection cycles. (**b**) Example of enrichment of subsequences corresponding to high-affinity sites for five different TFs during up to five SELEX selection cycles. Note that CAGGTGCA sequence enriches very similarly in two independent experiments for TCF4 (1st and 2nd). Adapted from [35]

SELEX can be used to solve binding specificities for proteins without previous knowledge of bound sites, and thus it has been used often to characterize affinities of new factors. To increase throughput, the method was further modified by increasing sequencing throughput first by adding an insert concatenation step (SELEX-SAGE) [40] and finally by using massively parallel sequencing [35–37]. Combining massively parallel sequencing with sample indexing using DNA barcodes allows hundreds of factors to be analyzed in a single sequencing reaction [35, 36].

As the SELEX reaction can be performed in liquid phase, and all DNA fragments can have a different sequence, the sequence space covered by HT-SELEX is 6–8 orders of magnitude larger than that of any other current TF binding specificity analysis method. The method is also very economical as the samples can be indexed for sequencing, and very small amounts of protein (nanograms) are needed [35]. HT-SELEX can also be easily automated using standard liquid handling equipment to analyze hundreds of TFs in parallel.

7.3.2 DIP and ChIP

DNA-immunoprecipitation (DIP) is essentially similar to SELEX but uses fragmented genomic DNA instead of synthetic random sequences. This is principally useful for very long binding specificities, as synthetic libraries are unlikely to contain such sequences. For example, a haploid human genome weighs approximately 3 pg, and in it any biologically relevant regulatory sequence should be present at least once. In contrast, a thousand times larger amount of completely random DNA includes a defined 26 bp subsequence on average only once. Thus, use of genomic libraries is very useful in identifying very long and highly specific binding sites, such as those suggested for proteins containing a large number of zinc fingers [41]. However, using genomic sequences to analyze TFs that bind to shorter sites is not generally sensible, as random libraries will cover the sequence space much more evenly and are easier to generate than genome-based libraries [31].

Biochemical binding specificity of a TF can also be estimated using in vivo based approach such as ChIP-chip or ChIP-seq, which are described in more detail in Chapter 8. Although ChIP-seq yields a highly reliable snapshot of genomic sites occupied by a given TF in a particular cell line or sample, it is not a reliable method for measuring biochemical affinity of a TF to different DNA sequences. This is due to several reasons, the most important of which is the fact that the occupancy of TF sites in the genome is not determined simply by the DNA-binding specificity of a given TF, but also strongly affected by nucleosome occupancy and higher order chromatin structure that affect accessibility of DNA sequences, secondary and tertiary protein–protein interactions, and co–operative interactions mediated by DNA bending, and/or unwinding. In addition, the genome is not a random sequence, and the accessible regions (e.g. promoters) are not similar in sequence to the whole genome, and thus devising a background model that corrects for these biases is non-trivial. Finally, in vivo generated binding profiles are less useful for certain types of analysis than in vitro profiles. In particular, they cannot be used to develop a predictive model of in vivo binding, as this leads to circular reasoning (explaining data with itself).

That said, ChIP-seq is an extremely important validation tool for the in vitro methods as it examines full-length TFs in their native environment. If the sites that are derived from ChIP-seq are substantially different from those found using the in vitro methods, and the differences cannot be explained by the factors described above, it is likely that the TF in question binds to DNA as a dimer with other factors, or contains additional determinants of specificity that are not present in the domain used for the in vitro analysis [35, 42, 43].

7.3.3 High-Throughput Bacterial One-Hybrid (B1H)

B1H is a method where the activity of a DBD-fusion protein is assessed in vivo inside living bacterial hosts. This method utilizes two plasmids, the first of which encodes a fusion protein combining the desired TF DBD with a subunit of bacterial RNA polymerase [33, 34, 44]. The other plasmid contains a complex library of randomized DNA-sequences in front of a minimal promoter driving expression of positive and negative selection markers. The size of the library is limited by bacterial plasmid transformation efficiency, and can practically be in the order of 10^6–10^8 independent clones.

Sequences that drive marker expression in the absence of the TF fusion protein are first depleted from the library by passage through bacteria grown under conditions selecting against expression of the negative marker. Subsequently, the library is transformed into bacteria that express a TF-fusion protein, and the bacteria are grown in media that requires expression of the positive selection marker. Binding specificity of the TF is then estimated by sequencing the randomized regions, which should be enriched in sites that bind to the TF tested [33].

This method can be easily adapted to massively parallel sequencing [34]. B1H has some benefits compared to the other methods. It does not require purification of proteins, and as the stringency of the selection can be easily adjusted, only one experimental selection cycle is needed. These benefits are however outweighed by multiple drawbacks of the method. Most importantly, in B1H the relationship between the measured variable (total plasmid copy number after selection) and the variable of interest (binding affinity) is not simple, making quantitative analysis of data very difficult if not impossible. Furthermore, analysis of mammalian full-length TFs is in most cases not possible as they express poorly in bacteria. B1H is also subject to position effects and saturation. For example, low affinity binding at an optimal position and orientation may drive expression of enough resistance marker to allow maximum growth. In addition, analysis of some TFs can be complicated by the depletion from the selection library of sequences that are bound by endogenous bacterial TFs. B1H does yield sites that qualitatively resemble binding sites determined by other methods, but in general, the results appear to be of lower quality and resolution than those obtained using other methods [34, 44, 45].

7.3.4 Protein-Binding Microarrays

Protein binding microarrays (PBMs) are microarrays containing spots of double-stranded DNA with different sequences [46, 47]. The arrays are incubated with TFs or DBDs to saturate the DNA with protein, and then washed extensively to remove weakly bound factors. Subsequently, fluorescent-labeled antibodies are used to detect the proteins and resulting spot intensities are interpreted computationally to yield an estimate of the binding specificity of the TF analyzed (Fig. 7.2a).

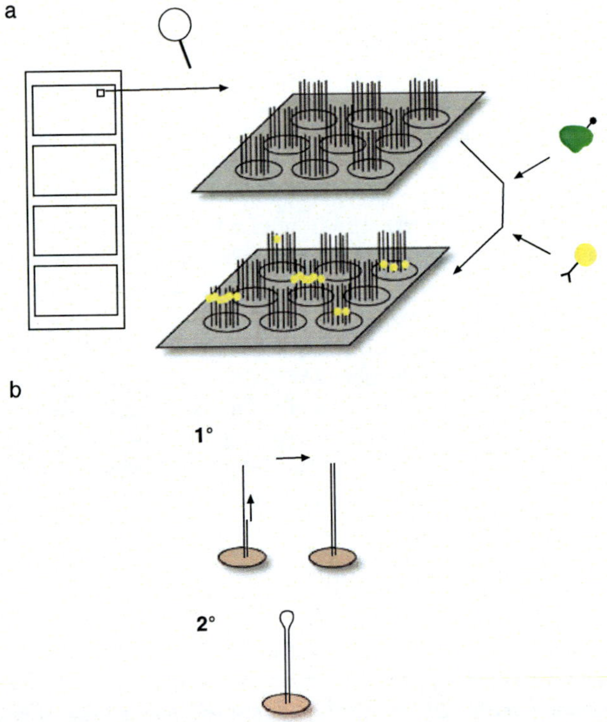

Fig. 7.2 Protein-binding microarrays (PBMs). (**a**) Glass slides containing spots of different DNA-sequences (*top*) are incubated with tagged TF (*green*) and fluorescent antibodies against the tag (*yellow*), followed by washing. Different DNA sequences bind to the TF with different affinities, resulting in different levels of fluorescent signal for different spots (*bottom*). Although all spots contain long sequences that can be divided to many different subsequences, the intensity data can be converted to estimates of affinity of the TF to all different subsequences by comparing fluorescence intensities in different spots that contain the same subsequence and different flanking regions. (**b**) Two strategies for in situ synthesis of DNA chips consisting of double-stranded oligonucleotides. *Top*: conversion of a ssDNA array to a dsDNA array by enzymatic synthesis of the complementary strands (primed by a common primer) [50]. *Bottom*: Direct chemical synthesis of self-complementary stem-loop sequences [51]

The early versions of PBMs had only tens to hundreds of different variants of target sequences or thousands of yeast intergenic sequences printed onto glass-slides [46, 48]. The most recent designs are based on synthetic DNA arrays, and contain ~ 44,000 different spots of 60 bp long dsDNA. Each 60 bp sequence contains a common primer binding site and a spot-specific subsequence from a B(4,10) De Bruijn sequence (i.e. a sequence that contains all 10 bp sequences exactly once). This results in an array that contains all possible 10 bp sequences, the large majority of which are in different spots [49, 50]. Initially the DNA is synthesized directly on the array as single stranded DNA, which is then enzymatically converted to dsDNA using the common primer (Fig. 7.2b). In an alternative design, the DNA is directly synthesized as a self-complementary hairpin. The hairpin design, known as Cognate-Site Identifier (CSI) has thus far been used primarily to characterize specificities of non-protein DNA binding molecules [51].

The PBM method is efficient and relatively economical, and has already been used to characterize binding specificities of more than 400 different TFs [28, 45, 52]. The main drawback of PBMs is that the arrays can contain only a limited number of sequences. As the number of possible DNA-sequences increases exponentially as a function of length, the universal PBMs are currently limited to analysis of sites that are less than ~12 bp long, although binding sites as wide as 14 bases have been inferred [48, 49, 52, 53]

The current system therefore functions well with small binding sites, and could presumably be used to model most of the known interactions between single TF and DNA in eukaryotes. Universal PBMs do not however work well to analyze TFs with long target sites such as dimeric RFX-proteins, or characterization of orientation and spacing preferences of heterodimeric TF pairs [35]. Custom arrays could be designed to analyze such cases, but the requirement for synthesis of a different array for each scientific question significantly increases cost and makes analysis of large numbers of factors using current universal PBM design not feasible.

7.4 Data Analysis

All of the methods described in Section 7.3 above can be used to generate motifs (e.g. PWMs) to search a large sequence space to identify sequences that are bound to a given TF (Table 7.1). In addition, the data can in general be used to rank individual sequences (e.g. 8-mers) in order of their binding affinities. However, none of the methods directly measures binding affinity of TFs to individual DNA sequences, and deriving relative binding affinity data from the raw data of any of the methods is not simple. The reasons for this are to some extent different for the different methods.

Occupancy of sites in ChIP-seq and B1H is determined by equilibrium that depends on concentration of the analyzed TF, which is generally not measured. In addition, number of plasmids bearing different sequences in B1H depends primarily on growth rate of the bacteria, which may not be linearly related to TF binding to DNA.

In SELEX, an equilibrium reaction is followed by washing, resulting in dissociation of some complexes. Although in an equilibrium reaction where TF is limiting binding is proportional to affinity, dissociation during washes results in exponential relationship between binding and affinity. This is because dissociation proceeds according to first order kinetics where the half-lives of the complexes are proportional to their affinities. In addition, some sequences are non-specifically carried over or bound to the TF. In addition, the ideal condition where TF is limiting with respect to the highest affinity site is in practice not achievable in most cases. If there are fewer high affinity sites than TF molecules, the high affinity sites can almost all be bound by the TF (i.e. the high affinity sites are approaching saturation). Thus, the high affinity sites cannot effectively compete against the lower affinity sites. This results in stronger enrichment of weaker sites than what would be expected from their relative affinity. Using multiple rounds of selection increases the concentration of preferred sites, but results in exponential enrichment of the high affinity sites. These opposite effects compensate for each other, and in practice enrichment after two to four cycles appears to be relatively close to that expected from the relative affinities [35].

Spot intensity data from PBM experiments is also not directly related to affinity. The signal in PBMs is generated by washing away weakly bound TFs, and thus the length of the wash affects the relative signal intensities due to the first order dissociation kinetics (see above). In addition, the DNA concentration is very high at the array surface, so there may be significant re-binding of dissociated TFs to the same spot. Such consecutive binding reactions make it more difficult to transform the data to affinities, as they increase signal at higher affinity spots relative to lower affinity spots in a way that is not very easy to model. Immobilization of the DNA and surface effects may also significantly affect the binding.

Correcting for the non-linear effects is not trivial for any of the methods, but can be accomplished by developing a model for the reactions, and subsequently defining the unknown variables (e.g. TF concentration) by finding the values that result in best fit of the observed data to predictions from the model (see for example [34, 37]). In order to have more measurements than unknown variables, this approach requires either making some assumptions about how the measured values relate to each other, or performing additional measurements using another method.

Making the common assumption that effects of substitutions at some DNA positions are independent of each other allows defining unknown variables by modeling [34, 37]. In general, this is a reasonable assumption, as in most studied cases, most DNA positions do appear to act independently [40, 54], and indeed, the commonly used model for TF DNA binding, the position weight matrix, assumes complete independence at all positions. However, as clear dependencies between bases have been observed for some factors (e.g HNF4 [52]), it is preferable to not to use the complete independence assumption in such models.

The best method to define the unknown variables in such models is to directly measure the relative affinities for some target sequences with different affinities using simpler methods, some of which are described below.

7.5 Modern Methods that Directly Measure TF-DNA Affinity

Several methods have been developed that can be used to determine actual absolute and/or relative affinities of TFs to DNA. As these methods currently allow only approx. 100 different sequences to be tested, they cannot be efficiently used to find novel binding sequences, or define binding affinity landscapes. However, they are very useful for refining binding specificities assuming that most or all substitutions act independently, for confirming binding specificities determined using other methods, and for defining the values for the unknown variables that are required for converting the data derived using the other methods to relative affinities.

7.5.1 *Plasmon Resonance (SPR) Arrays*

SPR is an optical method that can be used to monitor binding of molecules (e.g. TFs) from solution to surfaces coated with other molecules (e.g. DNA) quantitatively, in real time and without labels. This technology is highly suitable for measurement association and dissociation kinetics. While the method has been traditionally limited to observations of only one binding reaction at a time, it can be adapted to arrays by using SPR imaging (reviewed in [57]). In principle, this would allow parallel kinetic measurements of TF binding to 120 different target sequences [58].

7.5.2 *Microfluidic Methods*

MITOMI (Mechanically Induced Trapping of Molecular Interactions) devices can be used to measure tens to hundreds of TF–DNA interactions in parallel. This system is based on microfabricated array of chambers that are conjugated together by operable gates. The chambers are coated with antibodies that can be used to capture the desired TF. Each chamber can then be loaded with a different double stranded fluorescently labeled DNA-fragment, and the reactions allowed to equilibrate. The initial level of fluorescence is measured to determine the concentrations of the labeled DNAs. Subsequently, mechanical displacement buttons are pressed against the bottoms of the chambers, removing the liquid and capturing the bound TF-DNA complex in a state that very closely resembles equilibrium. Using the method to measure signals for multiple different DNA-fragments it is possible to generate good quality binding models [59]. The main advantage of this method is the ability to remove unbound DNA extremely rapidly, allowing capture the equilibrium state. Thus, the complexes best suited for this method are ones that have very rapid dissociation kinetics, which may include TFs with very short target sites. A more traditional competition assay described below can be equally well used to determine affinities for most other TF-DNA complexes that dissociate relatively slowly.

7.5.3 Microwell-Based Competition Assay

Protein-DNA binding can also be analyzed using an approach [60] similar to competition-ELISA. In this method, a biotin-labeled dsDNA with a high-affinity to a given TF is incubated with a TF fused to a *Renilla* luciferase reporter enzyme. The dsDNA is captured using a streptavidin plate and the amount of TF bound measured using a luminometer. Competing this reaction with different dsDNA sequences allows simple measurement of the relative affinities of the different competitors. This method is quantitative but requires prior knowledge of one high-affinity binding site for the protein of interest [61]. As the method is based on standard 96-well plates, it can be efficiently automated.

7.6 Perspective

There have been very rapid advances in the past 5 years in methods that can be used to determine TF-DNA binding specificity. The advances have been largely driven by development of microarray and massively parallel sequencing technologies. All of the methods have their own advantages and drawbacks, and in general can be used to qualitatively determine TF binding specificity and/or rank different sequences in order of their affinities to the TF. However, determining true relative affinities from the raw data remains a challenge, and in general the bioinformatic methods for analysis of data lag significantly behind the progress that has been made in the wet-lab. Thus, development of computational algorithms for data analysis is centrally important for further development of the field.

In addition to developing methods to derive affinities from raw data, more work is needed to develop models that represent TF binding to all possible DNA sequences. These range from the position weight or position specific scoring matrices (PWMs or PSSMs) that assume complete independence of bases to a model assuming complete dependence, where each possible sequence encoded by the bases that contribute to TF binding is given a separate affinity or score. For some classes of factors the PWMs may be sufficient, whereas for others more complicated model is clearly needed. However, it is clear that complete dependence is not a reasonable assumption, and intermediate models between these extremes are clearly needed, and several such models have already been developed. These include k-mer/E-score approach that represents TF binding specificity as-scores for all subsequences of a given length (6–8 bp; see for example Ref. [45], Nitta et al., in preparation) and models that use kmer data to generate a more compact predictors [55]. TF binding can also be represented as series of different PWMs for the same factor [52], and by using a feature based model that represent binding as a set of sequence features [56]. However, it remains to be determined what model is best at capturing the specificity landscape of TFs with more complex binding modes without overfitting to incorporate noise that is inevitably included in the data.

Despite caveats associated with all of the new TF binding analysis methods, they all do appear to reliably capture the major features of the DNA-binding activity of the proteins analyzed. In a very recent review by Stormo and Zhao [34] three of the methods, PBM, HT-SELEX and B1H were compared to each other using data for the same DNA-binding domain (from the zinc-finger factor zif268). The models based on all methods were qualitatively similar. At least in this case, the best quality model was generated using HT-SELEX experiments, whereas the B1H model had the lowest resolution and seemed to lack some aspects of specificity identified by both HT-SELEX and PBM [34]. In the cases we have analyzed, HT-SELEX and PBM in general yield similar results when used to analyze DBDs that bind to relatively short target sites (Fig. 7.3 and Refs. [28, 35]).

It is likely that methods based on SELEX and PBM technologies will dominate the field in the future. SELEX can cover a larger sequence space than any other method and can be efficiently multiplexed making it the most economical of all the technologies. The PBM approach has the advantage that it yields quantitative information for each sequence, as opposed to simple counts in SELEX, ChIP-seq or B1H. A massively parallel sequencer can in principle be adapted to generate and analyze an array with approximately two to three orders of magnitude (10^8) more subsequences than the current universal PBM design. Although such an approach is not easily multiplexed for analysis of a very large number of factors in high throughput, it could be used to generate more information than what is possible for any other method. In addition, by using multiple fluorescence channels and live imaging, the PBM approach could be developed to analyze complex reactions and kinetic parameters, which are harder to study using the other methods.

In the future, DNA-binding specificity of TFs, including the dependencies between the individual bases could also be determined based on structural data and physical modelling, and the rules that determine binding specificity could ultimately

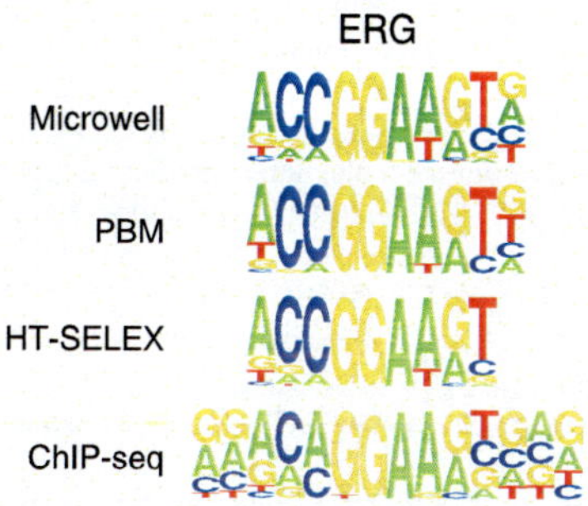

Fig. 7.3 Comparison of binding specificity models for the ETS-family TF ERG generated using four different methods. The methods used were microwell-based competition assay [28], protein binding microarrays [28], HT-SELEX [35] and ChIP-seq followed by MEME analysis [28]. Note that all in vitro methods generate very similar profiles, and that the ChIP-seq derived profile is broadly similar to the in vitro models. The differences are likely caused by the presence of a large number of GGAAGGAA repeats in the human genome, resulting in underrepresentation of C and T at some positions of the ChIP-seq profile. All in vitro analyses were performed using human ERG DBD, except PBM, which used mouse ERG DBD

be based on protein primary structure. Some progress in this area has already been made in the study of zinc finger TFs (reviewed in [62]). In addition to modeling TF–DNA interactions in a static setting, understanding how TFs find their target sites requires also dynamic models that take into account both the non-sequence specific and sequence specific binding. Such dynamic analyses have been performed in vivo using single-molecule fluorescence in *E. coli* [63, 64], and in vitro using PRE-NMR [65].

As transcription is controlled by combinatorial binding of many TFs, another key area of further study includes analysis of co–operative interactions between TFs. Co-operativity between TFs can be due to direct protein–protein interactions [66], mediated by local conformational change of DNA induced by TF binding [67], or caused by higher order effects (e.g. displacement of nucleosomes). The two first cases can be analyzed by PBMs or HT-SELEX. As this analysis would preferentially be performed using full-length TFs, and the sequence space that needs to be searched is very large, it is likely that HT-SELEX with tandem purification will be the method of choice for unbiased analysis of spacing and orientation preferences of TF pairs.

References

1. Vaquerizas JM, Kummerfeld SK, Teichmann SA, Luscombe NM (2009) A census of human transcription factors: function, expression and evolution. Nat Rev Genet 10:252–263
2. Fulton DL, et al. (2009) TFCat: the curated catalog of mouse and human transcription factors. Genome Biol 10:R29
3. Bryne JC, et al. (2008) JASPAR, the open access database of transcription factor-binding profiles: new content and tools in the 2008 update. Nucleic Acids Res 36:D102–6
4. Newburger DE, Bulyk ML (2009) UniPROBE: an online database of protein binding microarray data on protein–DNA interactions. Nucleic Acids Res 37:D77–82
5. Tuupanen S, et al. (2009) The common colorectal cancer predisposition SNP rs6983267 at chromosome 8q24 confers potential to enhanced Wnt signaling. Nat Genet 41:885–890
6. Jiang J, Levine M (1993) Binding affinities and cooperative interactions with bHLH activators delimit threshold responses to the dorsal gradient morphogen. Cell 72:741–752
7. Jen-Jacobson L (1997) Protein-DNA recognition complexes: conservation of structure and binding energy in the transition state. Biopolymers 44:153–180
8. Berg OG, von Hippel PH (1987) Selection of DNA binding sites by regulatory proteins. Statistical-mechanical theory and application to operators and promoters. J Mol Biol 193:723–750
9. von Hippel PH, Berg OG (1989) Facilitated target location in biological systems. J Biol Chem 264:675–678
10. Moss T, Leblanc B (2009) DNA–protein Interactions: Principles and Protocols (Humana Press, New York, NY)
11. Galas DJ, Schmitz A (1978) DNAse footprinting: a simple method for the detection of protein-DNA binding specificity. Nucleic Acids Res 5:3157–3170
12. Tullius TD, Dombroski BA, Churchill ME, Kam L (1987) Hydroxyl radical footprinting: a high-resolution method for mapping protein-DNA contacts. Methods Enzymol 155:537–558
13. Connaghan-Jones KD, Moody AD, Bain DL (2008) Quantitative DNase footprint titration: a tool for analyzing the energetics of protein–DNA interactions. Nat Protoc 3:900–914
14. Jain SS, Tullius TD (2008) Footprinting protein-DNA complexes using the hydroxyl radical. Nat Protoc 3:1092–1100

15. Hughes TR, Weilbaecher RG, Walterscheid M, Lundblad V (2000) Identification of the single-strand telomeric DNA binding domain of the *Saccharomyces cerevisiae* Cdc13 protein. Proc Natl Acad Sci U S A 97:6457–6462
16. Hellman LM, Fried MG (2007) Electrophoretic mobility shift assay (EMSA) for detecting protein-nucleic acid interactions. Nat Protoc 2:1849–1861
17. Gille J, Swerlick RA, Caughman SW (1997) Transforming growth factor-alpha-induced transcriptional activation of the vascular permeability factor (VPF/VEGF) gene requires AP-2-dependent DNA binding and transactivation. EMBO J 16:750–759
18. Fried M, Crothers DM (1981) Equilibria and kinetics of lac repressor–operator interactions by polyacrylamide gel electrophoresis. Nucleic Acids Res 9:6505–6525
19. Garner MM, Revzin A (1981) A gel electrophoresis method for quantifying the binding of proteins to specific DNA regions: application to components of the *Escherichia coli* lactose operon regulatory system. Nucleic Acids Res 9:3047–3060
20. Fried MG, Liu G (1994) Molecular sequestration stabilizes CAP-DNA complexes during polyacrylamide gel electrophoresis. Nucleic Acids Res 22:5054–5059
21. Hu S, et al. (2009) Profiling the human protein-DNA interactome reveals ERK2 as a transcriptional repressor of interferon signaling. Cell 139:610–622
22. Deplancke B, et al. (2006) A gene-centered *C. elegans* protein–DNA interaction network. Cell 125:1193–1205
23. Kadonaga JT, Tjian R (1986) Affinity purification of sequence-specific DNA binding proteins. Proc Natl Acad Sci U S A 83:5889–5893
24. Mittler G, Butter F, Mann M (2009) A SILAC-based DNA protein interaction screen that identifies candidate binding proteins to functional DNA elements. Genome Res 19: 284–293
25. Ong SE, et al. (2002) Stable isotope labeling by amino acids in cell culture, SILAC, as a simple and accurate approach to expression proteomics. Mol Cell Proteomics 1:376–386
26. Markljung E, et al. (2009) ZBED6, a novel transcription factor derived from a domesticated DNA transposon regulates IGF2 expression and muscle growth. PLoS Biol 7:e1000256
27. Dejardin J, Kingston RE (2009) Purification of proteins associated with specific genomic Loci. Cell 136:175–186
28. Wei GH, et al. (2010) Genome-wide analysis of ETS-family DNA-binding in vitro and in vivo. EMBO J 29:2147–2160
29. Berger MF, Bulyk ML (2006) Protein binding microarrays (PBMs) for rapid, high-throughput characterization of the sequence specificities of DNA binding proteins. Methods Mol Biol 338:245–260
30. Ren B, et al. (2000) Genome-wide location and function of DNA binding proteins. Science 290:2306–2309
31. Liu X, Noll DM, Lieb JD, Clarke ND (2005) DIP-chip: rapid and accurate determination of DNA-binding specificity. Genome Res 15:421–427
32. Johnson DS, Mortazavi A, Myers RM, Wold B (2007) Genome-wide mapping of in vivo protein–DNA interactions. Science 316:1497–1502
33. Meng X, Brodsky MH, Wolfe SA (2005) A bacterial one-hybrid system for determining the DNA-binding specificity of transcription factors. Nat Biotechnol 23:988–994
34. Stormo GD, Zhao Y (2010) Determining the specificity of protein-DNA interactions. Nat Rev Genet 11:751–760
35. Jolma A, et al. (2010) Multiplexed massively parallel SELEX for characterization of human transcription factor binding specificities. Genome Res 20:861–873
36. Zykovich A, Korf I, Segal DJ (2009) Bind-n-Seq: high-throughput analysis of in vitro protein–DNA interactions using massively parallel sequencing. Nucleic Acids Res 37:e151
37. Zhao Y, Granas D, Stormo GD (2009) Inferring binding energies from selected binding sites. PLoS Comput Biol 5:e1000590
38. Oliphant AR, Brandl CJ, Struhl K (1989) Defining the sequence specificity of DNA-binding proteins by selecting binding sites from random-sequence oligonucleotides: analysis of yeast GCN4 protein. Mol Cell Biol 9:2944–2949

39. Tuerk C, Gold L (1990) Systematic evolution of ligands by exponential enrichment: RNA ligands to bacteriophage T4 DNA polymerase. Science 249:505–510
40. Roulet E, et al. (2002) High-throughput SELEX SAGE method for quantitative modeling of transcription-factor binding sites. Nat Biotechnol 20:831–835
41. Urrutia R (2003) KRAB-containing zinc-finger repressor proteins. Genome Biol 4:231
42. Gangwal K, et al. (2008) Microsatellites as EWS/FLI response elements in Ewing's sarcoma. Proc Natl Acad Sci U S A 105:10149–10154
43. Hollenhorst PC, et al. (2009) DNA specificity determinants associate with distinct transcription factor functions. PLoS Genet 5:e1000778
44. Noyes MB, et al. (2008) Analysis of homeodomain specificities allows the family-wide prediction of preferred recognition sites. Cell 133:1277–1289
45. Berger MF, et al. (2008) Variation in homeodomain DNA binding revealed by high-resolution analysis of sequence preferences. Cell 133:1266–1276
46. Bulyk ML, Huang X, Choo Y, Church GM (2001) Exploring the DNA-binding specificities of zinc fingers with DNA microarrays. Proc Natl Acad Sci U S A 98: 7158–7163
47. Bulyk ML, Gentalen E, Lockhart DJ, Church GM (1999) Quantifying DNA–protein interactions by double-stranded DNA arrays. Nat Biotechnol 17:573–577
48. Mukherjee S, et al. (2004) Rapid analysis of the DNA-binding specificities of transcription factors with DNA microarrays. Nat Genet 36:1331–1339
49. Berger MF, et al. (2006) Compact, universal DNA microarrays to comprehensively determine transcription-factor binding site specificities. Nat Biotechnol 24:1429–1435
50. Berger MF, Bulyk ML (2009) Universal protein-binding microarrays for the comprehensive characterization of the DNA-binding specificities of transcription factors. Nat Protoc 4: 393–411
51. Puckett JW, et al. (2007) Quantitative microarray profiling of DNA-binding molecules. J Am Chem Soc 129:12310–12319
52. Badis G, et al. (2009) Diversity and complexity in DNA recognition by transcription factors. Science 324:1720–1723
53. Mintseris J, Eisen MB (2006) Design of a combinatorial DNA microarray for protein–DNA interaction studies. BMC Bioinformatics 7:429
54. Benos PV, Bulyk ML, Stormo GD (2002) Additivity in protein–DNA interactions: how good an approximation is it? Nucleic Acids Res 30:4442–4451
55. Agius P, Arvey A, Chang W, Noble WS, Leslie C (2010) High resolution models of transcription factor-DNA affinities improve in vitro and in vivo binding predictions. PLoS Comput Biol 6:e1000916
56. Sharon E, Lubliner S, Segal E (2008) A feature-based approach to modeling protein–DNA interactions. PLoS Comput Biol 4:e1000154
57. Scarano S, Mascini M, Turner AP, Minunni M (2010) Surface plasmon resonance imaging for affinity-based biosensors. Biosens Bioelectron 25:957–966
58. Shumaker-Parry JS, Aebersold R, Campbell CT (2004) Parallel, quantitative measurement of protein binding to a 120-element double-stranded DNA array in real time using surface plasmon resonance microscopy. Anal Chem 76:2071–2082
59. Maerkl SJ, Quake SR (2007) A systems approach to measuring the binding energy landscapes of transcription factors. Science 315:233–237
60. Hallikas O, et al. (2006) Genome-wide prediction of mammalian enhancers based on analysis of transcription-factor binding affinity. Cell 124:47–59
61. Hallikas O, Taipale J (2006) High-throughput assay for determining specificity and affinity of protein–DNA binding interactions. Nat Protoc 1:215–222
62. Negi S, Imanishi M, Matsumoto M, Sugiura Y (2008) New redesigned zinc-finger proteins: design strategy and its application. Chemistry 14:3236–3249
63. Li GW, Elf J (2009) Single molecule approaches to transcription factor kinetics in living cells. FEBS Lett 583:3979–3983

64. Elf J, Li GW, Xie XS (2007) Probing transcription factor dynamics at the single-molecule level in a living cell. Science 316:1191–1194
65. Iwahara J, Clore GM (2006) Detecting transient intermediates in macromolecular binding by paramagnetic NMR. Nature 440:1227–1230
66. Passner JM, Ryoo HD, Shen L, Mann RS, Aggarwal AK (1999) Structure of a DNA-bound Ultrabithorax-Extradenticle homeodomain complex. Nature 397:714–719
67. Panne D, Maniatis T, Harrison SC (2004) Crystal structure of ATF-2/c-Jun and IRF-3 bound to the interferon-beta enhancer. EMBO J 23:4384–4393
68. Robertson G, et al. (2007) Genome-wide profiles of STAT1 DNA association using chromatin immunoprecipitation and massively parallel sequencing. Nat Methods 4:651–657
69. Hampshire AJ, Rusling DA, Broughton-Head VJ, Fox KR (2007) Footprinting: a method for determining the sequence selectivity, affinity and kinetics of DNA-binding ligands. Methods 42:128–140

Chapter 8
Identification of Transcription Factor–DNA Interactions In Vivo

Duncan T. Odom

Abstract Recent technological developments have revolutionized our understanding of transcriptional regulation by providing an unprecedented ability to interrogate in vivo transcription factor binding. The combination of high-throughput sequencing with chromatin precipitation of transcription factors and specifically labeled histones has allowed direct protein-DNA contacts to be visualized across genomes as large and complex as mammals at base-pair resolution. This chapter reviews the developments that led to these insights, with particular focus on examples of early protein-DNA localization experiments using genomic microarrays in mammals and yeast. Four state-of-the-art research directions are highlighted as examples of previously unimaginable frontiers now under active investigation.

8.1 The Biochemistry and Regulation of the Genome

Living organisms use nucleic acids to store genetic information compactly and reliably. Although ribonucleic acids (RNA) were likely used in the earliest life forms to store genetic information [1], virtually all present-day organisms from prokaryotes to mammals use deoxyribonucleic acid (DNA) to store the instructions for creating proteins and RNAs. DNA is a highly negatively charged polymer that self-assembles into a stacked double helix of unsaturated nitrogenous bases (Fig. 8.1a). This helix has a major groove with substantial potential for hydrogen bonding and van der Waals interactions to other molecules, with a somewhat less information-rich minor groove; structural implications for site-recognition were in part recognized in the initial reports of the DNA double-helical structure [2–4] (Fig. 8.1b).

The proper deployment of the instructions found in DNA is controlled by a specialized class of proteins called transcription factors (TFs), numbering from scores in bacteria to thousands in vertebrates (reviewed in [5]) and higher plants [6, 7] (see Chapters 2 and 3 for description and enumeration of TFs in bacteria and eukarya). These proteins have evolved specialized amino acid surfaces to interact most often

D.T. Odom (✉)
Cancer Research UK, Li Ka Shing Centre, University of Cambridge, Cambridge, UK
e-mail: duncan.odom@cancer.org.uk

T.R. Hughes (ed.), *A Handbook of Transcription Factors*, Subcellular Biochemistry 52,
DOI 10.1007/978-90-481-9069-0_8,

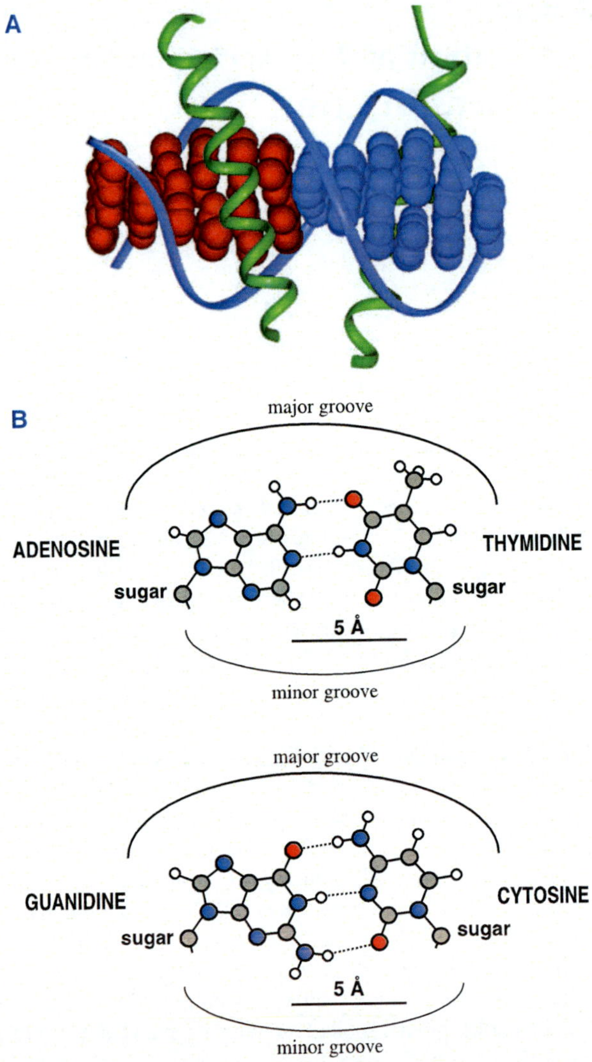

Fig. 8.1 The structure of deoxyribonucleic acid (DNA). **a** A schematic model of the stacked B-form 12 base sequence (*blue* and *red*, indicating the two halves of a palindromic sequence) is shown with the DNA binding domain of a leucine zipper protein (*green*) inserted into the major groove symetrically. **b** The adenine-thymine (A–T) and cytosine-guanine (C–G) base pairings proposed in the original 1953 Nature report from Watson and Crick are shown. The hydrogen-bonding potential in the major groove (upper surface of the base pairs) is suggested by the nitrogen-hydrogen bond in the A–T pair, as well as the (implied) lone pairs directed from the nitrogens and oxygens in both A–T and C–G (reprinted by permission from [4])

with the DNA major groove (an example is shown in Fig. 8.1a). Proteins accomplish this by using highly positively charged residues, directed hydrogen bonding, and architectural van der Waal interactions to select for and bind to specific biochemical structures formed by the exposed surfaces of stacked base-pairs within the DNA major groove [8]. These major groove interactions can be augmented by specific contacts in the minor groove (for instance, polyamides [9]). The DNA sequence preferences of individual TFs can be determined in vitro (see Chapter 7), but in vivo, the precise locations where TFs bind can be influenced by a variety of other factors also present in the cell, including nucleosomes and other TFs. This is presumably due to competition and cooperation, which can be either direct or indirect (see Chapter 9). Therefore, examining the genomic locations bound by TFs is critical to understanding the functions of the individual TFs as well as the operational principles of genomes.

8.2 Early Approaches to Identifying TF Binding Locations: Nuclease Sensitivity

One of the earliest methodologies used to identify where TFs bind DNA exploits the protection that a protein-DNA contact affords DNA when exposed to an endonuclease, in particular DNAse I [10]. A bound TF physically protects stretches of DNA of varying base length from digestion, depending on the context and other possible neighboring TFs bound to the same promoter region.

This procedure has been used extensively to explore the structure of the hepatitis viral promoter, beta-globin [11], interleukins [12], many liver-specific gene classes [13, 14] (and others), and many prokaryotic genes. The utilization of this method is reviewed also in Chapter 7.

Originally, this technique referenced digestion results against a combination of Maxim-Gilbert sequencing reactions to identify the sites protected in vitro against digestion. The same general approach has been adapted recently in in vivo work with isolated mammalian nuclei to identify regions protected from DNAse digestion, methods that have been coupled with the methods described below for TF binding assays to obtain global views of open versus closed chromatin [15]. When combined with bioinformatic analysis of the protected sequences, these methods can suggest which TFs bind to a particular regulatory region (for instance, [16]).

8.3 Antibodies Specifically Targeting DNA-Binding Proteins Allowed Identification of In Vivo Binding Events

The use of formaldehyde to crosslink proteins to nucleic acids was reported in the 1960s for ribonucleotides [17], and continues to be used in numerous RNA-protein identification protocols [18–20]. The ability of formaldehyde to reversibly crosslink proteins with DNA gradually evolved from work with SV40 minichromosomes and nucleosomes (for example [21, 22]).

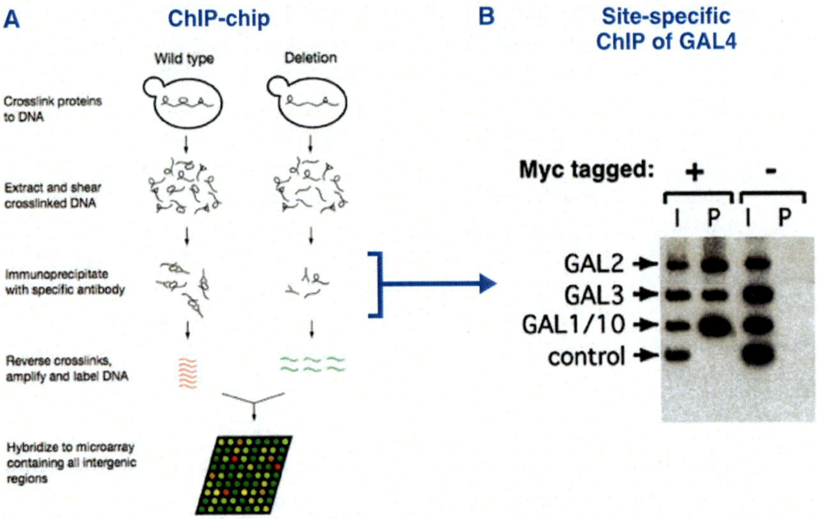

Fig. 8.2 Identification of protein-DNA contacts using chromatin immunoprecipitation. **a** Microarrays that contain the genetic sequence of promoter regions can be used to interrogate the complete set of nucleic acids enriched by antibody binding to TF-DNA complexes (reprinted by permission from [28]). **b** Primer sets can also be used to interrogate or confirm limited subsets of these binding events (reprinted by permission from [29])

By using antibodies against specific TFs of interest, the DNA regulatory regions bound in vivo can be isolated as nucleic acids, and then further interrogated, a process known as chromatin immunoprecipitation (ChIP). Historically, the enrichment of particular TFs at specific sites has been established using pairs of oligonucleotide primers at pre-selected promoter region(s) (Fig. 8.2). Direct comparison can be made of the number of copies of a potentially bound region versus random and unbound regions in the genome by simultaneous amplification of these regions, followed by gel electrophoresis and quantitation of the nucleic acid bands.

8.4 Microarrays First Allowed the Genome-Wide Determination of TF Binding in the Yeast *Saccharomyces cerevisiae*

In the early 2000s, ChIP experiments were combined with the then-nascent technology of microarrays. The most popular method for gene expression microarray synthesis in the first years of the technology was to PCR-amplify mRNA sequences, print them onto glass slides, and fix chemically. Gene expression arrays had been successfully reported to interrogate yeast [23, 24] and mammalian gene expression in tissues [25, 26] and in response to stimuli [27] (and many, many other publications on gene expression). Since the early days of gene expression microarray analysis, scores of species have had microarrays designed to interrogate gene expression.

In principle, ChIP experiments such as those described above simultaneously isolate and enrich all promoter regions that are bound by a protein, even if only a small subset are interrogated for ChIP enrichment using specific primers. This fact led to a number of groups realizing that one possible method for obtaining genome-wide information on TF-DNA binding would be to create promoter-sequence microarrays, as opposed to coding-sequence gene expression microarrays (Fig. 8.2). Synthesis of these microarrays was combined with methods to fluorescently label ChIP DNA one color and input (or a mock ChIP experiment performed without the specific antibody) DNA a second color, followed by co-hybridization against the promoters present on the promoter microarrays. The creation of promoter microarrays that tile the noncoding regions of the yeast genome was greatly facilitated by the extraordinarily dense yeast genome. In contrast to higher eukaryotes, yeast has few repeated sequences. When combined with ChIP experiments, the use of a microarray to identify TF binding become a technique that quickly gained the name *ChIP-chip*.

The first reports of genome-wide TF binding in vivo were both reported in yeast, essentially simultaneously [28, 29]. Three major genetics research groups were active in this then-nascent field, and used as a proof-of-principle TFs that had been well-studied by yeast transcriptional biologists for years. Richard Young's laboratory at the Whitehead Institute used Ste12 and Gal4, both tagged with a myc epitope and induced with either mating hormone (Ste12) or galactose (Gal4), to perform genome-wide location analysis using an anti-myc antibody [29]. In addition, to showing that this technology yielded results consistent with site-specific analysis, these authors were able to identify a number of novel components of the carbon metabolic pathways involved in galactose utilization, as well as new mating genes regulated by Ste12.

The Snyder and Brown laboratories, at Yale and Stanford respectively, collaboratively performed ChIP-chip experiments against two TFs active at the G1/S transition in yeast cell cycle [28]. On a technical basis, their experiments demonstrated that in vivo tagging of yeast TFs (here, SBF and MBF) afforded the same results as antibodies raised specifically against the TF proteins themselves. In addition, their study demonstrated for the first time the power of using TF binding and systematic gene expression perturbations as independent methodologies to deconstruct on a genome-wide basis the direct versus indirect effects of TF binding. As noted below, and in other chapters in this book, the relationships between TF binding and expression output can be complex. Notably, similar approaches in mammals have revealed that transcription factor binding rarely if ever follows the intuitively predicted expectation of direct regulation (Box 8.1) – nevertheless, few researchers, even in the field of transcription, appear to have absorbed this lesson.

Box 8.1

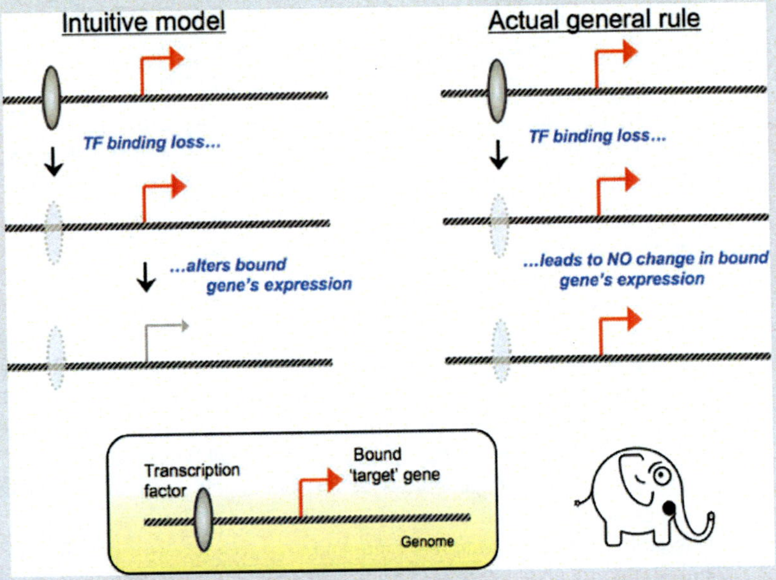

The elephant in the room: TFs rarely regulate the expression of bound genes simplisitically or intuitively. A major finding of the simultaneous comparison of gene expression and TF binding experiments in the same tissues and cells has been that our early and overly-simplistic model of TF binding leading directly to transcriptional regulation (*left*) is rarely true in complex eukaryotes like mammals. It is common for a TF to bind tens or even hundreds of thousands of locations in the mammalian genome, yet only a few hundred transcripts noticeably change their level in response to removal via siRNA or genetic deletion of this TF – and few of these transcriptionally altered genes are bound directly by the TF (*right*). One (controversial) model pioneered by part of the fly transcriptional community that may help account for this disparity is that the regional concentration of *all* protein–DNA interactions may control transcription [63]. This model could also help explain the surprisingly rapid divergence of mammalian TF binding sites seen in recent experiments [61]

Shortly thereafter, the Brown laboratory also reported the genome-wide binding of the Rap1 protein [30]. Among their notable observations was the large-scale binding of this protein to the telomeric regions of yeast, as well as to the ribosomal protein genes. This link between a site-specific TF that regulates a large number of coding regions in the genome with regions containing noncoding RNA was among the first of many major surprises in investigations utilizing genome wide TF binding.

8.5 Combining Multiple TF Binding Experiments on the Same Platform Allows Network Elaboration

The first systems-biology application of the ChIP-chip technology was to deconstruct the combinatorial transcriptional networks active in the model organism *Saccharomyces cerevisiae*. It has been long known that TFs at specific promoters bind to and regulate gene expression combinatorially. Two publications reporting large numbers of TF binding experiments exploiting a carefully matched yeast genetic system were reported within a year and half of the first report of ChIP-chip as a technique [31, 32].

One publication pioneered a comprehensive systems-wide characterization of the transcriptional regulatory networks active in yeast by controlling the growth condition (YPD media), genetic background (yeast strain), microarrays, and experimetnal protocols and reagents [32]. Over a hundred independent yeast strains containing epitope-tagged TFs were created and grown, followed by ChIP-chip analysis. Two crucial observations made in the simplest of eukaryotes indicated that (1) on a genome-wide basis, TFs bind to promoters in a highly combinatorial manner, and (2) that regulators not only bound transcriptional regulators in the same functional class, but that the interconnections between TFs and the promoters of other TFs often crosses functional classes (Fig. 8.3). For instance, cell cycle regulators were found to bind to the promoters of TFs that had well-characterized metabolic functions.

A simultaneous publication pioneered a different strategy of characterizing regulatory networks combinatorially by identifying other regulators downstream of a known TF's binding (MBF), followed by characterization of their genome-wide binding and further analysis [31]. The G1/S transition was carefully dissected by identifying the TFs downstream of MBF, tagging these with a biochemical handle, and performing matched ChIP-chip experiments at the same point of the cell cycle.

8.6 The Vast Size and Repeat Content of Mammalian Genomes Provoked Four Different Approaches to Microarray Analysis of TF Binding

Mammalian genomes like human and mouse can be over a hundred times the size of yeast genomes, as well as being composed of over 50% repeated sequences found at high density in introns, promoters, enhancers, gene deserts – basically everywhere outside coding regions. These facts greatly complicated the original assemblies of the human and mouse genomes [33, 34].

When considering strategies to interrogate TF binding, the enormous size and complex, repeat-laden sequence content were also very complicating. Four microarray strategies were simultaneously pioneered to interrogate on a genome-wide scale human TF binding and the histone mark patterns obtainable from ChIP

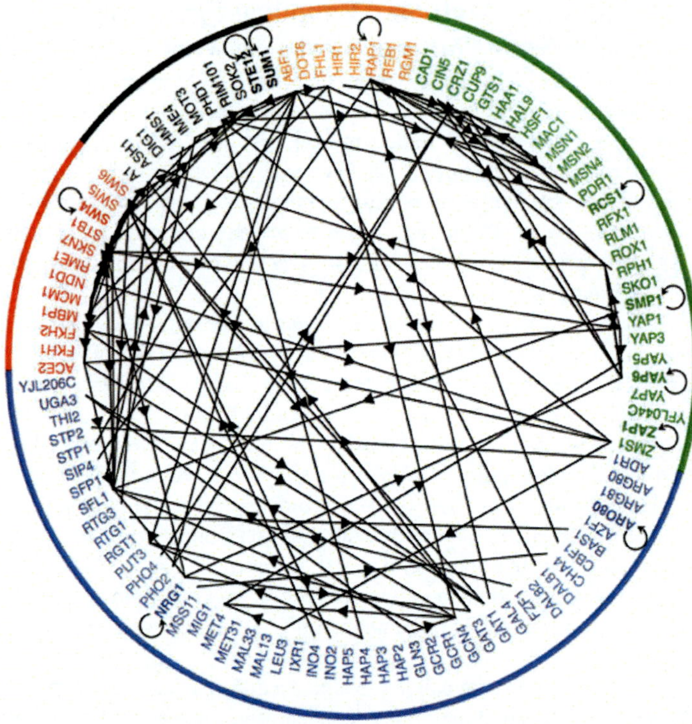

Fig. 8.3 Transcriptional regulators often bind to the genes coding for other transcriptional regulators in eukaryotes. Yeast transcriptional regulators are shown in a *circle* and organized by their functional categories. Potential regulatory linkages are shown as *arrows* between one TF and the promoter of another that is bound; *lines* between categories are very common (reprinted by permission from [32])

experiments: (1) CpG island microarrays, (2) PCR-based promoter microarrays, (3) PCR-based chromosome tiling microarrays, (4) short oligonucleotide chromosomal tiling microarrays.

8.6.1 CpG Island Microarrays

Many mammalian regulatory regions have an enrichment in CpG dinucleotides, and as such are named CpG islands. One strategy to capture on a genome-scale TF binding in mammals was to PCR amplify specifically the regions of the mammalian genome that contain CpG islands for printing onto a microarray platform. Such a microarray was first reported in combination with E2F4 ChIP experiments to reveal scores of newly discovered binding events across the human genome [35].

8.6.2 Promoter Region PCR Microarrays

A substantial fraction of transcriptional regulation has long been known to occur within the first few kilobases of transcription start sites in mammals. This fact was exploited to create PCR products against non-repeated sequence found within the first 2 kb of potentially regulatory sequence upstream of transcription start sites across the human genome, which were then used to create microarrays. The first report using this form of microarray characterized the cell-cycle promoter regions bound by E2F1 in in vitro human fibroblast cells [36]. A somewhat later report used a slightly larger platform that represented the promoters from most human genes, yet reported for the first time ChIP experiments performed against disease-related TFs in primary human tissues, here liver and pancreatic islets [37].

8.6.3 Regional and Chromosomal Tiling PCR Microarrays

The first microarrays designed to represent entire regions [38] and chromosomes [39] were created by using PCR to amplify overlapping, or nearly overlapping DNA regions, followed by printing onto microarrays. The first chromosome-wide interrogation of TF binding was performed against NFkB sites across human chromosome 22, and was the first genome-scale experimental evidence that TFs bind extensively far from gene loci. This observation led eventually to similar pioneering discoveries in tissue-specific transcriptional regulation, including with Estrogen Receptor [40] and Androgen Receptor [41].

8.6.4 Short Oligonucleotide Chromosomal Tiling Microarrays

Finally, it was quickly realized that commercially available Affymetrix microarrays designed to tile human chromosomes for other purposes could be adapted to ChIP studies. The first report exploiting this described the comparison of histone marks in human and mouse lung fibroblasts across microarrays that tiled human chromosomes 21 and 22, finding strong similarities in the histones found in a few selected human and mouse orthologous regions [42]. This study was among the first to focus on understanding histone marks; since then, similar studies have rapidly become far more common than genome-wide TF binding studies, due to commercial availability of high affinity antisera.

8.6.5 Limitations of Microarrays

Though the ability to print DNA sequences at high density onto slides was quickly developed, thus overcoming a major roadblock towards having a genome on a single microarray, other hurdles remained. Most important of these is the requirement to omit repeated regions from microarray designs, because of their extremely high cross-hybridization with multiple regions in the mammalian genome. In effect, typically half of a mammalian genome cannot be interrogated with microarrays regardless of the microarray technology being used, and this is an unavoidable design limitation. Finally, microarrays are by their nature designed to be species-specific, which is a significant limitation if a researcher wishes to investigate non-standard model organisms. For instance, how large could a market be for microarrays to interrogate butterfly gene expression [43]?

8.7 High-Throughput Sequencing Has Revolutionized Interrogation of In Vivo Binding in Complex Higher Genomes

A variety of methods have been developed to sequence nucleic acids (reviewed in [44]), and the most widely used method until recently was Sanger sequencing [45]. This methodology sequences stretches of hundreds of bases, which were painstakingly computationally sewn together using chromosomal maps to assemble the human [33] and mouse [34] genomes. Most placental mammals sequenced more recently have relied heavily on these early, carefully annotated mammalian genomes to facilitate the assembly of genomes (for instance [46–48]), as synteny is widespread.

The recent development and rapid widespread adaptation of technology that uses millions of parallel sequencing reactions [49] is currently displacing these methods in techniques as diverse as genome resequencing, mRNA sequencing, and characterization of TF binding globally by directly sequencing nucleic acids bound by TFs (ChIP-seq), instead of microarray hybridization (ChIP-chip) (HTP sequencing is reviewed in [50], and different platforms that can be used as of 2008 is reviewed in [51]).

To interrogate TF binding globally, oligonucleotides between 20 and 100 bases are read from the end of the ChIP-isolated DNA, aligned to the genome, and compiled into a global picture of TF binding. Typical reactions use >10 million reads per experiment, and, importantly, resolve up to 90% of the mammalian genome by covering sufficient nonrepeat region to allow unambiguous mapping of the reads to occur. The speed of data acquisition, and tremendous coverage of the mammalian genome, has revolutionalized the frontiers of modern genomic biology.

8.8 Major New Frontiers in Mammalian Transcription Are Possible with Genome-Wide Interrogation of In Vivo TF Binding

The ability to identify the global binding of mammalian TFs has created new opportunities to revisit questions that have been raised over the last 100 years in biology. How does one genome create multiple tissues? What mechanistically directs this process? How do intra-species variations and evolution of different species interface? I present here three nascent research areas that exploit how HTP sequencing enhances our ability to answer ambitious research questions.

8.8.1 Variation of TF Binding Within Species

Genetic differences drive phenotypic differences within individuals of a species. How these genetic differences impact the gene expression and transcriptional regulatory differences between species has been a long-standing question, and a number of microarray based experiments have characterized gene expression differences between different individual humans [52] and mice [53]. How the regulation of these differences is driven has more recently been addressed using human cell lines [54, 55].

The variation among humans in matched cell lines for the same polymerase machinery and TFs has been found to be remarkably high, on the order of 10–25% variation between individual human genotypes [54]. Closer inspection of inter-allelic regulation indicates that up to one in ten active chromatin sites can differ between the two chromosome copies within the same individual [55]. Taken together, these results suggest that the natural assumption of large-scale similarity between individuals within a species has important limitations. If up to a quarter of genomic regulation can vary between one individual and another, then assigning a consensus regulatory pattern becomes problematic; this observation has clear relevance to the ENCODE project's objectives, described below.

8.8.2 Evolution of TF Binding Among Mammals

Using microarrays, my laboratory showed that TF binding has diverged significantly between human and mouse in matched tissues [56]. We showed using ChIP-chip on specially designed proximal promoter microarrays that a set of liver-specific TFs (FOXA2, HNF1A, HNF4A, HNF6) whose function, amino acid sequence, and targeted binding motif are highly conserved throughout mammals, change both their potentially targeted genes and also their binding locations globally. However, our early study was limited to the proximal promoters around 4,000 transcription start sites in human and mouse, purely due to the technical limitations imposed on our experiments by the extant microarray densities.

By performing similar genome-wide binding studies using ChIP-seq for two TFs (CEBPA and HNF4A) in five vertebrates, we unambiguously identified how each species has tens of thousands of binding events that are unique to each evolutionary lineage (Fig. 8.4) [57]. The role of repeat elements could now be addressed, and it was shown that substantial numbers of binding events in liver were found on repeat elements in each lineage. Similar results have been found also for the OCT4 and

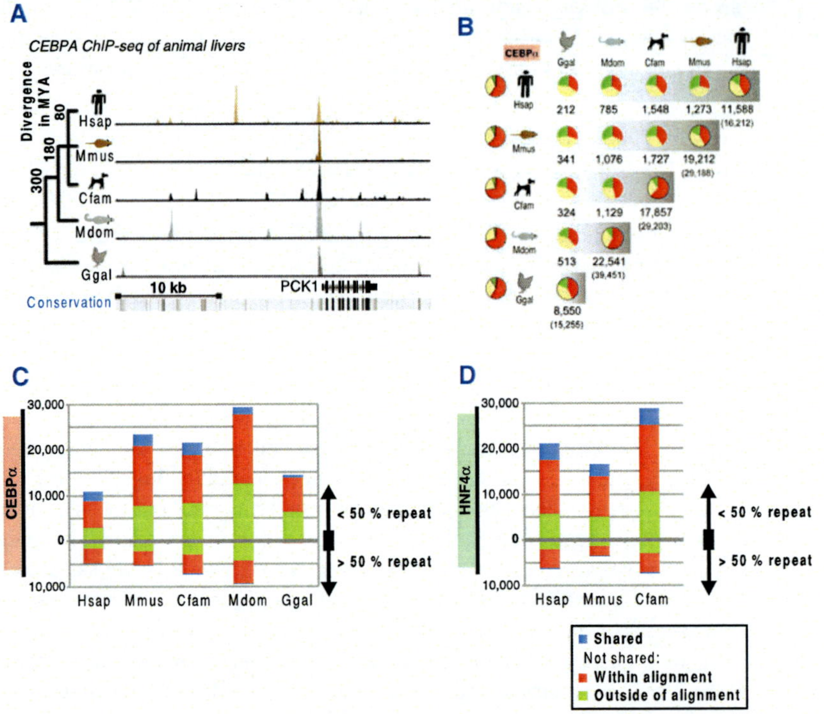

Fig. 8.4 Tissue-specific TF binding evolves rapidly among mammals. **a** The in vivo binding of CEBPA is shown in five species near the PCK1 locus. **b** The pairwise overlap between two of five vertebrates is shown. The pie chart indicates the location in the genome of the total set of binding events (*red*: intergenic; *green*: promoter [TSS±3 KB]; *yellow*: intronic). The *left* most piechart is the whole genome, the *right* most diagonal shows where the set of all binding events fall. (*c* & *d*) A sizable fraction of binding events are found in regions that are >50% repeat for CEBPA (*c*) and HNF4A (*d*). Reprinted by permission from [57]

NANOG transcriptional regulators in mouse and human in embryonic stem cells [58, 59], incidating that in mammals, rapid evolution of transcriptional regulation is the general rule.

8.8.3 Comprehensive Analysis of Multiple Cell Lines from One Species – The ENCODE Project

Inspired by the success of the human, mouse, and other mammalian genome projects, the National Human Genome Research Institute have undertaken an effort to fully annotate all functional elements in the human genome [60] (see also http://www.genome.gov). The ENCyclopedia Of DNA Elements (ENCODE) pilot project began by exhaustive characterization of 1% of the human genome using different microarray platforms and many of the techniques reviewed in this and other chapters in this book. More recently, with the advent of HTP sequencing, genome-wide analysis has displaced the more limited views obtainable by microarrays. Scores of laboratories have been funded to characterize a handful of human cell lines via TF binding profiles, gene expression, small RNA expression, chromatin structure, histone marks, proteomics, and CpG methylation, among other techniques.

8.8.4 Widespread Adaptation of ChIP-Seq Experiments in the Transcriptional Community

Many laboratories have now adapted sequencing to interrogate the genome-wide binding of transcription factors in cultured cell lines and in vivo. Virtually all are (re-)discovering that (i) tens of thousands of transcription factor binding events exist in the mammalian genome, (ii) that TFs bind combinatorially in tissue-specific patterns, and (iii) that TF binding rarely controls gene expression in a direct or intuitive manner. By community agreement, most datasets are freely available from the European Nucleotide Archive at the EMBL EBI (http://www.ebi.ac.uk/ena/) and from the Gene Expression Omnibus in America at the NCBI (http://www.ncbi.nlm.nih.gov/geo/). The sequencing data in ChIP-seq experiments is literally growing exponentially, and will continue to pose unprecedented challenges in both the storage of such massive amounts of data, and even more so in the analysis required to biologically interpret such truly genome-wide data.

8.9 Future Directions

What will the future hold for understanding how transcriptional regulation operates in the vast mammalian genome? Efforts like the ENCODE consortium will clearly produce enormous insight into how one species can deploy a single genome to create hundreds to thousands of distinct cell types. Combining TF binding and chromatin status for all possible proteins and post-transcriptional modifications may allow the development of models with true predictive value. Though simple, this goal has eluded computational and experimental science for decades. Intriguingly, one species of placental mammal (mouse) can readily and accurately interpret the genetic instructions embedded in the sequence of another mammal (human) [61] (Fig. 8.5), suggesting that an understanding of the underlying grammar is indeed an obtainable goal.

Nevertheless, the most likely result, as is always the case for high impact research areas, will be more questions. Many of these questions are predictable and clear extrapolations of the trends in genomics and proteomics. For instance, how does the three dimensional genome interact with the proteomic components of the rest of the cell? Other predictable directions may include controlled engineering of development: tailoring phenotypic traits using a soon-to-be-known regulatory toolbox in

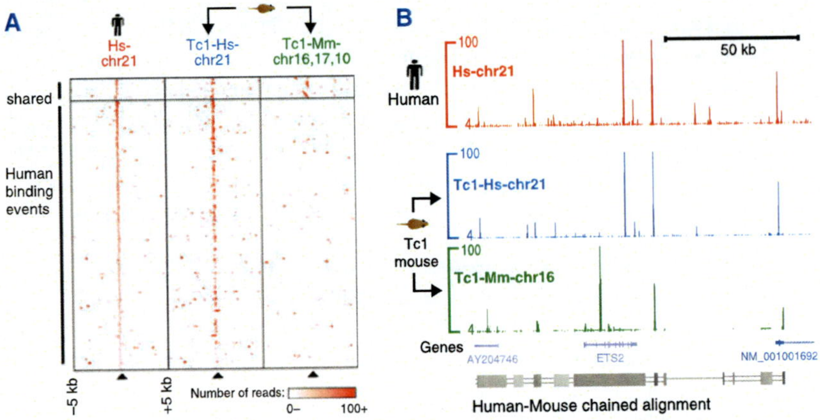

Fig. 8.5 A mouse nucleus can accurately read human genetic instructions to place the CEBPA TF in adult liver. **a** The chromosome-wide binding events for CEBPA in human liver were sorted first by whether they were shared with mouse, and then within these two classes by intensity of binding. The Tc1-Hs-chr21 lane is the signal obtained from the human chromosome in a mouse environment, which is largely indistinguishable from the native human case. **b** The CEBPA binding events near the ETS2 locus reflect the largely identical signal in human or mouse for this chromosome. Reprinted by permission from [57]; see also [61]

mammals. Many vertebrates can regenerate limbs using ancient developmental systems driven by TFs [62], which may be applicable to higher mammals that may have lost these functions.

Other questions will certainly be driven by unpredictable discoveries, much like the realization in the last years of the twentieth century that small RNAs can and do play integral regulatory roles in mammalian transcription. The ability to identify TF binding across the entire mammalian genome will be an important tool; one whose value will become fully apparent only after it is as routine and reliable as gene expression microarrays.

References

1. Gilbert W (1986) Evolution of antibodies. The road not taken. Nature 320(6062):485–486
2. Wing R, et al. (1980) Crystal structure analysis of a complete turn of B-DNA. Nature 287(5784):755–758
3. Watson JD, Crick FH (1953) Genetical implications of the structure of deoxyribonucleic acid. Nature 171(4361):964–967
4. Watson JD, Crick FH (1953) Molecular structure of nucleic acids; a structure for deoxyribose nucleic acid. Nature 171(4356):737–738
5. Vaquerizas JM, et al. (2009) A census of human transcription factors: function, expression and evolution. Nat Rev Genet 10(4):252–263
6. Riechmann JL, et al. (2000) *Arabidopsis* transcription factors: genome-wide comparative analysis among eukaryotes. Science 290(5499):2105–2110
7. Riano-Pachon DM, et al. (2007) PlnTFDB: an integrative plant transcription factor database. BMC Bioinformatics 8:42

8. Jordan SR, Pabo CO (1988) Structure of the lambda complex at 2.5 A resolution: details of the repressor–operator interactions. Science 242(4880):893–899
9. Chenoweth DM, Dervan PB (2009) Allosteric modulation of DNA by small molecules. Proc Natl Acad Sci U S A 106(32):13175–13179
10. Galas DJ, Schmitz A (1978) DNAse footprinting: a simple method for the detection of protein-DNA binding specificity. Nucleic Acids Res 5(9):3157–3170
11. Hardison RC, et al. (1979) The structure and transcription of four linked rabbit beta-like globin genes. Cell 18(4):1285–1297
12. Stranick KS, et al. (1997) Identification of transcription factor binding sites important in the regulation of the human interleukin-5 gene. J Biol Chem 272(26):16453–16465
13. Kardassis D, et al. (1990) Characterization of the promoter elements required for hepatic and intestinal transcription of the human apoB gene: definition of the DNA-binding site of a tissue-specific transcriptional factor. Mol Cell Biol 10(6):2653–2659
14. Kardassis D, Zannis VI, Cladaras C (1992) Organization of the regulatory elements and nuclear activities participating in the transcription of the human apolipoprotein B gene. J Biol Chem 267(11):7956
15. Shibata Y, Crawford GE (2009) Mapping regulatory elements by DNaseI hypersensitivity chip (DNase-Chip). Methods Mol Biol 556:177–190
16. Quitschke WW, et al. (2000) Differential effect of zinc finger deletions on the binding of CTCF to the promoter of the amyloid precursor protein gene. Nucleic Acids Res 28(17): 3370–3378
17. Perry RP, Kelley DE (1966) Evidence for specific association of protein with newly formed ribosomal subunits. Biochem Biophys Res Commun 24(3):459–465
18. Economidis IV, Rousseau GG (1985) Association of the glucocorticoid hormone receptor with ribonucleic acid. FEBS Lett 181(1):47–52
19. Licatalosi DD, et al. (2008) HITS-CLIP yields genome-wide insights into brain alternative RNA processing. Nature 456(7221):464–469
20. Wang Z, et al. (2009) CLIP: construction of cDNA libraries for high-throughput sequencing from RNAs cross-linked to proteins in vivo. Methods 48(3):287–293
21. Solomon MJ, Varshavsky A (1985) Formaldehyde-mediated DNA-protein crosslinking: a probe for in vivo chromatin structures. Proc Natl Acad Sci U S A 82(19): 6470–6474
22. Varshavsky AJ, Sundin O, Bohn M (1979) A stretch of "late" SV40 viral DNA about 400 bp long which includes the origin of replication is specifically exposed in SV40 minichromosomes. Cell 16(2):453–466
23. Gasch AP, et al. (2000) Genomic expression programs in the response of yeast cells to environmental changes. Mol Biol Cell 11(12):4241–4257
24. Causton HC, et al. (2001) Remodeling of yeast genome expression in response to environmental changes. Mol Biol Cell 12(2):323–337
25. Ross DT, et al. (2000) Systematic variation in gene expression patterns in human cancer cell lines. Nat Genet 24(3):227–235
26. Alizadeh AA, et al. (2000) Distinct types of diffuse large B-cell lymphoma identified by gene expression profiling. Nature 403(6769):503–511
27. Boldrick JC, et al. (2002) Stereotyped and specific gene expression programs in human innate immune responses to bacteria. Proc Natl Acad Sci U S A 99(2):972–977
28. Iyer VR, et al. (2001) Genomic binding sites of the yeast cell-cycle transcription factors SBF and MBF. Nature 409(6819):533–538
29. Ren B, et al. (2000) Genome-wide location and function of DNA binding proteins. Science 290(5500):2306–2309
30. Lieb JD, et al. (2001) Promoter-specific binding of Rap1 revealed by genome-wide maps of protein-DNA association. Nat Genet 28(4):327–334
31. Horak CE, et al. (2002) Complex transcriptional circuitry at the G1/S transition in *Saccharomyces cerevisiae*. Genes Dev 16(23):3017–3033

32. Lee TI, et al. (2002) Transcriptional regulatory networks in *Saccharomyces cerevisiae*. Science 298(5594):799–804
33. Gilad Y, Wiebe V, Przeworski M, Lancet D, Pääbo S (2004) Finishing the euchromatic sequence of the human genome. Nature 431(7011):931–945
34. Waterston RH, et al. (2002) Initial sequencing and comparative analysis of the mouse genome. Nature 420(6915):520–562
35. Weinmann AS, et al. (2002) Isolating human transcription factor targets by coupling chromatin immunoprecipitation and CpG island microarray analysis. Genes Dev 16(2):235–244
36. Ren B, et al. (2002) E2F integrates cell cycle progression with DNA repair, replication, and G(2)/M checkpoints. Genes Dev 16(2):245–256
37. Odom DT, et al. (2004) Control of pancreas and liver gene expression by HNF transcription factors. Science 303(5662):1378–1381
38. Horak CE, et al. (2002) GATA-1 binding sites mapped in the beta-globin locus by using mammalian chIp-chip analysis. Proc Natl Acad Sci U S A 99(5):2924–2929
39. Martone R, et al. (2003) Distribution of NF-kappaB-binding sites across human chromosome 22. Proc Natl Acad Sci U S A 100(21):12247–12252
40. Carroll JS, et al. (2005) Chromosome-wide mapping of estrogen receptor binding reveals long-range regulation requiring the forkhead protein FoxA1. Cell 122(1):33–43
41. Massie CE, et al. (2007) New androgen receptor genomic targets show an interaction with the ETS1 transcription factor. EMBO Rep 8(9):871–878
42. Bernstein BE, et al. (2005) Genomic maps and comparative analysis of histone modifications in human and mouse. Cell 120(2):169–181
43. Vera JC, et al. (2008) Rapid transcriptome characterization for a nonmodel organism using 454 pyrosequencing. Mol Ecol 17(7):1636–1647
44. Schuster SC (2008) Next-generation sequencing transforms today's biology. Nat Methods 5(1):16–18
45. Sanger F, Nicklen S, Coulson AR (1977) DNA sequencing with chain-terminating inhibitors. Proc Natl Acad Sci U S A 74(12):5463–5467
46. Gibbs RA, et al. (2004) Genome sequence of the Brown Norway rat yields insights into mammalian evolution. Nature 428(6982):493–521
47. Gibbs RA, et al. (2007) Evolutionary and biomedical insights from the rhesus macaque genome. Science 316(5822):222–234
48. Mikkelsen TS, et al. (2007) Genome of the marsupial Monodelphis domestica reveals innovation in non-coding sequences. Nature 447(7141):167–177
49. Shendure J, et al. (2005) Accurate multiplex polony sequencing of an evolved bacterial genome. Science 309(5741):1728–1732
50. Shendure J, Ji H (2008) Next-generation DNA sequencing. Nat Biotechnol 26(10):1135–1145
51. Mardis ER (2008) Next-generation DNA sequencing methods. Annu Rev Genomics Hum Genet 9:387–402
52. Schadt EE, et al. (2008) Mapping the genetic architecture of gene expression in human liver. PLoS Biol 6(5):e107
53. Yang X, et al. (2006) Tissue-specific expression and regulation of sexually dimorphic genes in mice. Genome Res 16(8):995–1004
54. Kasowski M, et al. (2010) Variation in transcription factor binding among humans. Science 328(5975):232–235
55. McDaniell R, et al. (2010) Heritable individual-specific and allele-specific chromatin signatures in humans. Science 328(5975):235–239
56. Odom DT, et al. (2007) Tissue-specific transcriptional regulation has diverged significantly between human and mouse. Nat Genet 39(6):730–732
57. Schmidt D, et al. (2010) Five-vertebrate ChIP-seq reveals the evolutionary dynamics of transcription factor binding. Science 328(5981):1036–1040
58. Bourque G (2009) Transposable elements in gene regulation and in the evolution of vertebrate genomes. Curr Opin Genet Dev 19(6):607–612

59. Kunarso G, et al. (2010) Transposable elements have rewired the core regulatory network of human embryonic stem cells. Nat Genet 42(7):631–634
60. Birney E, et al. (2007) Identification and analysis of functional elements in 1% of the human genome by the ENCODE pilot project. Nature 447(7146):799–816
61. Wilson MD, et al. (2008) Species-specific transcription in mice carrying human chromosome 21. Science 322(5900):434–438
62. Brockes JP, Kumar A (2005) Appendage regeneration in adult vertebrates and implications for regenerative medicine. Science 310(5756):1919–1923
63. MacArthur S, et al. (2009) Developmental roles of 21 *Drosophila* transcription factors are determined by quantitative differences in binding to an overlapping set of thousands of genomic regions. Genome Biol 10(7):R80

Chapter 9
How Transcription Factors Identify Regulatory Sites in Genomic Sequence

Yair Field, Eilon Sharon, and Eran Segal

Abstract Binding of transcription factors to functional sites is a fundamental step in transcriptional regulation. In this chapter, we discuss how transcription factors are thought to achieve specificity to their functional targets, despite their typically low concentrations and degenerate binding specificities, and the fact that in large genomes their functional binding sites must compete with their widespread alternative binding sites. We highlight the importance of the chromatin structure context of the binding sites in this process, and its dependency on the genomic DNA sequence.

9.1 Introduction

Coordinated binding of specific proteins to designated genomic locations is the basis for fundamental cellular processes and in particular for regulation of gene expression. The basic mechanism behind this process is the ability of DNA-binding proteins, mainly transcription factors, to specifically bind short DNA sequences, typically of ~6–20 basepairs in length. At a high-level view, genomes encode the regulatory binding sites by genomic matches to the recognized motifs, and they control the binding events in time by regulating the concentration and binding activity of the transcription factors. However, do these short motifs provide sufficient information for transcription factors to bind their functional target sites?

To answer this question, we need to consider the concentration of transcription factors against the competition with alternative binding. Occurrences of additional binding motifs within the genome will compete with a functional target site for factor binding. By chance, we expect larger genomes will have more motifs, with some motifs being more abundant than others (e.g., motifs present within highly repetitive sequences in the human genome [1]). Moreover, transcription factors have degenerate sequence specificities as they bind with variable but considerable efficiency to multiple different sequence motifs, with the level of degeneracy varying

Y. Field (✉)
Department of Computer Science and Applied Mathematics, Weizmann Institute of Science, Rehovot 76100, Israel
e-mail: yair.field@weizmann.ac.il

T.R. Hughes (ed.), *A Handbook of Transcription Factors*, Subcellular Biochemistry 52, DOI 10.1007/978-90-481-9069-0_9, © Springer Science+Business Media B.V. 2011 across transcription factors [2]. This degenerate specificity significantly increases the competition for a target site with alternative binding at decoy sites. In addition, transcription factors associate nonspecifically with DNA, and when this weak binding is integrated across the genome, it further enhances the competition with alternative binding [3, 4]. Thus, efficient binding to a regulatory site based solely on target sequence specificity requires that transcription factors would be present in high enough concentrations in order to overcome the widespread competition with alternative binding. Is this the case?

Theoretical studies suggest that the information of the single binding motif is typically sufficient to explain target specificity in the bacterium *E. coli*, but not in any eukaryote [2]. In fact, there is much evidence that in eukaryotes, most of the canonical binding sites for a transcription factor, based solely on its known sequence specificities, are not bound in vivo, or are infrequently bound [5–7]. In this chapter, we discuss additional determinants of transcription factor binding, highlighting the importance in eukaryotes of the chromatin context of binding sites.

9.2 General Importance of the Chromatin Structure Context of Binding Sites

In eukaryotes, a major determinant of transcription factor binding is chromatin structure, namely the state of the genome's packaging with specific structural proteins, mainly histones. Chromatin structure consists of several hierarchical levels of compaction and each level exists in multiple possible substructures, is dynamic and regulated, and affects transcription factor binding in several ways. Chromatin structure determines the physical proximity between genomic locations, thus affecting the ability of transcription factors to exert long-range protein–protein interactions. Moreover, at least for higher eukaryotes, the concentration of transcription factors within the nucleus is thought to be not uniform, and chromatin structure determines the proximity of regulatory regions to "transcription factories" – domains with elevated factor concentrations [8, 9]. In addition, chromatin structure serves as a direct template for binding and recruitment of proteins that are involved in regulation of transcription factor binding, such as proteins that bind the unstructured "tail" domains of histones with specific post-translational modifications [10, 11]. Finally, a central feature of chromatin is that several of its substructures render associated DNA inaccessible for transcription factor binding. Together, transcription factor binding in eukaryotes is highly dependent on the chromatin context of binding sites. Here, we focus on the prerequisite of transcription factors to gain access to their target sites.

9.3 Chromatin Structure and DNA Accessibility

The basic unit of chromatin is the nucleosome structure, in which a DNA segment of length 147 bp is tightly wrapped, in almost two super-helical turns, around an octamer of histone proteins [12]. In its most stable conformation, the nucleosome completely occludes its wrapped DNA from access to most transcription factors.

Since the nucleosome is dynamic and spontaneously undergoes transient unwrapping events, all parts of the nucleosomal DNA are in fact exposed to transcription factor binding, but with exponentially decreasing probability towards the middle of the wrapped DNA [13, 14]. This accessibility within the nucleosome is thought to be further regulated by factors that stabilize either the fully wrapped or partially unwrapped states. For example, binding of the linker histone H1 at the DNA entry/exit site of the nucleosome is thought to stabilize the inaccessible state, whereas binding of High Mobility Group (HMG) proteins to nucleosomes is thought to stabilize the partially accessible state [15]. In addition, ATP-dependent chromatin remodelers can actively expose nucleosomal DNA [16]. Thus, at the first level of chromatin compaction, that of nucleosomes separated by short linker DNA segments, the positions occupied by nucleosomes have a significantly reduced accessibility to transcription factor binding compared to linker DNA, and gaining high accessibility to these nucleosome occupied regions requires explicit regulation [17, 18]. In the next higher-order level of compaction, nucleosome arrays can fold into a more condensed chromatin structure known as the 30-nm fiber, which significantly reduces the accessibility of linker DNA [19, 20]. The basic requirement for local fiber condensation is an array of few (~6) consecutive H1-bound nucleosomes [21, 22]. In addition to its primary dependence on the linker histone H1, fiber folding is known to further depend, directly or indirectly, on several additional factors, including histone variants [23, 24] and modifications [25, 26], DNA-methylation [27], chromatin remodelers [28], and chromatin binding proteins such as HP1 [29] and HMG [30]. Notably, the positions of nucleosomes also affect fiber folding because H1 binding to the nucleosome requires a minimal linker length of ~20–40 bp [22, 31]. Thus, regions with very dense nucleosome organizations can only have sub-stoichiometric binding of H1 per nucleosome and should consequently form relatively relaxed fibers, with relatively accessible linker DNA and partially accessible H1-free nucleosomal DNA. In contrast, regions with sparse nucleosome organizations may have full H1 stoichiometry and consequently relatively condensed and inaccessible fiber structures (Field et al., unpublished).

Together, accessibility of DNA to transcription factor binding within chromatin depends on the presence of several chromatin binding proteins such as H1, HMG and HP1, the histone variant composition, the state of histone modifications and DNA methylation, the activity of chromatin remodelers, and the detailed positions of nucleosomes.

9.4 DNA Accessibility Preferences Encoded Through Nucleosome Positioning Signals

The nucleosome structure has DNA sequence preferences, which have a significant effect on the nucleosome organization of the genome [32]. As nucleosome organization is a major determinant of DNA accessibility within chromatin, genomes in practice encode explicit preferences for their DNA accessibility landscape through nucleosome positioning signals. However, in different organisms, accessibility is thought to be encoded by preferences for different nucleosome organizations, according to the typical context of higher-order chromatin structure (Fig. 9.1). Yeast

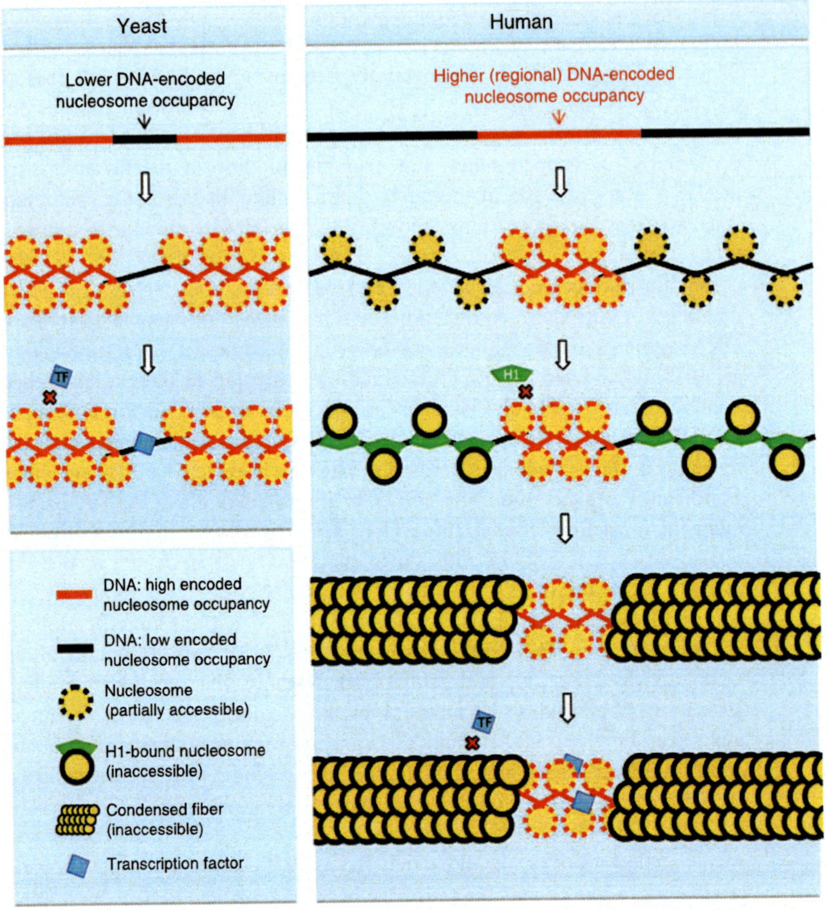

Fig. 9.1 Different DNA-encoded nucleosome organizations are thought to facilitate DNA accessibility in lower and higher eukaryotes. In yeast (*left panel*), whose H1 physiological concentration is relatively low and its fiber structures are relatively relaxed, local low DNA-encoded affinity to nucleosomes is thought to increase the potential that a region will be embedded within naked linker DNA, and thus be relatively accessible to transcription factors. In contrast, in human and other higher eukaryotes (*right panel*), which have relatively high H1 concentrations and relatively condensed fibers, increased potential to assemble into relatively accessible chromatin structures is thought to be achieved by high regional DNA-encoded affinity to nucleosomes, in opposite to the model in yeast. This is because such regions should be more likely to assemble into dense (closely spaced) nucleosome organizations that do not allow high binding stoichiometry of H1 per nucleosome, and consequently should have relatively relaxed and accessible fiber structures. Hence, low local affinity to nucleosomes in yeast and high regional affinity to nucleosomes in human, are thought to both facilitate relatively accessible chromatin structures in each genome. "X"'s under the transcription factor and H1 represent cases in which DNA accessibility for the factor or for H1 is relatively reduced by the chromatin structure

chromatin has largely relaxed fiber structures [33], as histone H1 is present in low stoichiometry [34] and the genome is highly gene-dense and mostly transcribed. Thus, in yeast, the nucleosome is the dominant inaccessible chromatin structure, and relatively high DNA accessibility is encoded by a preference for low nucleosome occupancy. In contrast, human chromatin typically has a high concentration of H1 [31] and much more condensed fiber structures, and this chromatin level is thought to have a dominant effect on DNA accessibility with respect to the nucleosome level [35]. As a result, we propose that high DNA accessibility in human may be encoded primarily by a preference for high regional nucleosome density, in order to facilitate H1 exclusion and formation of relaxed fiber structures (Field et al., unpublished).

In accordance, yeast genomes generally encode preferences for low nucleosome occupancy over their regulatory regions (local A/T rich sequences) [36], whereas higher eukaryotic genomes such as human [37] and fly [38] generally encode preferences for high regional nucleosome density over their regulatory regions (high regional G/C content). Hence, eukaryotic genomes generally demarcate their regulatory regions by setting their DNA-encoded preferences to relatively accessible nucleosome organizations.

The prerequisite of transcription factors to gain accessibility to their target sites does not imply that regulatory regions encoded for relatively inaccessible chromatin structures cannot be bound, but rather, that transcription factor binding in such cases requires explicit regulation to gain accessibility. Indeed, among yeast promoters, those that are encoded for relatively high nucleosome occupancy are generally associated with the presence of more transcription factor binding sites, more extensive regulation by chromatin remodelers, and more regulated gene expression patterns across different conditions [36]. Notably, this general trend is conserved across the evolution of yeast species, and even under expression divergence of genes between constitutive and regulated expression programs [39]. Similarly, regulatory regions in human that have relatively low regional DNA-encoded nucleosome occupancy (indicated by their relatively low G/C content) show more variability across different cell types in terms of their measured accessibility [40] and associated gene expression [41].

Thus, eukaryotic genomes utilize nucleosome positioning signals to encode functional preferences for local chromatin structure over their regulatory regions. This encoding for regulatory regions facilitates their demarcation compared to the bulk of the genome, and further determines their level of dependency on active regulation for DNA accessibility.

9.5 Combinatorial Binding of Transcription Factors

An additional important determinant of transcription factor binding is the crosstalk between multiple binding sites. Transcription factors can physically interact through protein–protein interactions, either directly or through intermediate co-factor proteins, and this may increase the binding probability of a target site [42]. These interactions can occur between adjacent binding sites, or between distal sites through formation of DNA loops [43]. The interactions between binding sites may also be negative, for example when binding sites overlap [44], or when transcription factors compete among themselves for interactions with additional proteins [45]. Cooperative interactions between sites, either positive or negative, can further occur without any physical interaction, for example through recruitment of chromatin remodelers or histone modifiers that regulate DNA accessibility within chromatin [18]. Moreover, unlike symmetric dependencies between binding sites due to protein–protein interactions or competition over overlapping sites, recruitment of DNA accessibility regulators is a general mechanism by which non–symmetric interactions are implemented. Finally, clustering of binding sites may impose implicit cooperativity between them even without a direct recruitment of accessibility regulators, just due to their collaborative competition for dissolving inaccessible chromatin structures [46]. For example, when one transcription factor overcomes a condensed fiber state to bind in linker DNA, this relieves the need for additional transcription factors to overcome this barrier for binding to adjacent sites. Together, crosstalk between binding sites is an important determinant of transcription factor binding.

9.6 A Modeling Framework for Transcription Factor Binding in Genomic Sequence

Quantitative models are being developed to test and validate our understanding of the determinants of transcription factor binding. These models can be further used to predict binding potential from genomic sequence and be integrated into more complex models to study and predict transcriptional regulation behaviors [47]. Recognition sequence specificity of transcription factors and nucleosomes is typically inferred statistically from high throughput binding measurements over a large pool of diverse sequences, either in vivo [36, 48] or in controlled in vitro systems [32, 49, 50], and less frequently, it is being predicted directly from structural considerations of the binding proteins and DNA [51]. A quantitative model of the recognition specificity can then be used to predict the binding affinity landscape of a genomic sequence. Many computational studies use such binding affinity landscapes to threshold for predicted high affinity sites, search for clusters of predicted sites, and consider the nucleosome affinity over these sites in order to explain and predict in vivo binding measurements [36, 52]. One recent and promising computational approach considers all the possible linear genomic binding arrangements (termed binding configurations) of the examined transcription factors and nucleosomes, in which binding events respect minimal steric hindrance constrains, and given parameter values for the concentrations of these factors computes the equilibrium distribution over all configurations [47]. This framework can model the competition between transcription factors on overlapping sites, the competition between transcription factors and nucleosomes, and the implicit cooperativity between transcription factors due to collaborate competition against nucleosome occupancy. This modeling approach was used to further model the effect of explicit cooperative protein–protein interactions [53], binding of linker histones to nucleosomes, and local fiber folding (Field et al., unpublished).

Whether the equilibrium assumption in this modeling framework is realistic or not, and what are the contributions of phenomena that are currently ignored in this modeling framework such as the non-linear arrangement of chromatin and long range interactions, are all open interesting questions; but regardless, this framework was proven useful to understand and predict binding behaviors. For example, studies have demonstrated that of the two conserved binding sites for the Pho4 transcription factor within the promoter region of the Pho5 gene in yeast, only the upstream site is bound effectively at low physiological Pho4 concentrations. In contrast, binding at the downstream site is effective only at high factor concentrations, because this site is typically occluded by a nucleosome, and its binding depends on recruitment of a remodeler by the bound Pho4 at the upstream site [18, 54]. Figure 9.2 illustrates the in vitro derived recognition specificity models of Pho4 [55] and the nucleosome [32], along with their predicted affinity landscapes over the Pho5 promoter region, and their predicted binding occupancy at equilibrium, over all possible configurations, for low and high concentration parameters of Pho4. These predictions, similar to a previous analysis [56], demonstrate that by this simple equilibrium competition model of Pho4 and nucleosomes, we can understand from DNA sequence alone the differential binding of Pho4 to these two studied sites, including their differential dependency on chromatin remodeling, which are both not explained by considering only the Pho4 affinity landscape. Improving this and other quantitative modeling frameworks is likely to play a central role in the ongoing study towards understanding transcription factor binding.

9.7 Summary and Perspective

The emerging view on the process of transcription factor binding to regulatory sites is that in the context surrounding a regulatory site, genomes encode the information that sets the ground for dynamic regulation of binding to this site. This encoding includes the sequence specificity at the target site and at additional sites involved in regulation of binding to the target site, and in eukaryotes, preferences for nucleosome organization and consequently, DNA accessibility. Over this encoded background, binding is dynamically controlled through global regulation of the concentration and activity of the target transcription factor, as well as of the auxiliary transcription factors and cofactors that interacts, directly or indirectly, with binding at the target site; and in eukaryotes, also through global regulation of effectors of chromatin structure. Transcription factor binding is thus a complex stochastic function that depends both on the encoding surrounding the binding site as well as on global *trans* regulation of factors.

The growing understanding of the determinants of transcription factor binding is increasingly allowing us to indentify regulatory regions and their characteristics. As discussed above, regulatory regions should typically include a cluster of binding

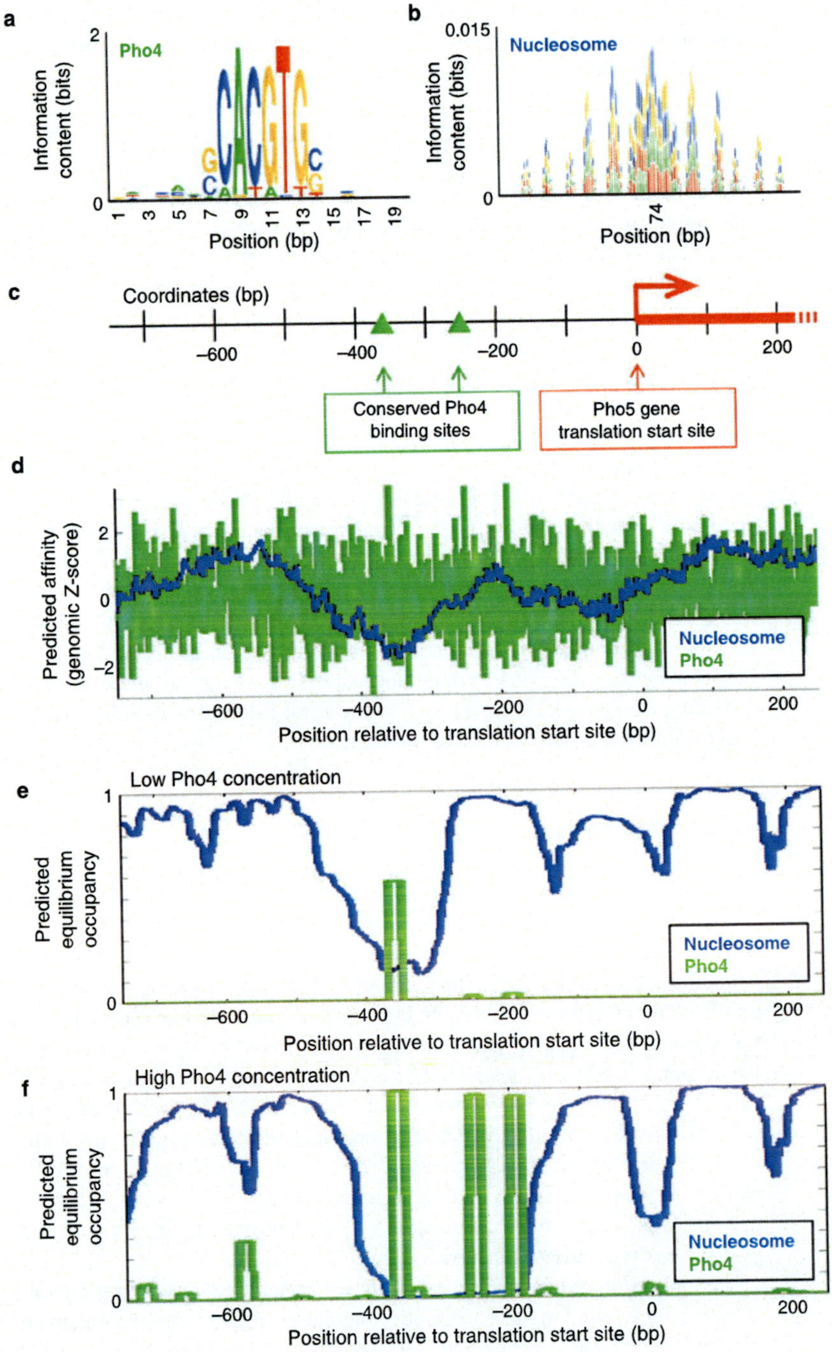

Fig. 9.2 (continued)

sites, and in eukaryotes, an encoding for relatively accessible nucleosome organization. However, the characteristics of regulatory regions should further depend on the functionality of these regions, and not only on determinants of binding. For example, a cluster of sites for the same transcription factor increases the probability that this factor would bind at the regulatory region, which may be functionally important even if these sites do not cooperate to increase the binding probability of the individual sites. Regulatory regions typically show other characteristics that can be useful tools for researchers in identifying them, but they are not the originating mechanistic signal by which cells recognize regulatory regions. One such example includes specific histone modifications [57, 58], and although some of them are known to affect DNA accessibility, others may be merely a consequence of the activity at the regulatory regions. As another example, sequence conservation across evolutionary related genomes, of the specificity to transcription factors [48] and the encoded preferences for an accessible nucleosome organization, may also be informative for indentifying regulatory regions, following the idea that sequences important for conserved functions are generally more conserved than nonfunctional sequences.

An interesting general trend that has been observed across evolution is that for more complex genomes, functionality of regulatory regions becomes more dependent on the combinatorial interactions between binding sites, rather than on the individual site. First, going from bacteria to yeast to human, regulatory regions typically have many more binding sites [2]. Second, the typical specificity of transcription factors becomes more degenerate as we go from bacteria to yeast to human [2]. Third, the average accessibility of regulatory regions is lower in yeast than in *E. coli*, due to the presence of nucleosomes, and is lower in human than in yeast, due to the additional level of the chromatin fiber, implying that functionality of regulatory regions in more complex genomes is more strongly dependent on regulation of DNA accessibility. This general trend makes sense if we expect the functionality of regulatory regions, and mostly the regulation of gene expression, to exhibit more complex regulation in more complex genomes. Notably, a similar trend seems to also hold within genomes, when considering the variability in functionality between

Fig. 9.2 A quantitative computational model for DNA dependent binding competition between transcription factors and nucleosomes. The sequence specificities of the transcription factor Pho4 (**a**) and the nucleosome (**b**) as measured by high throughput in vitro assays [32, 55], presented as the information content of the preferred distribution over nucleotides in each position along the recognition sequence. Notice the different scales of the y-axis, indicating much lower sequence specificity (per position) for the nucleosome compared with the transcription factor. (**c**) The Pho5 gene promoter region in yeast has two evolutionary conserved binding sites for Pho4: the downstream site is typically occluded by a nucleosome and is bound only at high factor concentrations and in dependence on remodeler recruitment; and the upstream site is typically accessible in linker region and is bound at lower factor concentrations [18]. (**d**) The predicted affinity landscape for Pho4 and nucleosomes over the Pho5 gene promoter region, presented as the number of standard deviations above the genomic mean affinity (Z-score). (**e, f**) Shown over the Pho5 promoter region are the predicted equilibrium probabilities of binding occupancy for Pho4 and nucleosomes, in a model of competition between Pho4 and nucleosomes. The concentration parameter for Pho4 in (**f**) is 100-fold higher than in (**e**) different regulatory regions. For example, in both yeast and human, promoters of more constitutive genes are generally encoded with preferences for more accessible nucleosome organizations, whereas more regulated genes are generally encoded for relatively less accessible organizations, implying that they are encoded for an additional dependency for regulation of DNA accessibility within chromatin [36]. In yeast, it has been further shown that more regulated promoters indeed associate with more binding sites [36, 59]. Thus, we propose that, in line with intuition, more complex regulation of gene expression may be generally coupled to a regulatory region architecture that has more binding sites and a DNA-encoded preference for less accessible chromatin structure.

In summary, although many of the related mechanisms are still poorly understood, we are beginning to understand the determinants of transcription factor binding, and how to further identify and characterize the functionality of regulatory regions from genomic sequence. The context of chromatin structure, its encoding and dynamic regulation, emerges as a major determinant of transcription factor binding in eukaryotes.

References

1. Polak P, Domany E (2006) Alu elements contain many binding sites for transcription factors and may play a role in regulation of developmental processes. BMC genomics 7:133
2. Wunderlich Z, Mirny LA (2009) Different gene regulation strategies revealed by analysis of binding motifs. Trends Genet 25:434–440
3. Kao-Huang Y, Revzin A, Butler AP, O'Conner P, Noble DW, von Hippel PH (1977) Nonspecific DNA binding of genome-regulating proteins as a biological control mechanism: measurement of DNA-bound *Escherichia coli* lac repressor in vivo. Proc Natl Acad Sci U S A 74:4228–4232
4. von Hippel PH, Revzin A, Gross CA, Wang AC (1974) Non-specific DNA binding of genome regulating proteins as a biological control mechanism: I. The lac operon: equilibrium aspects. Proc Natl Acad Sci U S A 71:4808–4812
5. Guertin MJ, Lis JT (2010) Chromatin landscape dictates HSF binding to target DNA elements. PLoS Genet 6:9
6. Lidor Nili E, Field Y, Lubling Y, Widom J, Oren M, Segal E (2010) p53 binds preferentially to genomic regions with high DNA-encoded nucleosome occupancy. Genome Res 20: 1361–1368
7. Yang A, Zhu Z, Kapranov P, McKeon F, Church GM, Gingeras TR, Struhl K (2006) Relationships between p63 binding, DNA sequence, transcription activity, and biological function in human cells. Mol Cell 24:593–602
8. Cremer T, Cremer C (2001) Chromosome territories, nuclear architecture and gene regulation in mammalian cells. Nat Rev Genet 2:292–301
9. Jackson DA, Hassan AB, Errington RJ, Cook PR (1993) Visualization of focal sites of transcription within human nuclei. EMBO J 12:1059–1065
10. Dhalluin C, Carlson JE, Zeng L, He C, Aggarwal AK, Zhou MM (1999) Structure and ligand of a histone acetyltransferase bromodomain. Nature 399:491–496
11. Fischle W, Wang Y, Jacobs SA, Kim Y, Allis CD, Khorasanizadeh S (2003) Molecular basis for the discrimination of repressive methyl-lysine marks in histone H3 by Polycomb and HP1 chromodomains. Genes Dev 17:1870–1881
12. Richmond TJ, Davey CA (2003) The structure of DNA in the nucleosome core. Nature 423:145–150

13. Anderson JD, Widom J (2000) Sequence and position-dependence of the equilibrium accessibility of nucleosomal DNA target sites. J Mol Biol 296:979–987
14. Polach KJ, Widom J (1995) Mechanism of protein access to specific DNA sequences in chromatin: a dynamic equilibrium model for gene regulation. J Mol Biol 254:130–149
15. Zlatanova J, Seebart C, Tomschik M (2008) The linker-protein network: control of nucleosomal DNA accessibility. Trends Biochem Sci 33:247–253
16. Lorch Y, Maier-Davis B, Kornberg RD (2010) Mechanism of chromatin remodeling. Proc Natl Acad Sci U S A 107:3458–3462
17. Iyer V, Struhl K (1995) Poly(dA:dT), a ubiquitous promoter element that stimulates transcription via its intrinsic DNA structure. EMBO J 14:2570–2579
18. Lam FH, Steger DJ, O'Shea EK (2008) Chromatin decouples promoter threshold from dynamic range. Nature 453:246–250
19. Poirier MG, Bussiek M, Langowski J, Widom J (2008) Spontaneous access to DNA target sites in folded chromatin fibers. J Mol Biol 379:772–786
20. Zlatanova J, Leuba SH, Yang G, Bustamante C, van Holde K (1994) Linker DNA accessibility in chromatin fibers of different conformations: a reevaluation. Proc Natl Acad Sci U S A 91:5277–5280
21. Graziano V, Gerchman SE, Ramakrishnan V (1988) Reconstitution of chromatin higher-order structure from histone H5 and depleted chromatin. J Mol Biol 203:997–1007
22. Routh A, Sandin S, Rhodes D (2008) Nucleosome repeat length and linker histone stoichiometry determine chromatin fiber structure. Proc Natl Acad Sci U S A 105:8872–8877
23. Braunschweig U, Hogan GJ, Pagie L, van Steensel B (2009) Histone H1 binding is inhibited by histone variant H3.3. EMBO J 28:3635–3645
24. Clausell J, Happel N, Hale TK, Doenecke D, Beato M (2009) Histone H1 subtypes differentially modulate chromatin condensation without preventing ATP-dependent remodeling by SWI/SNF or NURF. PloS ONE 4:e0007243
25. Robinson PJ, An W, Routh A, Martino F, Chapman L, Roeder RG, Rhodes D (2008) 30 nm chromatin fibre decompaction requires both H4-K16 acetylation and linker histone eviction. J Mol Biol 381:816–825
26. Wang X, He C, Moore SC, Ausio J (2001) Effects of histone acetylation on the solubility and folding of the chromatin fiber. J Biol Chem 276:12764–12768
27. Karymov MA, Tomschik M, Leuba SH, Caiafa P, Zlatanova J (2001) DNA methylation-dependent chromatin fiber compaction in vivo and in vitro: requirement for linker histone. FASEB J 15:2631–2641
28. Corona DF, Siriaco G, Mcclymont SA, Armstrong JA, Snarskaya N, McClymont SA, Scott MP, Tamkun JW (2007) ISWI regulates higher-order chromatin structure and histone H1 assembly in vivo. PLoS Biol 5:e232
29. Fan JY, Rangasamy D, Luger K, Tremethick DJ (2004) H2A.Z alters the nucleosome surface to promote HP1alpha-mediated chromatin fiber folding. Mol Cell 16:655–661
30. Postnikov Y, Bustin M (2010) Regulation of chromatin structure and function by HMGN proteins. Biochim Biophys Acta 1799:62–68
31. Woodcock CL, Skoultchi AI, Fan Y (2006) Role of linker histone in chromatin structure and function: H1 stoichiometry and nucleosome repeat length. Chromosome Res 14:17–25
32. Kaplan N, et al. (2009) The DNA-encoded nucleosome organization of a eukaryotic genome. Nature 458:362–366
33. Dekker J (2008) Mapping in vivo chromatin interactions in yeast suggests an extended chromatin fiber with regional variation in compaction. J Biol Chem 283:34532–34540
34. Downs JA, Kosmidou E, Morgan A, Jackson SP (2003) Suppression of homologous recombination by the *Saccharomyces cerevisiae* linker histone. Mol Cell 11:1685–1692
35. Kornberg RD, Lorch Y (1999) Twenty-five years of the nucleosome, fundamental particle of the eukaryote chromosome. Cell 98:285–294
36. Field Y, Kaplan N, Fondufe-Mittendorf Y, Moore IK, Sharon E, Lubling Y, Widom J, Segal E (2008) Distinct modes of regulation by chromatin encoded through nucleosome positioning signals. PLoS Comput Biol 4:e1000216

37. Tillo D, et al. (2010) High nucleosome occupancy is encoded at human regulatory sequences. PLoS ONE 5:e9129
38. Li X, et al. (2008) Transcription factors bind thousands of active and inactive regions in the *Drosophila* blastoderm. PLoS Biol 6:e27
39. Field Y, Fondufe-Mittendorf Y, Moore IK, Mieczkowski P, Kaplan N, Lubling Y, Lieb JD, Widom J, Segal E (2009) Gene expression divergence in yeast is coupled to evolution of DNA-encoded nucleosome organization. Nat Genet 41:438–445
40. Xi H, et al. (2007) Identification and characterization of cell type-specific and ubiquitous chromatin regulatory structures in the human genome. PLoS Genet 3:e136
41. Landolin JM, Johnson DS, Trinklein ND, Aldred SF, Medina C, Shulha H, Weng Z, Myers RM (2010) Sequence features that drive human promoter function and tissue specificity. Genome Res 20:890–898
42. Giniger E, Ptashne M (1988) Cooperative DNA binding of the yeast transcriptional activator GAL4. Proc Natl Acad Sci U S A 85:382–386
43. Merika M, Orkin SH (1995) Functional synergy and physical interactions of the erythroid transcription factor GATA-1 with the Krüppel family proteins Sp1 and EKLF. Mol Cell Biol 15:2437–2447
44. Zhang Z, Fuller GM (1997) The competitive binding of STAT3 and NF-kappaB on an overlapping DNA binding site. Biochem Biophys Res Commun 237:90–94
45. Darieva Z, Clancy A, Bulmer R, Williams E, Pic-Taylor A, Morgan BA, Sharrocks AD (2010) A competitive transcription factor binding mechanism determines the timing of late cell cycle-dependent gene expression. Mol Cell 38:29–40
46. Polach KJ, Widom J (1996) A model for the cooperative binding of eukaryotic regulatory proteins to nucleosomal target sites. J Mol Biol 258:800–812
47. Segal E, Widom J (2009) From DNA sequence to transcriptional behaviour: a quantitative approach. Nat Rev Genet 10:443–456
48. Harbison CT, et al. (2004) Transcriptional regulatory code of a eukaryotic genome. Nature 431:99–104
49. Bulyk ML, Gentalen E, Lockhart DJ, Church GM (1999) Quantifying DNA–protein interactions by double-stranded DNA arrays. Nat Biotechnol 17:573–577
50. Maerkl SJ, Quake SR (2007) A systems approach to measuring the binding energy landscapes of transcription factors. Science (New York, NY) 315:233–237
51. Morozov AV, Fortney K, Gaykalova DA, Studitsky VM, Widom J, Siggia ED (2009) Using DNA mechanics to predict in vitro nucleosome positions and formation energies. Nucleic Acids Res 37:4707–4722
52. Sinha S, Adler AS, Field Y, Chang HY, Segal E (2008) Systematic functional characterization of cis-regulatory motifs in human core promoters. Genome Res 18:477–488
53. Segal E, Raveh-Sadka T, Schroeder M, Unnerstall U, Gaul U (2008) Predicting expression patterns from regulatory sequence in *Drosophila* segmentation. Nature 451:535–540
54. Venter U, Svaren J, Schmitz J, Schmid A, Hörz W (1994) A nucleosome precludes binding of the transcription factor Pho4 in vivo to a critical target site in the PHO5 promoter. EMBO J 13:4848–4855
55. Zhu C, et al. (2009) High-resolution DNA-binding specificity analysis of yeast transcription factors. Genome Res 19:556–566
56. Raveh-Sadka T, Levo M, Segal E (2009) Incorporating nucleosomes into thermodynamic models of transcription regulation. Genome Res 19:1480–1496
57. Ernst J, Kellis M (2010) Discovery and characterization of chromatin states for systematic annotation of the human genome. Nat Biotechnol 28:817–825
58. Heintzman ND, et al. (2007) Distinct and predictive chromatin signatures of transcriptional promoters and enhancers in the human genome. Nat Genet 39:311–318
59. Tirosh I, Barkai N (2008) Two strategies for gene regulation by promoter nucleosomes. Genome Res 18:1084–1091

Chapter 10
Transcription Factor Binding Sites and Other Features in Human and *Drosophila* Proximal Promoters

Charles Vinson, Raghunath Chatterjee, and Peter Fitzgerald

Abstract Eukaryotic promoters determine transcription start sites (TSSs), and are often enriched for transcription factor binding sites (TFBSs), which presumably play a major role in determining the location and activity of the TSS. In mammalian systems, proximal promoters are enriched for the CpG dinucleotide. The TFBSs that are enriched in proximal promoters (–200 bps to the TSS) are CCAAT, ETS, NRF1, SP1, E-Box, CRE, BoxA, and TATA. Only TATA occurs in a DNA strand dependent manner. In *Drosophila*, proximal promoters are AT rich and many putative TFBSs are enriched in proximal promoters. These sequences are different from those that occur in human promoters, except for TATA and E-Box, and many occur on a single strand of DNA giving directionality to the promoter. Thus, fundamental differences have arisen as promoters evolved in metazoans.

10.1 Introduction

The regulation of eukaryotic gene expression is a complex process involving many different control mechanisms, including chromatin structure and DNA sequences bound by specific proteins termed transcription factors (TFs). An important paradigm in gene expression studies is that TFs bind specific DNA sequences termed Transcription Factor Binding Sites (TFBSs) to control transcription. These TFBSs often localize near the Transcriptional Start Site (TSS) in an area termed the promoter, and specific locations elsewhere in the genome termed enhancers. These TFBSs are bound by TFs that recruit additional proteins to either activate or repress gene expression. Because TFBSs tend to be composed of defined short stretches of DNA (typically 6–12 base pairs), a simple search of the DNA sequence within a large genome therefore finds large numbers of matching sequences. A major question in current research is whether these potential binding sites are functional (for binding TFs, and for regulating transcription) and under what circumstances. The

C. Vinson (✉)
Laboratory of Metabolism, NCI, NIH, Bethesda, MD 20892, USA
e-mail: Vinsonc@mail.nih.gov

T.R. Hughes (ed.), *A Handbook of Transcription Factors*, Subcellular Biochemistry 52,
DOI 10.1007/978-90-481-9069-0_10, © Springer Science+Business Media B.V. 2011

picture is further complicated by the fact the most TFBSs are defined by a consensus sequence that contains ambiguous bases. Thus, the identification of DNA sequences that are biologically relevant TFBS is challenging.

We and others have focused on identifying DNA sequences that preferentially localize in the proximal promoter as a method to find TFBSs that are likely to be biologically important [1–6] and to help understand what characterizes and defines eukaryotic promoters. This strategy has been facilitated by methods that identify the TSS by determining the 5-CAP site of mRNA [7]. A complication in identifying the TSS for a given gene is that RNA Polymerase II (RNAP) does not always initiate mRNA synthesis from a unique nucleotide. Many tissue specific transcripts have a unique TSS; however, housekeeping genes, which often contain a CpG island in the promoter region, typically have a more variable TSS with mRNA synthesis starting over a 50–100 bp range [8]. Another complication is that mechanisms of both promoter definition and gene expression regulation are far from uniform across all eukaryotes. Even within the narrow region of the proximal promoter, we see major differences in the TFBSs used by different organisms, even within the same clade.

This book chapter will compare the promoter architecture of Human and *Drosophila* promoters and then discuss in detail the DNA sequences that preferentially localize in human proximal promoters. The analyses suggest that human promoters are embedded in CpG rich regions while *Drosophila* promotes are in A and T rich regions.

10.2 General Similarities Between *Drosophila* and Human Dinucleotide Content

We will first consider the simple sequence content of promoter (and non promoter) sequence. Comparing the dinucleotide frequency of the *Drosophila* and human genomes shows general similarity (Fig. 10.1a). For example, the AA/TT dinucleotide is the most abundant in each genome. The dinucleotide content is not completely explained by base content, presumably due to the differential expansion of simple repeat sequences. The most notable difference between these two genomes is that the human genome is depleted for the CpG dinucleotide. However, in the human, but not the *Drosophila* genome, the CpGs often occur in clusters (Fig. 10.1b), and these clusters are frequently, but not always, found in and around the proximal promoters of genes. This clustering of CpGs in mammals was noticed 25 years ago and these clusters were termed “CpG islands” that often occur in the promoters of “housekeeping” genes [9, 10]. In fact, all CpG islands may be associated with a TSS. An explanation for the depletion of the CpG dinucleotide in mammalian genomes follows from the observation that, in mammals, CpG dinucleotides that are not in CpG islands are methylated in early development. It is thought that the CpGs in CpG islands are not methylated during the wave of methlyation that occurs during early development because they are bound by TFs expressed at this time in development which includes primarily essential genes involved in housekeeping functions of the cell and not tissue specific genes that will become

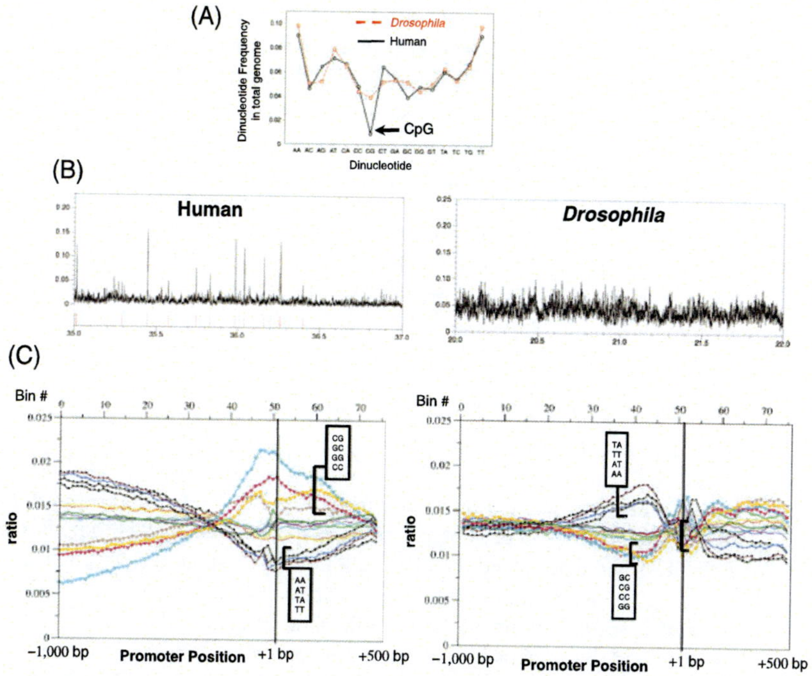

Fig. 10.1 **a** Dinucleotide frequency in the entire Human and *Drosophila* genome. **b** CpG density across 2 MB of the Human and *Drosophila* genome. Observe that in the human genome, CpGs on average are rarer than in *Drosophila* but they do occur in clusters called CpG islands which is not observed in *Drosophila*. The *red dashes* in the human trace are CpG islands as defined on the UCSC genome browser. **c** Dinucleotide density across promoters from −1,000 to +500 bps for *Drosophila* and humans using a 20 bp window. The CA dinucleotide peak that occurs exactly at the TSS in both Human and *Drosophila* promoters is not observed because we are using a 20 bp window for this calculation. Note that in *Drosophila* promoters, the dinucleotides containing T & A are enriched just upstream of the TSS while in human promoters, the dinucleotides containing C & G are enriched at the TSS

activated later in development [11]. Elsewhere in the genome, where CpGs are rare (including the promoters of tissue specific genes) the CpGs are not bound because the tissue specific TFs that bind them are not expressed. These unbound CpGs are methylated because they are accessible to the CpG methylation machinery. Mutation of methylated CpGs is due to their chemical property in which the methyl cytosine spontaneously deaminates to thymine, which in effect depletes CpG containing sequences throughout the genome except in CpG islands where the methylation levels are low. This solves a vexing problem of selecting against TFBSs that arise by mutation throughout the genome: natural selection does not need to select against these spontaneous arising TFBSs, because chemistry selects against these sites. For these reasons, when we examine DNA sequences that localize in mammalian promoters, we divide them into two classes, those with a CpG dinucleotide and those without a CpG dinucleotide. When thinking about CpG containing sequences, we need to keep in mind that they may be methylated, which may enhance or diminish the DNA binding of any protein that binds the sequence. CpG methylations in the promoters are generally transcriptionally repressive as occurs with X-chromosome inactivation and imprinting [11]. CpG methylation both recruits repressive complexes [11] and prevents the DNA binding of many transcription factors (TFs) [12]. In some cancers, methylation of tumor suppressor gene promoters is associated with gene repression [13]. Contrary to that, however, genomic analyses have identified low CpG promoters that are both methylated and transcriptionally active [14, 15], but the mechanism underlying the activation of methylated promoters remains unclear.

10.3 *Drosophila* and Humans Have Different Promoter Architecture

Figure 10.1c presents the dinucleotide frequencies from –1,000 to +500 bps for *Drosophila* and human promoters. For each species we aligned promoter sequences to the TSS and determined the distribution of DNA sequences throughout the promoter region. When we examine the distribution of dinucleotides, we observe that *Drosophila* promoters are enriched for the 4 dinucleotides AA, TT, AT, & TA which are over 50% more abundant at –200 bps than at –1,000 bps. Human promoters, in contrast, are enriched for the four dinucleotides CG, GC, GG, & CC, with the CpG dinucleotide being over three times more abundant at –200 bps compared to –1,000 bps. This fundamental difference in promoter architecture has far reaching consequences for the mechanisms of regulated gene expression in these two species, primarily because the CpG rich proximal promoters observed in human tend to be nucleosome binding site in vitro, but not in vivo, as will be discussed later. In addition, as we shall see, the stereotypic spatial arrangements of TFBSs are also different in *Drosophila* and human (with the notable exception of the E-Box (CANNTG) and TATA sequences), as are their spatial arrangements relative to both the TSS and each other.

10.4 DNA 8-Mers that Localize in Human Promoters

When we examine the distribution of 8-mers in human promoters, we observe that some sequences are preferentially localized near the TSS [2]. Our assumption is that these sequences may be TFBSs. An important issue to understand with this approach is that we can only identify abundant TFBSs: if a TFBS occurs in a limited number of promoters, we will not be able to identify it using this approach of examining all promoters because the signal may be too far diminished relative to the genomic background. Figure 10.2a shows the distribution of CGGAAGTG, an ETS motif that is the most preferentially localized DNA sequence in human promoters. When we count the occurrence of 8-mers in promoters, we have chosen to

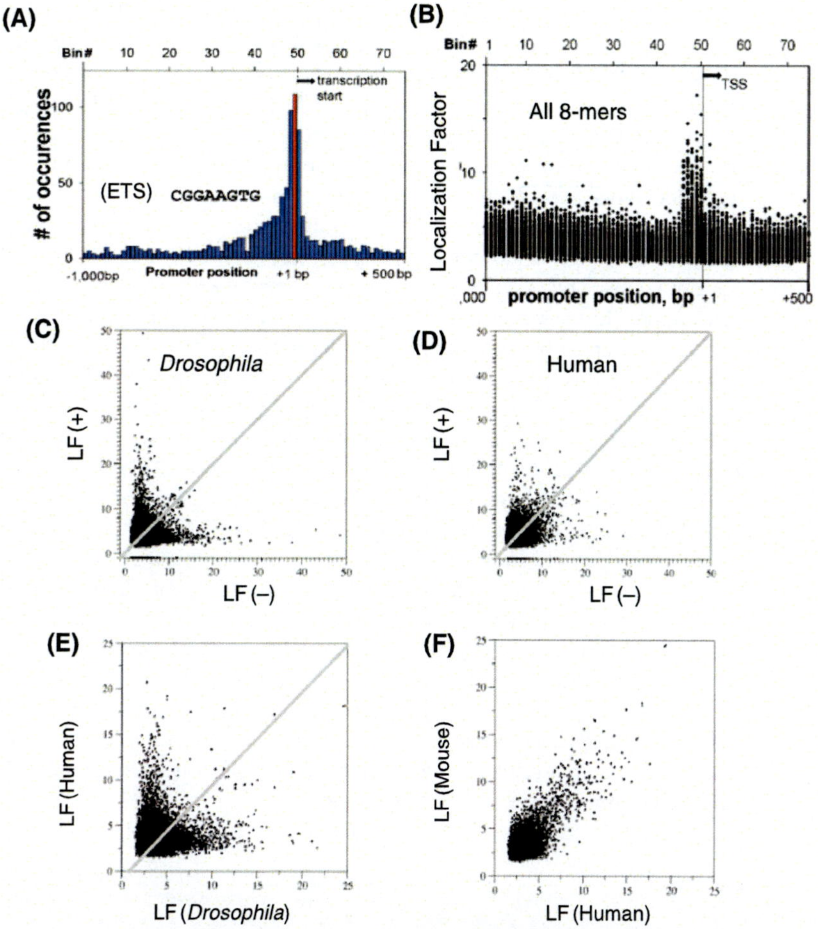

Fig. 10.2 **a** The most preferentially localized 8-mer in human promoters is CGGAAGTG, an ETS sequence. The number of occurrences of the CGGAAGTG sequence and its complement is counted in 13,010 promoters aligned to the TSS using 20 bp windows (bins) from –1,000 to +500 bps. From this distribution, a measure of non-random distribution or preferential localization termed Localization Factor (LF) is calculated. **b** Localization of all 8-mers in promoters region. For each 8-mer and its complement, a measure of non-random distribution is calculated and plotted in the most abundant window. Note that most of the preferentially localized 8-mers occur just upstream of the TSS. **c** Localization of 8-mers that occur on one strand compared to the opposite strand in 10,914 *Drosophila* promoters [3]. Here, many sequences are off the diagonal indicating that these sequences occur preferentially on one strand and not the other strand. **d** Localization in human promoters of 8-mers that occur on one strand compared to the opposite strand. Note that most sequences are on the diagonal indicating that these sequences occur on either DNA strand. **e** Localization of 8-mers in human compared to *Drosophila* promoters. The sequences off the diagonal indicate that different sequences localize in promoters in these two species. **f** Localization of 8-mers in human compared to mouse promoters (20,328 promoters). Most sequences are near the diagonal indicating that the same sequences localize in the promoters of these two species

use a 20 bp window or bin in this counting process for the following reasons. On average, in 13,010 promoters over a range from –1,000 to +500 bps, one would expect that each 8-mer would occur ((13,000×1,493)/65,536) = ~300 times. 1,493 is the number of 8-mers in 1,500 bp of DNA, 13,010 is the number of promoters we have examined, and 65,536 is the number of possible 8-mers. In each 20 bp bin, on average, each 8-mer would occur 4 times. To increase the statistical power of our calculations, we add the occurrences of an 8-mer together with the occurrences of its complementary sequence. This essentially doubles the number of occurrences of an 8-mer making it easier to evaluate the distribution across the promoter region for any non-random distribution properties. This raises a problem for this type of calculation; do we count the palindromic sequences (there are 256 palindromic 8-mers) as a single occurrence or as two occurrences, one on each strand? It should be appreciated that palindromic sequences have the property that they are identical on each strand of DNA, allowing a TF to bind to either strand, which essentially doubles their concentration compared to non-palindromic sequences that need to be recognized by a TF on only one strand of DNA. Figure 10.2b presents a measure of the non-random distribution for all 8-mers where we combine an 8-mer and its complement, resulting in 32,896 8-mers (32,640 non-palindromic 8-mers and 256 palindromic 8-mers). Previously, we used "clustering factor" as the name for this non-random distribution of an 8-mer in the promoter region [2, 3]. Now, we prefer the name Localization Factor (LF) because it more accurately captures what we are measuring. To determine if a DNA sequence localized, the mean ($\bar{x}$) and standard deviation (σ) were determined based on its abundance in each of the 75 bins (each 20 bp). Those bin values that were $\geq$ 2 SD above the mean were considered to be part of the cluster and a new mean ($\bar{x}'$) and standard deviation (σ') were calculated excluding these bin values. A localization factor (LF) was then calculated based on this corrected mean and standard deviation,

$$LF = \frac{x_{\max} - \bar{x}'}{\sigma'}$$

We have plotted this Localization Factor in the bin where the DNA sequence is the most abundant. In Fig. 10.2b, we observe that most of the 8-mers with high LF localize just upstream of the TSS, with some localizing just downstream of the TSS.

10.5 Comparing DNA 8-Mers that Localize in Human, Mouse, and *Drosophila* Promoters

The next several panels compare the 8-mers that localize in *Drosophila*, human, and mouse promoters (Fig. 10.2c–f). In *Drosophila* promoters, we also observe that some sequences preferentially localize in the proximal promoters. A startling difference between human and *Drosophila* promoters is revealed when one examines the DNA strand dependence of the localization of 8-mers in the proximal promoter. We calculated the distribution of all 65,536 8-mers and plotted the localization factor of an 8-mer vs. its complement. In *Drosophila* promoters, many 8-mers localize on one strand of DNA but not on the other strand, which imparts directional information to the promoter (Fig. 10.2c). In human promoters (Fig. 10.2d), the strand dependence of the localization factor of an 8-mer is much less strong (the exceptions are TATA sequences) suggesting that preferentially localized DNA sequences do not contain information that imparts direction information. When we examine the sequences that are preferentially localized in *Drosophila* and human promoters, there is little overlap (Fig. 10.2e) indicating that the DNA sequences that regulate promoter function are different between these two species, a result we found surprising having been taught that *Drosophila* is a good model organism to learn about humans, an assumption that is true for many aspects of biology but apparently not for understanding promoter sequences. In contrast, we observe that similar sequences are preferentially localized in human and mouse promoters, suggesting similar promoter architecture between these two species (Fig. 10.2f).

Additional differences between human and *Drosophila* promoters are identified when we examine the localization of discontinuous 8-mers that contain two 4-mers separated by an insert with the aim of identifying either dimeric motifs or wide TFBSs that are preferentially localized in promoters (Fig. 10.3). We examined insert length of 1–60 base pairs. In human promoters, we only identify sequences that localize and have a short insert length. Examination of these sequences shows they are primarily versions of the continuous 8-mers that localize in promoters. The exceptions are combinations of ETS:ETS and ETS:CRE sequences we will discuss later in this chapter. In *Drosophila*, we identify pairs of 4-mers that are separated by 20–30 bps that localize in promoters. These pairs of sequences are combinations of TATA and INR sequences identified previously [3] and additional sequences that we are currently studying. The general conclusion from this analysis is that human

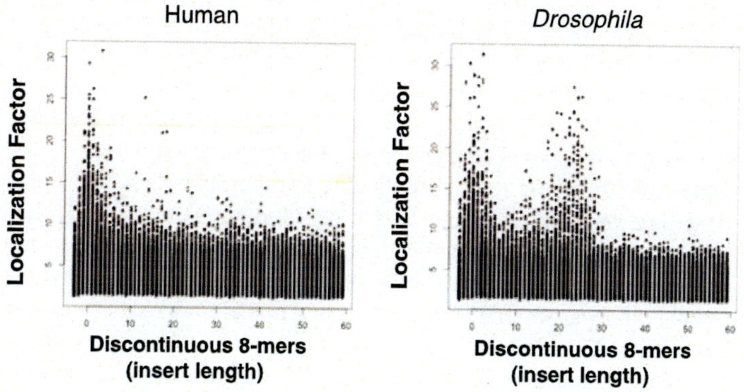

Fig. 10.3 Localization factor for 8-mers composed of two 4-mers separated by a variable insert length between the two 4-mers. Note the difference between *Drosophila* and human promoters. Individual pairs of 4-mers that localize in proximal promoters are noted

proximal promoters are comprised of continuous sequences with no fixed positioning information among different sequences in the promoters. In contrast, *Drosophila* promoters have strand specific sequences that are often uniquely positioned relative to other sequences in the promoter.

These differences in human and *Drosophila* promoter organization suggest that these two species use different mechanisms to regulate gene expression. First, the different frequency and distribution of mononucleotides and dinucleotides in promoters correlates with nucleosome positioning or occupancy. *Drosophila* promoters are A&T rich with a peak of A&T dinucleotides between –200 bp and the TSS (Fig. 10.1), a region that experimentally is known to be nucleosome free, particularly for active genes [16]. A similar correlation is observed in the yeast genome where the promoter regions between –200 and the TSS are A&T rich and devoid of nucleosomes [17]. This model of promoter organization in *Drosophila* has an appealing simplicity. The promoter region is accessible and is bound by multiple TFs that bind TFBSs that occur on a single strand of DNA and are uniquely positioned relative to each other. In contrast, in humans, there are usually CpG islands at promoters. These CpG island sequences experimentally bind nucleosomes because of their C&G content [18] but are devoid of nucleosomes because they are instead bound by TFs. The competition between TFs and nucleosomes is evident at inducible promoters where the induction of DNA hypersentitive sites is observed. Going forward, we are particularly interested in experimentally examining how CpG methylation can shift the equilibrium between TF binding and nucleosome binding. This scheme of a competition would allow for a DNA regulatory sequence to be repressed by nucleosome binding and activated by the displacement of the nucleosome and the binding of TFs. Additionally, the same sequences that are TFBSs are also nucleosome binding sites [19]. This switch mechanism theoretically allows more control over gene expression.

The dramatic difference in promoter organization between *Drosophila* and human indicates that the TFBSs that delineate promoters and that control expression of coordinately regulated genes have changed over evolutionary time. For example, between *Drosophila* and humans, the ribosomal proteins are conserved but the TFBS that regulate their expression are different [20]. It is hard to image how this could happen if one imagines that evolution is an incremental process with selection acting on each mutation. This idea of gradual change in genome structure does not explain the global change in ribosomal promoter structure that is observed. An alternative image is that, episodically, transposable elements ravage the genome, inserting during meiosis into active genes, these would include the housekeeping genes that are active during this time. The transposable elements could degrade over time with only the relevant TFBS remaining resulting in a dramatic increase in the number of a particular TFBS (those found in the transposable elements) in proximal promoters. This process of the housekeeping genes being ravaged by transposable element insertions could repeat itself over time resulting in the evolution of CpG islands. This image of promoter evolution could explain how TFs that are conserved in both *Drosophila* and humans have dramatically different occurrences in proximal promoters. This idea of promoter evolution is supported by the observation that some mammalian TFBS are derived from repetitive elements [20]. Some other mechanisms are reviewed recently [21].

10.6 8-Mers that Preferentially Localize in Human Proximal Promoters

Previously, we have taken the 150 most localizing sequences in human promoters and grouped them into 8 related sequences (Table 10.1) [2]. These DNA sequences are all known TFBSs. We have now updated this grouping of TFBS as we have gained more insight and have included this table of 150 8-mers that localize in promoters because we expect that even closer examination will reveal that the groups we have generated are overly simplistic. Both CpG and non-CpG sequences localize in proximal promoters. This process of grouping different sequences is fraught with complications. If two 8-mers are different by a single base pair, do we conclude that they are variants of the same TFBS or are they different TFBSs? We do not know the answer to this question. We have taken the approach of grouping sequences together that may be related, but further understanding is likely to result in a refinement of these groupings. The number of 8-mers in each TFBS group is variable. The TFBS with the most 8-mers is CCAAT, while the Box-A TFBS is observed in only one 8-mer. Given the fact that we are only examining 8-mers sequences, two extreme possibilities could explain the large number of 8-mers within the CCAAT group. The first possibility is that the multiple 8-mers containing CCAAT could represent a single TFBS consensus that is 8 or more bps long and each base pair is significant but variable. The second possibility is that the CCAAT TFBS is 5 bps long and the remaining 3 bases in the 8-mer represent unconstrained surrounding sequences. When we align the 31 8-mers placed into the CCAAT group, we observe an invariant 5 bp central core, surrounded by variant but constrained sequences, giving rise to a 9 bps consensus sequence. Several of the TFBS groups appear very consistent, e.g. CCAAT and ETS while others appear more varied, e.g. SP1 and NRF1 [2].

10.6.1 The 8 Consensus Sequences Representing TFBS

We have divided the TFBSs into two groups: the non-palindromic sequences which are bound by a protein monomer, and the palindromic sequences which are bound by protein dimers. We present the distribution of the TFBSs, their relevant variants, and an X-ray crystal structure of the protein bound to DNA if it exists in the literature to help understand the length of the TFBS (Fig. 10.4). The majority of these TFBSs preferentially occur in the promoters of housekeeping genes presumably reflecting the abundance of these types of promoters. The notable exception is that TATA preferentially occurs in tissue specific genes. GO term analysis of the genes whose promoters contain these different TFBSs reveals that individual TFBSs preferentially occur in the promoters of specific kinds of genes. This general conclusion

Table 10.1

CCAAT

47	690	CAATGGGA	11.3
47	601	CAATCAGC	13.1
46	708	CAATCAGA	14.4
45	310	CCAATCGG	8.1
46	871	CCAATCCC	8.0
48	620	CCAATCAC	11.7
47	1061	CCAATCAG	23.6
47	306	CCAATCGC	13.3
47	770	CCAATGGG	31.1
47	896	GCCAATCA	22.5
46	361	GCCAATAG	9.2
48	357	GCCAATCG	12.4
46	578	GCCAATGA	17.0
47	775	GCCAATGG	26.8
46	553	GCCAATCC	9.7
47	537	TCCAATCA	7.0
47	220	ACCAATCG	14.7
47	469	ACCAATGG	17.8
46	583	ACCAATCA	17.4
47	384	GACCAATG	9.8
47	400	GACCAATC	19.2
47	893	AGCCAATC	19.3
46	748	AGCCAATG	13.8
47	680	GGCCAATG	11.7
48	658	GGCCAATC	24.0
47	547	GAGCCAAT	10.2
47	324	GGACCAAT	8.8
47	483	GGGCCAAT	12.4
48	509	CGGCCAAT	10.9
47	1039	CAGCCAAT	31.4
47	774	TCAGCCAA	10.5
47	1036	GCAGCCAA	7.0

SP1

48	1332	GCCACGCC	15.7
48	8136	GCCCCGCC	25.2
48	3078	CGCCCCTC	7.3
48	5248	CGCCCCGC	13.7
48	3141	CGCCCCCT	7.4
48	7055	CCGCCCCC	18.1
47	2106	CCGCCCAC	8.1
48	5783	CCGCCTCC	7.0
47	5204	CCGCCCCG	16.6
48	3688	CCGCCCCT	12.6
48	10767	CCCGCCCC	28.3
48	1170	ACGCCCCC	15.4
48	829	ACGCCCCG	7.9
48	1639	CACGCCCC	13.9
48	2890	CCCGCCCT	8.9
47	2334	CCCGCCCA	10.8
48	2462	TCCGCCCC	8.4
48	4767	CCCGCCTC	18.8
48	3366	CTCCGCCC	11.8
48	11029	CCCCGCCC	31.3
48	3190	CCCCGCCT	12.5
49	918	TTCCGCCC	17.8
48	2673	GCTCCGCC	7.2
49	1213	CTTCCGCC	7.9
48	4947	GGCCCCGC	7.1
47	5139	CCTCCCTC	8.1
48	7985	CCCCTCCC	7.4

Box A

48	432	TCTCGCGA	10.6

E-Box

49	755	CACGTGAC	9.0
48	294	TCACGTGA	9.4
49	582	TCACGTGG	9.0

CRE

50	484	TGACGTCA	18.4
49	282	ATGACGTC	8.5
50	503	CTGACGTC	9.3
48	635	GTGACGTC	13.5
50	313	GTGACGCA	7.4
49	345	AGTGACGT	9.4
49	294	CGTGACGC	8.0
49	280	CGTGACGT	10.2
48	379	GGTGACGT	7.1
50	264	TGTGACGT	11.4
49	241	ACGTGACG	10.3
49	472	ACGTGACC	8.4

ETS:CRE

49	345	AGTGACGT	9.4
49	332	AAGTGACG	23.9
50	769	GAAGTGAC	10.4
49	1324	GGAAGTGA	16.2

TATA

49	486	CCTATAAA	9.3
49	571	GCTATAAA	7.1
49	496	CTATAAAG	10.1
49	809	TTATAAAG	10.9
49	861!	TATAAAAG	11.4
49	417	TATATAAG	9.7
49	542!	TATAAAGG	28.0
49	860!	ATAAAAGG	17.1
49	630	TAAAAGGC	9.9

NRF-1

50	1240	TGCGCCTG	11.9
50	2300	GCGCCTGC	12.3
50	1767	CGCCTGCG	11.6
50	2154	GCCTGCGC	7.8
48	1205	GCGTGCGC	7.4
50	1041	CCTGCGCA	12.9
50	903	ACTGCGCC	8.0
50	572	TGCGCATG	8.5
49	386	CGCGCATG	11.1
50	1179	GCGCATGC	18.5
50	463	CGCATGCG	15.5

ETS

49	1546	AGGAAGTG	7.6
49	923	GGAAGTGC	11.9
50	1892	GGAAGTGG	7.5
49	284	CGGAAGTA	23.1
50	484	CGGAAGCA	13.8
50	426	CGGAAGTC	24.8
51	402	CGGAAGTT	8.0
50	991	CGGAAGTG	29.5
51	356	CGGAAATG	7.8
49	567	CGGAAGCT	8.4
50	824	CGGAAGCG	19.2
49	1150	CCGGAAGC	20.9
50	1030	CCGGAAGT	31.9
51	459	CCGGAAAC	13.1
50	600	ACCGGAAG	40.6
50	1096	GCCGGAAG	23.2
49	1224	CCCGGAAG	20.1
51	603	ACCCGGAA	7.8
50	382	CACCGGAA	12.9
49	401	GACCGGAA	7.4
49	556	AGCCGGAA	8.7
50	600	GCGGAAGT	33.6
50	541	CGCCGGAA	24.9

ETS:ETS

51	820	GCGGAAGC	7.9
50	712	AGCGGAAG	18.5
50	433	AAGCGGAA	15.9

YY1

51	1018	CAAAATGG	9.8
51	1048	AAAATGGC	16.8
51	436	AAATGGCG	23.4
51	414	AATGGCGG	12.9

ATG/KOZAK

52	960!	CCAAGATG	7.5
52	617!	GCAAGATG	13.7
51	543	GCGCCATG	9.3
53	688!	GCACCATG	9.9
52	1152	CAGCCATG	11.1
53	1005	CACCATGG	8.6
52	426	CGCCATGC	9.0
52	931	CGCCATGG	9.4
52	1081!	CAAGATGG	39.6
52	1202!	AAGATGGC	36.9
52	881!	AGATGGCG	40.2
51	654	ACATGGCG	13.5
52	1026!	GATGGCGG	27.2
52	920	CATGGCGG	18.4
54	291	CATGGCGT	11.1
51	583	ATGGCGCC	23.6
52	1125!	ATGGCGGC	27.7
52	619	ATGGCGGG	8.2
52	468!	ATGGCGGA	16.0
52	966	ATGGCTGC	15.8

Protein coding?

54	791!	CCAGGTAA	7.1
56	307!	CGCAGGTA	8.2
51	443	CGCAGTCT	8.1
55	1638!	GGTGAGTG	7.6
53	848!	TGGTGAGT	7.9
52	1414	GAGAGCTG	7.4
53	3887!	CTGCTGCT	9.1
53	3570!	TGCTGCTG	8.0

Consensus sequences

SP1

44-50	8.8	CCCCGCCC	3424
44-50	8.3	GCCCCGCC	2687
44-50	8.7	CCCGCCCC	2257

CCAAT

42-49	10.0	RRCCAATSR	1170

ETS

44-51	13.1	VCCGGAARY	1031
48-51	11.6	RGCGGAAGY	260

TATA

48-49!	7.7	TATAAAD	472
48-49!	2.4	TATATAD	349
48-50!	5.5	TATAAGD	217

Box A

43-51	8.2	TCTCGCGA	211

NRF-1

46-51	7.4	CGCCTGCG	512
45-50	5.8	CGCGTGCG	220
46-51	9.0	CGCATGCG	186

CRE

45-50	9.5	TGACGTCA	190
45-51	5.1	TGATGTCA	125
46-50	7.1	TTGCGTCA	48

E-Box

46-50	7.3	CCACGTGA	123
47-51	7.6	TCACGTGA	89

lends support to the general proposition that one can unravel the function of a gene by knowing its promoter sequence.

10.6.1.1 Non-palidromic Sequences

SP1 is found in 21% of promoters. Twenty one 8-mers have been placed in this group (Fig. 10.4). The most abundant localizing sequence is the sequence CCCCGCCC bound by the SP1 family of 3-zinc finger motif proteins [22]. This sequence contains a CpG and methylation decreases binding [12]. Extended sequences also peak including the 8-mer GCCCCGCC and the 9-mer CCCCGCCCC which is the length of DNA that a 3-zinc finger protein could bind. Many 8-mers were placed in this group and it is not obvious that these sequences represent a unique TFBS. The KLF family of C2H2 zinc finger proteins is known to bind to the CCCCTCCC variant. There are many C2H2 zinc finger family members and one presumes that they may bind to SP1 related sequences.

CAAT is found in 8% of promoters. Thirty one 8-mers contain an invariant 5-mer (CCAAT) termed CAAT which was one of the first specific DNA sequences identified that was critical for gene expression [23]. This TFBS does not contain a CpG and thus is immune to epigenetic regulation. Neighboring DNA sequences are constrained resulting in the consensus 9-mer (**RR**CCAAT**SR**) (Fig. 10.4). This sequence is the furthest from the transcription start site, peaking about at –100 bps. There are several TFs that can bind to this sequence. One is a trimeric protein called CBF or NF-Y [24] with homology to the yeast proteins HAP2 and HAP3. There is no X-Ray structure for this protein DNA complex. 8-mers in this group appear to represent a unique TFBS.

ETS is found in 8% of promoters. Nineteen 8-mers have a core consensus CCGGAA which is bound by the ETS family of TFs [25, 26]. These sequences contains a CpG and methylation decreases DNA binding [12]. The extended consensus is the 9-mer VCCGGAARY. This extended consensus is found in DNA binding site selection experiments using ETS proteins [26]. Six 8-mers contain a variant ETS sequence, the 6-mer GCGGAA, a single base change from the ETS consensus. The extension of this sequence is the 9-mer RGCGGAAGY found in 2% of promoters. DNA binding site selection experiments indicate that this ETS site variant is bound by the PEA-3 subfamily of ETS proteins [27, 28].

◄

Table 10.1 Grouping of DNA 8-mer sequences that localize in human promoters. 150 DNA sequences are grouped into related sequences and arranged by their peak position relative to the TSS. From the left the table contains: the most abundant bin, the number of times the sequence occurs in the distribution, the 8-mer sequence, and finally the probability (P) that the cluster occurs by chance. The end of the table contains consensus sequences. Here the leftmost numbers are the bins defining the peak, followed by the localization factor (LF), the consensus sequence, and finally the number of occurrences of the sequence in the bins that comprise the peak. Exclamation point (!) denotes sequences that are at least threefold more abundant in the maximum bin on the DNA strand presented in the table than on the opposite strand. IUPAC letters used to represent degenerate bases are: R (G,A), W (A,T), Y (T,C), K (G,T), V (G, C, A), D (G,A,T), N (A,T,G,C)

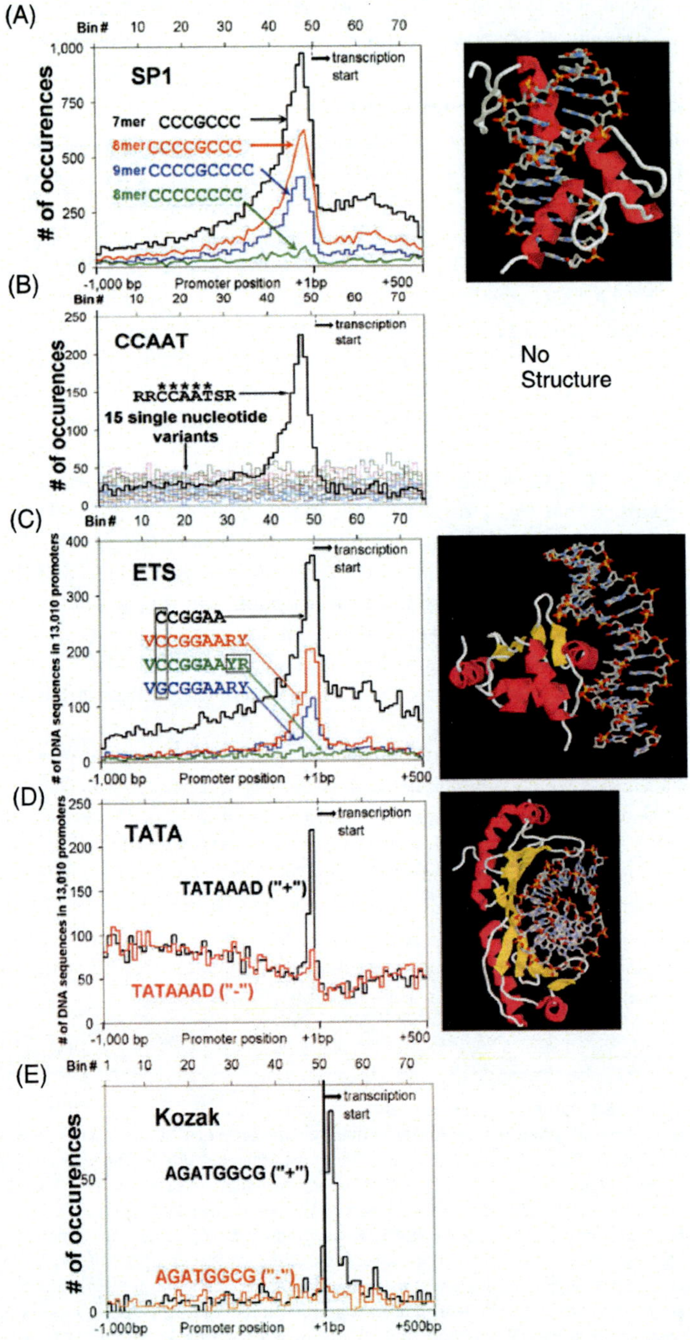

Fig. 10.4 (continued)

ETS:ETS and ETS:CRE pairs: Recent work from our group indicates that some of the discontinuous 8-mers that localize are combinations of ETS:ETS or ETS:CRE sites. We observe two continuous ETS sequences with the GCGGAA ETS variant always being a member of the ETS:ETS pair. This direct repeat is not what has been observed with biochemical selection experiments where inverted ETS sites are observed [29]. An 8-mer representing the pair of ETS sites is AAGCGGAA. An additional partner for ETS is observed in several 8-mers that overlap to produce **GGAAGTGACGT** that appear to be an ETS (CC**GGAAGTG**) and a CRE (**TGACGT**) site that overlap. An interesting aspect of these juxtapositions of two ETS sites and the ETS and CRE site is that the space between the two sites is invariant suggesting some structural constraint that would be exciting to examine.

TATA is found in 3% of promoters. Nine 8-mers contain the consensus 7-mer TATAAAD, a sequence bound by the TATA binding protein (TBP) [30] that recruits the basal machinery to initiate transcription [31]. This TFBS does not contain a CpG. The TATA sequence shows the sharpest peak but also has the highest background. This is the only TF binding site that localizes and occurs in a DNA strand specific manner (Fig. 10.3). TATA also localizes in a strand specific manner in *Drosophila*. TATA occurs in only a few percent of promoters when you restrict the analysis to around –30 bps [2, 32].

Box-A is found in 1% of promoters. Only one 8-mer contains this TFBS (TCTCGCGA). This TFBS is involved in the regulation of the ribosomal genes but the TF that binds this sequence is not known [33]. This TFBS has two CpGs allowing methylation to potentially modulate DNA binding.

Kozak: Downstream of the TSS we observe the Kozak sequence that contains the initiating ATG where protein synthesis initiates from the mRNA. As expected, this sequence is strand specific. It is sometimes difficult to observe the strand specific properties of the Kozak sequence because the sequence can be palindromic.

YY1: Previously, we grouped all ATG containing sequences that occur downstream of the TSS as Kozak sequences. Closer examination suggests that they are bound by YY1, a zinc finger protein [34].

Fig. 10.4 Distribution of non-palindromic TFBS in promoters. We include both the distribution of the TFBS and the X-ray crystal structure if it exists. **a** SP1 sequences (CCCGCCC, CCCCGCCC, CCCCGCCCC) and a non-peaking single base variation (CCCCCCCC). Crystal structure of a three zinc finger protein bound to DNA. **b** The CCAAT consensus RRCCAATSR and the 15 single base variants of the central CCAAT. Note the 5-mer CCAAT is needed for there to be any localization in the proximal promoter. No crystal structure is available. **c** ETS core (CCGGAA), consensus sequence (VCCGGAARY), and a peaking (VGCGGAARY) and non-peaking VCCGGAAYR variant. Crystal structure of ETS bound to DNA **d** Strand specific localization of the TATAAAD sequence. Note both the high background and the sharpness of the peak. Crystal structure of TATA bound to DNA. This is the only protein DNA complex presented here without an α-helix in the major groove of DNA. **e** Kozak sequence (AGATGGCG) on the plus strand (+) and minus strand (–). Again, note the DNA strand dependence of the localization of this sequence

Protein Coding: We observe multiple 8-mers downstream of the TSS that occur on a single strand and appear to be protein coding. They can translate into hydrophobic amino acids that occur at the 5′ end of proteins as a transmembrane signal.

10.6.1.2 The Palindromic Sequences

Three sequences that localize in promoters are palindromic (Fig. 10.5). The proteins that binding these palindromic sequences are dimeric raising the possibility that heterodimers can form and bind variants of the consensus sequence. This is known for the B-ZIP and B-HLH-ZIP proteins that bind the CRE and E-Box respectively. The crystal structures of dimer B-ZIP and B-HLH-ZIP protein help rationalize why these proteins bind palindromic sequences.

NRF1 is found in 6% of promoters. The palindromic CGCATGCG sequence is the most localizing 8-mer. This TFBS contains two CpGs. NRF-1 is the only member of the family and activates the expression of nuclear genes that function in the mitochondrion and helps to link general cellular respiration with other cellular functions including cell growth [35]. Unfortunately, no crystal structure exists. When we vary each bp, we identify two additional sequences that localize resulting in the consensus CGCVTGCG. We have grouped several C & G rich 8-mers into this TFBS group but these 8-mers may represent binding sites for other TFs.

E-Box is found in 1.5% of promoters. The palindromic 8-mer TCACGTGA and the related 8-mer, CCACGTGA, localize in proximal promoters. This sequence contains a CpG and methylation could affect DNA binding. These sequences are bound by the USF family of dimeric B-HLH-ZIP proteins [36, 37]. The core of this sequence is the E box sequence 6-mer CANNTG that is bound by B-HLH proteins [38]. Varying each base pair in this consensus does not identify additional DNA sequences that cluster. Keeping one half of the palindrome constant and varying the other half (**NNNN**GTGA) does not identify additional DNA sequences that localize. This is one of the two sequences that localize in both human and *Drosophila* promoters (the other is the TATA element). There are over 100 B-HLH-ZIP proteins and many are known to heterodimerize, e.g. E12 and MyoD heterodimerize and bind the E-Box sequence. A more comprehensive examination of the DNA binding of heterodimers using new comprehensive techniques is an exciting issue to examine.

CRE is found in 2.4% of promoters. The palindromic 8-mer TGACGTCA sequence is known as the cAMP responsive element (CRE) [39] [40, 41]. The CRE is bound by a variety of B-ZIP proteins homodimers including CREB, ATF1, and Oasis and by heterodimers including FOS|JUN and ATF2|JUN [42]. CpG methylation attenuated CREB binding to the CRE [43] but less is known about how CpG methyation affects the binding of other B-ZIP proteins to the CRE. We varied each base of the CRE TFBS and identified the TGATGTCA sequence that localizes in promoters. This sequence has the CG in the CRE changed to a TG as would be expected if the methyl CpG deaminates to TG. Thus this sequence cannot be regulated by CpG methylation. We identified an additional sequence that clusters when we keep one half of the palindrome constant and let the second half vary

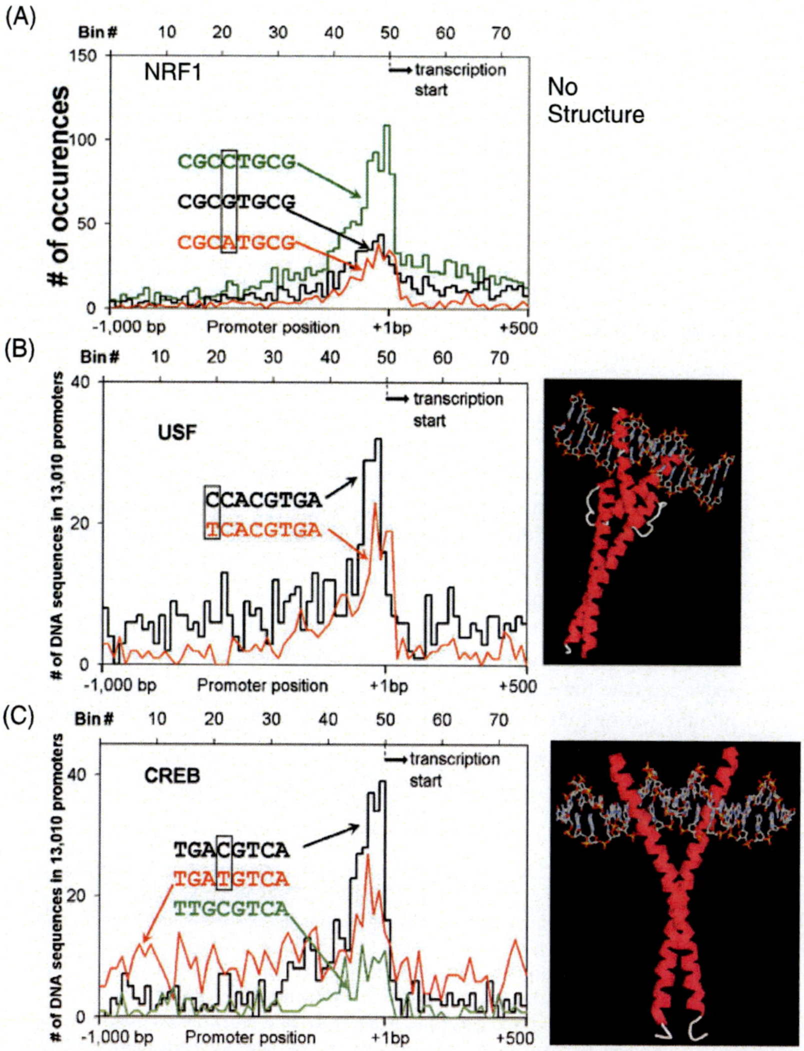

Fig. 10.5 Distribution of palindromic TFBS in promoters. We include both the distribution of the TFBS and the X-ray crystal structure if it exists. **a** NRF-1 sequence (CGCCTGCG, CGCGTGCG, CGCATGCG). No X-ray structure exists. **b** E-Box sequences (TCACGTGG, TCACGTGA). Crystal structure of USF bound to E-Box sequence [38]. **c** CRE-like sequences (TGACGTCA, TGATGTCA, TTGCGTCA). Crystal structure of CREB bound to the CRE sequence [46]

(NNNNGTCA). This sequence is TTGCGTAC that contains C/EBP and CREB half sites and can be bound by a C/EBP|ATF4 [44] or C/EBP|ATF2 heterodimer [45]. Twelve 8-mers contain the 5 bp sequence GTCAC which is observed in both the CRE and E-Box TFBSs. It could be that there is a competition for a B-ZIP or a B-HLH-ZIP protein to bind this sequence.

10.6.2 Additional DNA Sequences that Localize in Proximal Promoters

The analysis presented here highlights what can be gleaned from an examination of DNA sequences that preferentially localize in all promoters. Presently, we are examining subsets of promoters with similar properties to identify additional sequences that localize in proximal promoters. For example, when we examine the E2F binding site (TTTCGCG), a sequence known to localize in promoters of cell cycle genes, it does not appear when we examine all promoters but does when we examine promoters that are well bound by RNA polymerase II. This strategy will allow one to identify more DNA sequences in proximal promoters with biological function.

10.7 Conclusion

Ultimately, gene expression is controlled by the DNA sequence of the genome. It has been very challenging to unravel this code because of the difficulty of identifying the DNA sequences that are functional TFBSs. The analysis of the localization of DNA sequences in promoters has allowed us to begin to define DNA sequences that are important in regulating gene expression. As we learn more about the sequences that occur in the promoters of different organisms, we will be able to observe the changes that have occurred between humans and *Drosophila*. Is ETS a more ancient sequence than the CRE? The answer to these types of questions will give us insight into the wiring hierarchy that has occurred as promoters evolve in metazoans.

References

1. Ohler U, Liao GC, Niemann H, Rubin GM (2002) Computational analysis of core promoters in the *Drosophila* genome. Genome Biol 3:RESEARCH0087
2. FitzGerald PC, Shlyakhtenko A, Mir AA, Vinson C (2004) Clustering of DNA sequences in human promoters. Genome Res 14:1562–1574
3. Fitzgerald PC, Sturgill D, Shyakhtenko A, Oliver B, Vinson C (2006) Comparative genomics of *Drosophila* and human core promoters. Genome Biol 7:R53
4. Marino-Ramirez L, Spouge JL, Kanga GC, Landsman D (2004) Statistical analysis of over-represented words in human promoter sequences. Nucleic Acids Res 32:949–958
5. Bina M, et al. (2004) Exploring the characteristics of sequence elements in proximal promoters of human genes. Genomics 84:929–940
6. Bina M, et al. (2009) Discovering sequences with potential regulatory characteristics. Genomics 93:314–322
7. Suzuki Y, Yamashita R, Sugano S, Nakai K (2004) DBTSS, DataBase of Transcriptional Start Sites: progress report 2004. Nucleic Acids Res 32:D78–81
8. Zhang MQ (1998) A discrimination study of human core-promoters. Pac Symp Biocomput 3:240–251
9. Bird AP (1986) CpG-rich islands and the function of DNA methylation. Nature 321:209–213
10. Gardiner-Garden M, Frommer M (1987) CpG islands in vertebrate genomes. J Mol Biol 196:261–282
11. Bird A (2002) DNA methylation patterns and epigenetic memory. Genes Dev 16:6–21

12. Rozenberg JM, et al. (2008) All and only CpG containing sequences are enriched in promoters abundantly bound by RNA polymerase II in multiple tissues. BMC Genomics 9:67
13. Jones PA, Baylin SB (2007) The epigenomics of cancer. Cell 128:683–692
14. Eckhardt F, et al. (2006) DNA methylation profiling of human chromosomes 6, 20 and 22. Nat Genet 38:1378–1385
15. Weber M, et al. (2007) Distribution, silencing potential and evolutionary impact of promoter DNA methylation in the human genome. Nat Genet 39:457–466
16. Mito Y, Henikoff JG, Henikoff S (2005) Genome-scale profiling of histone H3.3 replacement patterns. Nat Genet 37:1090–1097
17. Yuan GC, et al. (2005) Genome-scale identification of nucleosome positions in *S. cerevisiae*. Science 309:626–630
18. Tillo D, Hughes TR (2009) G+C content dominates intrinsic nucleosome occupancy. BMC Bioinformatics 10:442
19. Tillo D, et al. (2010) High nucleosome occupancy is encoded at human regulatory sequences. PLoS One 5:9129
20. Polavarapu N, Marino-Ramirez L, Landsman D, McDonald JF, Jordan IK (2008) Evolutionary rates and patterns for human transcription factor binding sites derived from repetitive DNA. BMC Genomics 9:226
21. Weirauch MT, Hughes TR (2010) Dramatic changes in transcription factor binding over evolutionary time. Genome Biol 11:122
22. Pavletich NP, Pabo CO (1991) Zinc finger-DNA recognition: crystal structure of a Zif268-DNA complex at 2.1 A. Science 252:809–817
23. Dynan WS, Tjian R (1985) Control of eukaryotic messenger RNA synthesis by sequence-specific DNA-binding proteins. Nature 316:774–778
24. Sinha S, Maity SN, Lu J, de Crombrugghe B (1995) Recombinant rat CBF-C, the third subunit of CBF/NFY, allows formation of a protein-DNA complex with CBF-A and CBF-B and with yeast HAP2 and HAP3. Proc Natl Acad Sci U S A 92:1624–1628
25. Sharrocks AD (2001) The ETS-domain transcription factor family. Nat Rev Mol Cell Biol 2:827–837
26. Graves BJ, Petersen JM (1998) Specificity within the ets family of transcription factors. Adv Cancer Res 75:1–55
27. Brown TA, McKnight SL (1992) Specificities of protein–protein and protein–DNA interaction of GABP alpha and two newly defined ets-related proteins. Genes Dev 6:2502–2512
28. Wei GH, et al. (2010) Genome-wide analysis of ETS-family DNA-binding in vitro and in vivo. Embo J 29:2147–2160
29. Jolma A, et al. Multiplexed massively parallel SELEX for characterization of human transcription factor binding specificities. Genome Res 20:861–873
30. Kim Y, Geiger JH, Hahn S, Sigler PB (1993) Crystal structure of a yeast TBP/TATA-box complex. Nature 365:512–520
31. Geiger JH, Hahn S, Lee S, Sigler PB (1996) Crystal structure of the yeast TFIIA/TBP/DNA complex. Science 272:830–836
32. Kim TH, et al. (2005) A high-resolution map of active promoters in the human genome. Nature 436:876–880
33. Perry RP (2005) The architecture of mammalian ribosomal protein promoters. BMC Evol Biol 5:15
34. Kim J, Kim J (2009) YY1's longer DNA-binding motifs. Genomics 93:152–158
35. Scarpulla RC (2006) Nuclear control of respiratory gene expression in mammalian cells. J Cell Biochem 97:673–683
36. Bendall AJ, Molloy PL (1994) Base preferences for DNA binding by the bHLH-Zip protein USF: effects of MgCl2 on specificity and comparison with binding of Myc family members. Nucleic Acids Res 22:2801–2810
37. Boyd KE, Farnham PJ (1999) Coexamination of site-specific transcription factor binding and promoter activity in living cells. Mol Cell Biol 19:8393–8399

38. Ferre-D'Amare AR, Prendergast GC, Ziff EB, Burley SK (1993) Recognition by Max of its cognate DNA through a dimeric b/HLH/Z domain. Nature 363:38–45
39. Montminy M (1997) Transcriptional regulation by cyclic AMP. Annu Rev Biochem 66: 807–822
40. Shaywitz AJ, Greenberg ME (1999) CREB: a stimulus-induced transcription factor activated by a diverse array of extracellular signals. Annu Rev Biochem 68:821–861
41. Mayr B, Montminy M (2001) Transcriptional regulation by the phosphorylation-dependent factor CREB. Nat Rev Mol Cell Biol 2:599–609
42. Vinson C, et al. (2002) Classification of human B-ZIP proteins based on dimerization properties. Mol Cell Biol 22:6321–6335
43. Iguchi-Ariga SM, Schaffner W (1989) CpG methylation of the cAMP-responsive enhancer/promoter sequence TGACGTCA abolishes specific factor binding as well as transcriptional activation. Genes Dev 3:612–619
44. Vinson CR, Hai T, Boyd SM (1993) Dimerization specificity of the leucine zipper-containing bZIP motif on DNA binding: prediction and rational design. Genes Dev 7:1047–1058
45. Shuman JD, Cheong J, Coligan JE (1997) ATF-2 and C/EBPalpha can form a heterodimeric DNA binding complex in vitro. Functional implications for transcriptional regulation. J Biol Chem 272:12793–12800
46. Schumacher MA, Goodman RH, Brennan RG (2000) The structure of a CREB bZIP.somatostatin CRE complex reveals the basis for selective dimerization and divalent cation-enhanced DNA binding. J Biol Chem 275:35242–35247

Chapter 11
Interactions of Transcription Factors with Chromatin

Harm van Bakel

Abstract Sequence-specific transcription factors (TFs) play a central role in regulating transcription initiation by directing the recruitment and activity of the general transcription machinery and accessory factors. It is now well established that many of the effects exerted by TFs in eukaryotes are mediated through interactions with a host of coregulators that modify the chromatin state, resulting in a more open (in case of activation) or closed conformation (in case of repression). The relationship between TFs and chromatin is a two-way street, however, as chromatin can in turn influence the recognition and binding of target sequences by TFs. The aim of this chapter is to highlight how this dynamic interplay between TF-directed remodelling of chromatin and chromatin-adjusted targeting of TF binding determines where and how transcription is initiated, and to what degree it is productive.

11.1 Introduction

The basic principles of transcriptional regulation are similar between prokaryotes and eukaryotes and involve the binding of TFs to specific DNA sequences at target genes, where they recruit and stabilize the general transcriptional machinery required for gene expression [1, 2]. Despite these general similarities, transcription initiation in eukaryotes is considerably more complex, which is likely related to the increased genome size and greater need for organization compared to prokaryotes. One key difference is that DNA in eukaryotes is not readily accessible, but tightly packaged by architectural proteins into chromatin. The basic unit of this packaging is the nucleosome, which consists of ~147 bp of DNA wrapped around an octamer of histone proteins [3, 4]. Nucleosomes play an important role in condensing DNA, thereby allowing the large eukaryotic genome to fit into the nucleus. Perhaps not surprisingly, this compaction also negatively affects transcription initiation in vitro

H. van Bakel (✉)
Banting and Best Department of Medical Research, University of Toronto, Toronto, ON, Canada M5S 3E1
e-mail: hvbakel@gmail.com

T.R. Hughes (ed.), *A Handbook of Transcription Factors*, Subcellular Biochemistry 52, DOI 10.1007/978-90-481-9069-0_11,

[5, 6] and in vivo [7], as it forms an impediment to the binding of TFs and the formation of a preinitiation complex (PIC) [8, 9]. To initiate transcription, TFs and the PIC must first overcome the physical barrier posed by nucleosomes; however, the stability of nucleosomes means that direct competition for DNA access is inefficient. A host of coactivators therefore exist that can be recruited to regulatory regions by TFs to facilitate transcription initiation. These coactivators typically consist of (or recruit) chromatin modifier (CM) complexes that either displace or evict nucleosomes or covalently modify histones to loosen their interactions with DNA. CMs can also function as corepressors by effecting a more closed chromatin conformation. Consequently, the recruitment of coregulators that affect chromatin structure is now recognized as a major mechanism by which TFs can regulate gene expression.

Knowledge of general chromatin architecture has greatly expanded in recent years due to the broad application of classical and novel techniques to map TF binding sites, histone modifications, and chromatin accessibility. Mapping of TF binding sites and histone modifications is typically done using chromatin immunoprecipitation (ChIP) or related techniques such as DamID, which are discussed in more detail in Chapter 8. Most of the techniques to map chromatin accessibility make use of the fact that regulatory sites and the short DNA linkers connecting nucleosomes are more sensitive to nuclease digestion by micrococcal nuclease (MNase) or DNase I, each of which has distinct cleavage patterns that provide a different view of chromatin structure [10]. MNase cuts preferentially in linker regions between nucleosome and it is therefore typically used to map the positions of nucleosomes. On the other hand, DNaseI also cuts DNA associated with nucleosomes, when used at higher concentrations, and its cleavage pattern therefore typifies general chromatin accessibility. Another approach to identify regions of open chromatin, formaldehyde-assisted isolation of regulatory elements (FAIRE), has also been described [11]. This method exploits the property that fragmented DNA that is highly crosslinked to histones after formaldehyde treatment (i.e. closed chromatin) can be separated from DNA with a low degree of crosslinking (i.e. open chromatin) by phenol extraction.

Advances in microarray and sequencing technology have made it possible to apply these various methods to create genome-wide maps of nucleosome occupancy [12–15], potential regulatory sites [16, 17], as well as patterns of histone modifications and TF binding [18–23]. A common observation in these studies is that active promoters and distal regulatory elements such as enhancers are associated with regions of open chromatin and enriched for bound TFs and their coregulators, underscoring that transcriptional regulation is universally linked to chromatin remodelling. These studies have also provided an unprecedented view of the higher-order structure of the genome, where broad domains of more accessible chromatin (i.e. euchromatin) alternate with regions that are less accessible to the transcription machinery (i.e. heterochromatin). It should be noted, though, that these techniques provide only a snapshot of the chromatin structure at the time of fixation and while many regulatory regions may appear stable, several lines of evidence suggest that remodelling is in fact a highly dynamic and continuously ongoing process. For example, nucleosomes found in yeast promoters exchange more rapidly than nucleosomes located in gene bodies [24, 25] and FRAP (fluorescence recovery after photobleaching) studies suggest that many TFs only transiently interact with DNA in vivo, even at active promoters [26–29]. Thus, chromosomal domains and regulatory regions with apparently stable chromatin are likely in a dynamic equilibrium between competing forces, the balance of which ultimately determines the degree of DNA accessibility [8].

Following a brief introduction into the types of CM involved in chromatin remodelling, this chapter will highlight how TFs can regulate gene expression by recruiting these coregulators to orchestrate changes in the chromatin state, and in turn, how chromatin can affect TF target recognition and binding. Then, I will discuss how these dynamic and antagonistic forces may be coordinated to organize chromatin and direct transcription at specific locations in the genome. Other recent reviews that consider these and related topics include [30–33], as well as Chapters 10 and 12 in this volume, which specifically consider TF–nucleosome interactions, and the auxiliary domains of TFs that mediate many of these functions, respectively. This chapter also contains a Glossary at the end which provides an overview of key terminology used throughout.

11.2 An Overview of Coregulators that Effect Changes in Chromatin Structure

A broad distinction can be made between two types of CMs, based on their mechanism of action: histone modifiers and ATPase nucleosome remodelling complexes. Histone modifiers are responsible for the wide variety of covalent modifications found on histone proteins, in particular on their unstructured N-terminal tails (Reviewed in [34, 35]). At least eight different types of histone modifications and their associated enzymes have been identified, with the number of distinctly modified residues currently standing at well over a hundred [34]. It has been proposed that combinations of these modifications constitute a "histone code" that is read by proteins that interact with specific modifications [36], allowing for an organized association of proteins with different stages of transcription. Indeed, the different modifications can serve as interaction sites for other coregulators, such as ATPase remodelers, that can direct further changes to chromatin structure (see examples below). The ultimate effect of histone modifications on chromatin structure – be it compacting or unwrapping – is therefore presumably to a large degree determined by the type of proteins that interact with them. Another way that histone modifications can affect chromatin structure is by changing the electrostatic properties of nucleosomes. For example, the acetylation of histone tails by histone acetyl transferases (HATs) neutralizes positive charges that would otherwise interact with negatively charged DNA [37], facilitating nucleosome unwrapping and mobility (Fig. 11.1a). It is unclear whether other modifications similarly affect chromatin through effects on the chemical properties of nucleosomes, but it has been suggested that phosphorylation may, like acetylation, reduce chromatin compaction through its effects on nucleosome charge [34].

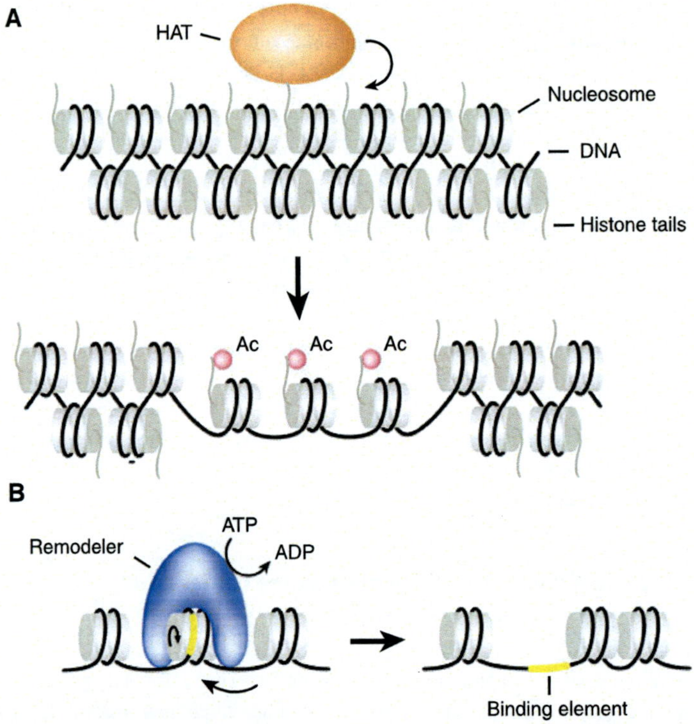

Fig. 11.1 Effects of chromatin modifiers on chromatin structure. **a** Acetylation of histone tails by histone acetyl transferases (HATs) results in a more open chromatin conformation. **b** Model for nucleosome sliding by ATPase remodelers based on studies of the ACF complex [273]. In this model, the ATPase remodeler draws in DNA from the linker region (*bottom arrow*), resulting in the formation of a small DNA loop at the nucleosome entry site, which then propagates over the nucleosome, resulting in a lateral displacement along the DNA. The illustration shows one possible effect of remodelling at regulatory regions, namely the exposure of TF binding sites that would otherwise be rendered inaccessible by nucleosomes

Genome-wide studies have revealed that the occurrence of most modifications is tightly coupled to the location and activity of genes and their regulatory regions, in a manner that reflects their effects on chromatin structure. For example, acetylation marks are predominantly found at the beginning of active genes in yeast [22, 38–41] and at promoters and CpG islands in higher eukaryotes [42–45], although activation has also been linked to *decreased* acetylation of lysine residue 16 on histone H4 (i.e. H4K16ac) [38, 46, 47]. In contrast, methylation patterns differ depending on the residue that is modified, and distinct methylation states can be associated with either repression or activation [31, 34]. Classical examples include H3K4me and H3K27me, which mark regions of active and silent chromatin, respectively. The difference between acetylation and methylation patterns is mirrored in the specificity of their enzymes: HATs typically act indiscriminantly on multiple histone residues

[34], whereas methyltransferases are restricted to a single residue on one histone type [48]. Some effects of HATs on chromatin may also be mediated through other targets, as it has become increasingly clear that they can acetylate many non-histone proteins, including TFs [49–51]. For other modifications, the relation to the transcriptional state is less well characterized, but in general, phosphorylation appears to correlate with activation [52, 53], while sumoylation has been associated with repression [54, 55]. Ubiquitination, like methylation, can be associated with either transcriptional state [56–58]. Extensive crosstalk between modifications presumably contributes to these complex patterns. For example, phosphorylation of H3S10 can stimulate acetylation of H3K14 [59, 60] and inhibit H3K19 methylation [61], while repression by sumoylation may be directly related to the fact that it competes for the same residues as acetylation and ubiquitination.

The second class of CMs, ATPase remodelers, can directly affect the degree of chromatin packing by repositioning or sliding nucleosomes along the DNA (Reviewed in [62]) (Fig. 11.1b). The primary driving force behind this motion comes from a central catalytic subunit, which contains a conserved ATPase domain that provides the energy to move nucleosomes by rewinding the DNA around them. This process involves breaking and reforming most histone–DNA interactions, which likely explains the broad effects that remodelers can have on nucleosomal DNA accessibility [63, 64], nucleosome eviction [65–67] and histone exchange [68, 69]. Besides the ATPase domain, the catalytic subunits contain various additional domains that have been used to classify these remodelers into four major families: SWI/SNF, ISWI, CHD and INO80. Interestingly, with the exception of INO80 subunits, many of these additional domains mediate affinity to distinct histone modifications [70, 71], which are thought to confer different preferences for specifically modified chromatin structures to each family [72, 73]. SWI/SNF remodelers contain a bromodomain which binds acetylated histones [74], while the CHD family possesses chromodomains that can interact with methylated histone tails [75–78]. ISWI family proteins have a pair of SANT and SLIDE domains that are believed to form a module with affinity for unmodified histones [79], though it is as yet unclear to what degree this interaction may be affected by specific modifications.

The diversity of CMs is further increased through the association of the core catalytic subunits with different complements of additional proteins, which can vary even within families [62, 70, 80]. These accessory subunits can play a structural role, and can also contribute a variety of additional interaction domains and catalytic activities. Some complexes, such as NURD (nucleosome-remodelling and histone deacytelase), even combine ATPase remodeler and histone modifier activities [81]. As in the case of histone modifications and their associated enzymes, a broad classification can be made regarding the effects of the ATPase remodelers on gene expression. For example, recruitment of SWI/SNF complexes is predominantly associated with transcriptional activation, consistent with its preference for acetylated histones, while ISWI complexes typically function as repressors [82]. This distinction is by no means sharply defined, though, and most ATPase remodelers have been found to function as activators at some promoters and repressors at others. Thus the ultimate effect of remodelling can vary depending on the context in which this remodelling takes place.

11.3 TFs Play a Central Role in Targeting Chromatin Remodelling

Exactly how chromatin remodelling complexes are guided to their target regions remains an active area of investigation. One clearly established pathway is direct recruitment by TFs, with TFs providing the targeting component through their sequence-specific DNA binding domains. This recruitment typically involves transient interactions with the transactivation or effector domains of TFs, which are discussed in more detail in Chapter 12 of this volume. The intrinsic preferences for specific histone modifications found in many CMs, discussed above, do indicate that there are also alternative routes that do not involve direct recruitment by TFs. For example, the bromodomains in the yeast Swi2/Snf2 remodelers and Gcn5 HAT are sufficient to anchor their respective complexes to acetylated promoters in the absence of transcriptional activators [74]. Individual histone binding domains may in general not be sufficient for effective targeting, however, given the low binding affinities of the domains characterized to date [62]. Instead, the interaction domains could serve other purposes that do not involve recruitment, such as regulating remodeler ATPase activity [62]. Regardless, even if histone modifications indeed provide important targeting cues for CMs, the question remains as to how these modifications are established in the first place, given that histone-modifying enzymes generally do not posses intrinsic DNA sequence preferences. One possible answer comes from detailed studies of model genes in yeast (Reviewed in [83]), which have shown that the actions of histone modifiers in the early stages of transcription initiation are primarily guided by sequence-specific TFs. It is therefore likely that TFs play a central role in targeting chromatin remodelling, whether this is through direct interactions with remodelling complexes, or by guiding initial histone modifications and/or other coregulators that mediate these interactions indirectly. An overview of some of the key features of TF-mediated recruitment of CMs and their implications for gene regulation will be given in the following paragraphs; readers are referred to Chapter 12 for more details.

Individual TFs can interact with a surprisingly wide variety of modifier complexes and other coregulators. This promiscuity is in part due to the intrinsic characteristics of the TF transactivation domains (also discussed in greater detail in Chapter 12), which are generally unstructured and only become stabilized upon interacting with their binding partners [84, 85] property that may allow for some degree of flexibility in the selection of binding partners [86]. The diversity of TF partners is also increased through interactions with subunits that are shared between different CM complexes. For example, acidic activation domains such as those found in the yeast Gal4 TF can recruit both the SAGA and NuA4 HATs through interactions with the Tra1 subunit that is present in both these complexes [87–89]

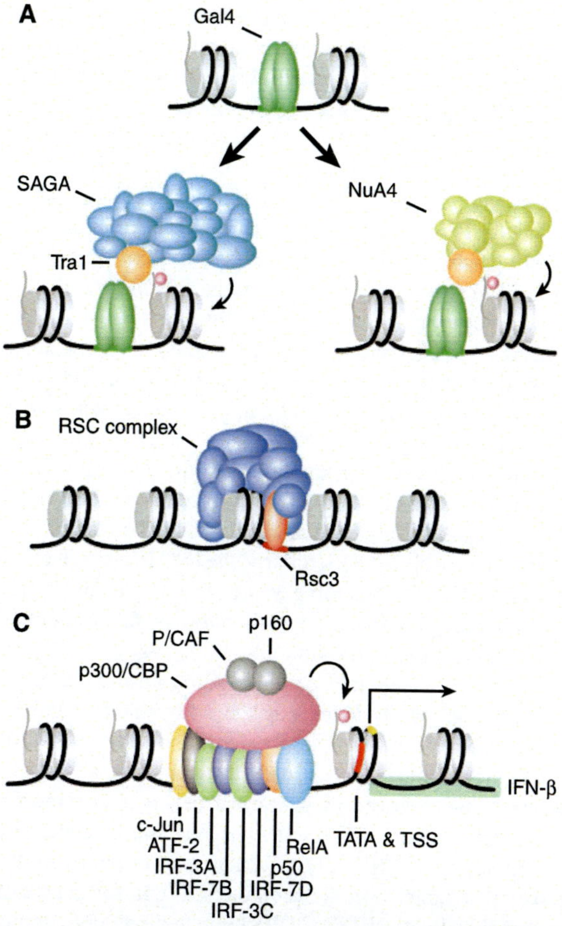

Fig. 11.2 Targeting of chromatin remodelling by TFs. **a** The diversity of TF interactions with CMs is increased through shared subunits in remodeler complexes, as illustrated here by the interaction between the Gal4 TF and the Tra1 subunit in the SAGA and NuA4 complexes. **b** Targeting of the RSC complex in *S. cerevisiae* by the Rsc3 TF subunit. **c** CBP hub function at the IFN-β enhanceosome. CBP interacts with the enhanceosome TFs, resulting in recruitment of the RNA polymerase II holoenzyme, PIC assembly and the initiation of transcription [274]

(Fig. 11.2a). The great diversity of TF binding partners may serve multiple purposes. First, it enables the same TF to participate in distinct mechanisms of transcription initiation at different genes, as has been described for the activation of transcription by Pho2 and Pho4 at the *PHO5* and *PHO8* promoters in budding yeast [83]. Second, the transient nature of TF interactions at individual regulatory regions [26–29] could allow for repeated cycles of TF binding to the same target site with different coregulators, enabling a TF to affect initiation in more than one way. The particular coregulator(s) recruited at each site likely depends on other elements such as local chromatin structure and interactions with other TFs.

In addition to mediating targeting through transient interactions, TFs can be integrated into CM complexes as stable components (Fig. 11.2b). The budding yeast TF Rsc3 is a subunit of the RSC chromatin remodelling complex [90], and was shown to promote nucleosome exclusion at promoters containing Rsc3 binding motifs [91], suggesting that it directs the RSC complex to these locations. Likewise, the Iec1 TF subunit of the INO80 complex is required for recruitment to target genes in fission yeast, and for associated histone remodelling [92]. Numerous putative DNA binding domains have also been identified in subunits of SWI/SNF remodelers in higher eukaryotes, including high mobility group (HMG) domains, C2H2 zinc fingers, and AT–rich interaction domains (ARIDs) [93]. The function of these domains is still largely uncharacterized and some, such as the HMG and ARID domains, are known to predominantly bind DNA in a sequence-independent manner and likely have structural roles [94, 95]. Nevertheless, it is possible that others will turn out to be important for targeting. Interestingly, the integration of sequence-specific TFs in remodelling complexes does not appear to be highly conserved between species. The RSC complex in higher eukaryotes lacks the specific DNA-binding determinants found in yeast [93, 96]; similarly, the INO80 component Iec1 is fungal-specific and has no ortholog in budding yeast. The stable integration of these particular TFs in remodelling complexes may therefore be the result of adaptations to specific selective pressures during evolution.

The multitude of subunits found in CMs means that they too can have many binding partners, greatly increasing their potential to regulate diverse targets. The subunit composition of complexes associated with each CM can also vary, such that different versions can pair with distinct sets of TFs. This enables individual complexes to be involved in gene- and cell type-specific functions, as exemplified by the mammalian SWI/SNF-type ATPases Brahma (BRM) and its paralog Brahma related gene 1 (BRG1), which are part of numerous chromatin remodelling complexes that target specific promoters to control gene expression [97]. BRG1 can be associated with WINAC (WSTF including nucleosome assembly complex), which can inhibit or activate target gene expression through subunit–specific interactions with the Vitamin D receptor [98]. Alternatively, when incorporated in the NUMAC (nucleosomal methylation activation) complex it can associate with estrogen receptor-responsive promoters to activate transcription [99]. Dynamic changes in CM subunit composition during development have also been shown to result in alterations in targeting by TFs. For example, the BRG1/BRM associated factors (BAFs) BAF45A and BAF53A in the SWI/SNF-type neuronal-progenitor-specific BAF complex (npBAF) are replaced by BAF45B and BAF53B upon differentiation, to form a neuron-specific complex (nBAF) [100]. The inclusion of BAF53B allows the nBAF complex to interact with the calcium-responsive transactivator (CREST) to regulate genes that are essential for dendritic outgrowth in the differentiated cells [101]. A similar requirement for specific BAF complex components has been observed in the differentiation of cardiomyocytes, where ectopic expression of the GATA4 and TBX5 TFs in combination with the BAF60C but not BAF60A subunits can induce the differentiation of mesoderm into contracting cardiomyocytes in developing mouse embryos [102]. Together, these observations indicate that TF binding can be interpreted differently in distinct cell types, depending on the complement of coregulators that is expressed. This modularity underscores the importance of combinatorial subunit assembly in establishing gene regulatory networks and reveals an additional layer of complexity that must be considered in our attempts to reconstruct these networks.

CM complexes can also be used as scaffolds for the assembly of different components of the transcriptional machinery. Indeed, the main catalytic function of CMs is sometimes dispensable altogether, as illustrated by the fact that SAGA-mediated activation of GAL genes does not require its HAT activity [103–105]. Instead, SAGA is believed to serve as a platform for the assembly of the PIC at GAL promoters. Similar functions have also been demonstrated for the general transcriptional coactivators CREB binding protein (CBP) and P300, two highly similar HATs with homologs in most multicellular organisms. In addition to the HAT domain, P300/CBP proteins contain other domains that mediate interactions with RNA polymerase II and a multitude of basal and gene-specific TFs [106, 107], allowing P300/CBP proteins to operate as hubs that can integrate signals from multiple TFs. This function has been most clearly described at the IFN-β enhanceosome, a stable complex of TFs and other nucleoproteins directly upstream of the IFN-β core promoter [108]. In this complex, CBP simultaneously interacts with multiple TFs bound across a 55 bp region, acting as a mediator for their synergistic activation of IFN-β transcription [108, 109] (Fig. 11.2c).

Consistent with their numerous interaction partners, P300/CBP have been linked to regulation of many genes, often acting at enhancers. Indeed, recent ChIP studies have identified P300/CBP binding as a key component of a wider signature of histone modifications and trans-acting factors that distinguish distal enhancers from gene promoters [20, 110–114]. Another component of this signature is H3K4 monomethylation, which peaks at enhancers but not promoters. Nevertheless, despite the predominance of P300/CBP at distal enhancers, both proteins can also be associated with proximal promoters and genes [115], underscoring their versatile roles in gene regulation.

11.4 Determinants of TF Access to Chromatin

A complicating factor for any model of chromatin remodelling based primarily on targeting by TFs is that they typically recognize small DNA motifs (~6–12 bp) that can occur randomly at high frequencies. For example, an 8-bp recognition sequence will appear 45,000 times in a human-sized genome with random sequence composition, and in reality this number will be dozens of times greater considering that TFs typically bind degenerate motifs in vitro [116]. Chromatin is believed to significantly increase TF specificity by reducing the accessibility of many spurious binding sites [117, 118]. This central role of chromatin in restricting where transcription initiation takes place is underscored by observations that failure to properly reconstitute nucleosomes in the body of transcribed yeast genes results in the appearance of cryptic transcripts, presumably initiated from exposed sequences that resemble promoters [119, 120]. Nonetheless, the packaging of DNA by nucleosomes is not the only means by which TF specificity is achieved in vivo. For example, TF–TF interactions, direct or indirect (e.g. through scaffold proteins or by outcompeting nucleosomes), can decrease the number of potential target sites due to the larger size of the combined binding specificity. Moreover, recognition sites are often clustered together in regulatory regions, allowing for further synergistic interactions between TFs [121–123]. A more in-depth overview of the various factors that play a role in TF target site selection can be found in Chapters 8 and 9.

The fact that nucleosomes can restrict access to DNA to prevent spurious transcription raises an important question: how can TFs bind their *bona fide* target sites to initiate the remodelling required for active transcription, given that much of the genome is covered by nucleosomes? Part of the answer to this question lies in the aforementioned fact that regulatory sites tend to be associated with open chromatin and nucleosome depleted regions (NDRs) [12–15, 124]. In yeast and *C. elegans*, there is strong evidence that the intrinsic DNA sequence preferences of nucleosomes play a key role in establishing these regions, and that these preferences are encoded in the genome sequence [125, 126]. Rigid DNA sequences such as poly(dA:dT) tracts are common in many eukaryotic promoters and have long been known to disrupt nucleosome–DNA interactions, increasing accessibility of nearby TF binding sequences [12, 127–130] (Fig. 11.3a). For example, the presence of a poly(dA:dT) tract in the *Candida glabrata AFT1* promoter destabilizes a well-positioned nucleosome containing a metal responsive element, enabling Aft1 to bind and autoactivate its gene expression [131–133]. Poly(dA:dT) tracts were also found to be major determinants of nucleosome exclusion in studies aimed at predicting in vivo nucleosome positions from DNA sequence features in a range of species [12, 134, 135]. Perhaps the most direct indication of the importance of intrinsic nucleosome sequence preferences in the establishment of NDRs at promoters has come from comparisons of in vivo yeast nucleosome occupancy patterns to those of nucleosomes reconstituted in vitro on purified yeast genomic DNA, which showed a high correlation between the two profiles [125, 136]. The importance of nucleosome disfavouring sequences in establishing NDRs is now widely accepted, though there is still some debate about the degree in which intrinsic sequence preferences dictate nucleosome positions outside these regions [137–140].

Despite their general applicability, models based on intrinsic nucleosome sequence preferences alone cannot fully explain the architecture of promoters and other regulatory sequences observed in living cells, even in yeast. An assessment of the influence of a wide range of sequence features on in vivo nucleosome positioning in budding yeast revealed additional strong nucleosome excluding elements that corresponded to binding motifs of sequence-specific TFs such as Reb1 and Abf1 [12]. The role of these factors in establishing NDRs was confirmed in Reb1 and Abf1 loss-of-function mutants that showed greatly increased nucleosome occupancy at hundreds of promoters containing their binding motifs [91, 141]. Moreover, the in vitro reconstituted nucleosome occupancy at Abf1 and Reb1 binding sites was

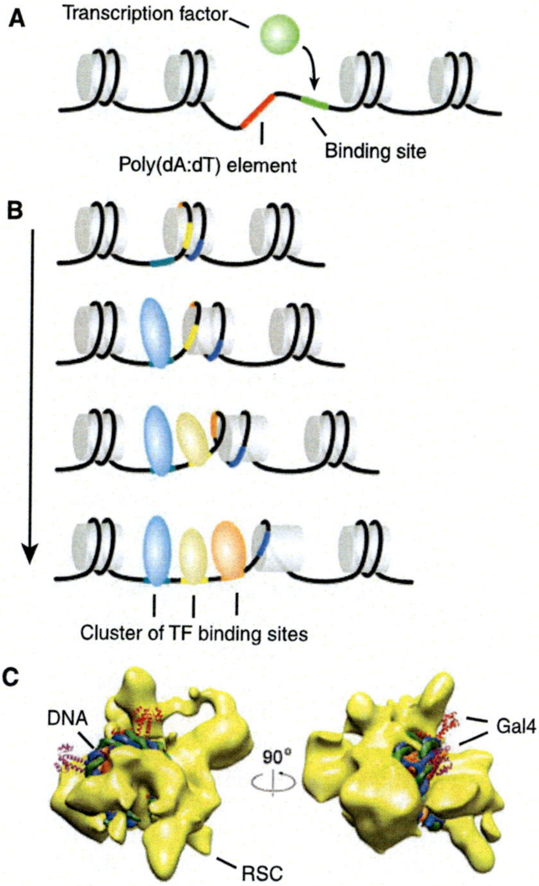

Fig. 11.3 Mechanisms of TF access to chromatin. **a** Rigid Poly(dA:dT) elements (*red*) are refractory to nucleosome assembly, allowing TFs to access nearby binding sites. **b** A model for progressive opening of chromatin by sequential binding of multiple TFs, as proposed by Polach and Widom [153]. **c** Presentation of the Gal4 UAS by RSC (Reproduced, with permission, from [63]). The model shows two exposed binding sites in the Gal4 UAS in the RSC/UASg/nucleosome complex and that Gal4 (*red* and *purple*) can access these sites without disrupting the complex. The structure of RSC/nucleosome complex (*yellow*) was determined by cryo-electronmicroscopy [275] and the position of the DNA helix is indicated in *green* and *blue*

higher than that measured in vivo [125]. Taken together, these data clearly indicate that TFs are capable of establishing NDRs at yeast promoters that lack intrinsic nucleosome-disfavouring sequences. Correspondingly, the concept of a universally encoded open promoter structure does not appear to apply to all genes: a subset of yeast genes that display highly variable expression levels have increased nucleosome occupancy in their promoters, consistent with predictions based on intrinsic sequence preferences [142]. It was proposed that the positioning of nucleosomes in these promoters plays a key role in the variable regulation of these genes.

The degree of basal nucleosome occupancy at promoters and other regulatory sequences also appears to vary between species. When applied to the human genome, models based on intrinsic nucleosome sequence preferences actually predict an overall increased occupancy at regulatory sites, in sharp contrast to most yeast promoters [143]. One explanation that was offered for this difference is that higher eukaryotes have greater requirements for variable gene expression, such as in the case of cell-type specific genes, and a constitutive open state might therefore not be desired [143]. Examples of TF binding to regions with high nucleosome occupancy have been described for the CCCTC-binding factor (CTCF) [144] and p53 [145], suggesting that the predicted increased nucleosome binding preferences in regulatory regions are relevant in vivo. Given these various observations it is evident that other mechanisms must exist to ensure TF access to DNA in regulatory regions that are occupied by nucleosomes. One model of TF binding to nucleosomal DNA that does not depend on external factors is based on in vitro observations that compacted DNA can undergo spontaneous transitions to more open states, allowing for brief windows of opportunity for TF access [146–148]. These movements can affect relatively small regions of DNA near the nucleosome entry sites, a process referred to as "nucleosome breathing", or involve the unwinding of DNA over longer stretches [147, 149]. The increased accessibility of DNA at nucleosome entry sites is consistent with observations that TF binding sites are, on average, enriched at these locations in vivo [150–152]. Given the need to prevent cryptic transcription initiation, the thermodynamic balance in cells is likely such that individual TF binding events are not sufficient to prevent rapid rewrapping of nucleosomal DNA; however, cooperative binding of multiple TFs may overcome this barrier. Polach and Widom proposed that the binding of one TF could lead to further unwinding of the DNA on a nucleosome, enabling other factors to bind to nearby sites in a stepwise process that could ultimately result in a stable TF-DNA complex [153] (Fig. 11.3b). This cooperative model of TF access to nucleosomal DNA has two major additional benefits. First, it enables TFs to interact with each other without direct protein-protein contacts, creating new opportunities for coregulated gene expression. Second, the requirement for multiple closely spaced TF binding sites ensures regulatory site specificity. Cooperative binding of TFs to nucleosomal DNA has been demonstrated both in vitro [154] and in vivo [154–157], though it remains difficult to assess how widespread this mode of regulation is across the genome.

There is also evidence that TFs can interact with DNA in a manner that involves additional direct contacts with nucleosomes. For example, FOXA1 (HNF3A) binds more strongly to nucleosomal DNA than to naked DNA [158]. The source of this unique behaviour can be traced to the protein structure of the FoxA family members. FOXA1-3 contain a C-terminal domain that interacts with the core histones H3 and H4, as well as a winged helix N-terminal *forkhead* DNA binding domain that structurally resembles that of linker histone H1 [159]. In stark contrast to H1 linker histones, which are known for their ability to stabilize nucleosomes and higher order chromatin structures [160, 161], FoxA factors have intrinsic chromatin opening activity [159, 162]. Interestingly, this activity does not require the action of CMs such as SWI/SNF. Because of their ability to open condensed chromatin, FoxA proteins have been proposed to function as "pioneer" TFs that facilitate the binding of other factors [159]. A similar pioneer function has also been described for the RAR and RXR members of the nuclear receptor family, due to their ability to bind a highly compacted chromatin fibre containing a *PEPCK* promoter in an in vitro system that recaptured the chromatin dynamics observed at this promoter in vivo [163]. In this system, the action of the RAR/RXR heterodimer together with CMs was required to disrupt the chromatin for subsequent binding of nuclear factor 1 (NF1), an essential coregulator for transcriptional activation of *PEPCK*. The requirement for additional coregulators in transcriptional activation by both FoxA and RAR/RXR may be essential to ensure that their actions do not result in spurious transcription at non-specific sites in the genome. In the case of FoxA, methylation patterns associated with repressive or active chromatin domains also further guide recruitment to specific sites [164].

Other TFs that are able to access condensed chromatin include the CAAT-box/enhancer binding protein (C/EBP), though its pioneering role may be limited to a subset of genes [165]. In yeast, the Reb1 and Abf1 TFs can clearly function as pioneers as well, as evidenced by their aforementioned ability to direct the formation of NDRs [91, 141]. Finally, Gal4 upstream activating sequences (UAS) are able form mini-promoters regardless of their location in the genome [166], indicating that Gal4 binding can also disrupt chromatin. The Gal4 UAS used in this study contained multiple Gal4 binding sites, suggesting cooperative binding as a possible mechanism underlying this effect. Alternatively, Gal4 access to nucleosomal DNA can also be aided by the actions of CMs in a manner that does not involve displacing nucleosomes away from binding sites, as it was recently shown that the RSC complex can envelop and partially unwind a nucleosome in the *GAL1/GAL10* promoter, with RSC essentially "presenting" this element for Gal4 binding [63] (Fig. 11.3c).

11.5 A Dynamic Regulatory Role for Chromatin

Up to this point, the relationship between TFs and chromatin has mainly been explored in terms of how TFs overcome the chromatin barrier to access DNA and facilitate further remodelling. However, the involvement of chromatin in gene expression goes beyond merely forming a passive impediment to TF binding. Indeed, there are many indications that CMs are causative for gene expression outputs, so presumably they must be both regulated and regulatory. In the remainder of this chapter I will examine some of the other roles of chromatin remodelling, such as effecting transcriptional repression and controlling the accessibility and activity of regulatory regions, as well as establishing higher-order chromatin organization. In all these cases the role of TFs will be highlighted in particular.

CMs are essential coregulators in TF-mediated repression of many target genes. A large number of these coregulators belong to a family of histone deacetylases (HDACs) [167, 168], which catalyze the removal of acetyl groups that are closely associated with a relaxed chromatin structure. Accordingly, they prevent initiation by maintaining chromatin in a condensed state that is inaccessible to the transcription machinery. Some of the effects of HDACs may also be mediated by deacetylation of proteins other than histones, such as TFs [169]. Like their HAT counterparts, HDACs typically operate as part of larger corepressor complexes that include other chromatin binding or remodelling activities, as has been described for the NURD [81] and NCoR (nuclear receptor corepressor) complexes [168, 170]. The importance of HDACs in transcriptional repression is reflected in the size of their family, which includes as many as 6 different members in yeast and 18 in human, distributed over four main classes [171]. In addition to HDACs, other CMs such as ATPase nucleosome remodelers have also been implicated in the formation of repressive chromatin structures. For example, the ISW2 complex can be recruited to a large variety of promoters by the Ume6 repressor in budding yeast, where it establishes a repressive chromatin environment as evidenced by decreased nuclease sensitivity [172]. SWI/SNF remodelers can also effect transcriptional repression, either directly [173–175], or as part of larger corepressor complexes that include deacetylase activities [81, 170, 176]. In contrast to HDACs, the mechanisms by which ATPase remodelers act to repress transcription are less well understood, but presumably involve chromatin compaction [172, 173] and/or the repositioning of nucleosomes to block important TF binding sites [177].

By condensing chromatin at promoters of repressed genes, CMs can place important restrictions on the actions of TFs, as illustrated by the effects of the Tup1-Cyc8 corepressor on Rap1-mediated gene activation in budding yeast [178]. The Tup1-Cyc8 complex was one of the first corepressors to be identified [179] and is targeted to promoters by a variety of sequence-specific TFs [180–183] where it recruits HDACs and the Isw2 remodeler complex to induce chromatin condensation [184, 185]. Among the Tup1-Cyc8 targets are promoters of genes that are bound by Rap1 in low- but not high-glucose conditions, despite the fact that Rap1 directs the expression of other genes encoding glycolytic enzymes and ribosomal protein subunits when glucose is present [186, 187]. The increased number of Rap1 targets in low-glucose is even more surprising given that global Rap1 levels actually decrease during a shift to low glucose medium [178]. The contradictory behaviour of Rap1 binding was explained by the actions of Tup1-Cyc8, which prevent Rap1 binding to low-glucose specific genes when glucose is present. The Tup1-Cyc8-mediated promoter compaction is only released upon glucose depletion, presumably through a mechanism that involves the release or inactivation of the TFs responsible for recruiting Tup1-Cyc8, allowing Rap1 to bind [178]. This example shows that chromatin remodelling can provide an additional level of regulation of gene expression by preventing activators from recognizing their binding sites in target promoters.

An unexpected finding has been that the actions of chromatin-targeting corepressors are not just limited to transcriptionally silent regions. Genome-wide ChIP experiments have revealed that HDACs are also associated with active promoters [188, 189]. Even more surprising, the degree of HDAC recruitment was positively correlated with transcription levels. To explain this paradox, it was proposed that the presence of HDACs at active promoters was needed to reset the chromatin state between subsequent rounds of initiation [189, 190], which suggests that histone acetylation – like TF and nucleosome interactions – may be inherently transient. Indeed, the dynamic nature of TFs interactions with DNA in vivo may well be directly connected to negative feedback from CMs. For example, the human glucocorticoid receptor can be actively removed from promoter templates by SWI/SNF remodelers [26, 191] and Rsc2 can speed up the release of Ace1 from non-specific binding sites in yeast [27]. Nevertheless, the presence of remodelling complexes associated with repression at active promoters does not necessarily have to be associated with returning these promoters to their basal state. The yeast SWI/SNF ATPase Mot1 is a global repressor known for its role in removing TBP from DNA [192], and like HDACs, its presence at promoters is positively correlated with transcript levels [193]. However, in this particular case it was shown that Mot1 can actually make a positive contribution to PIC assembly at active promoters by releasing a transcriptionally inert TBP complexed with the NC2 inhibitor, thereby allowing entry of free TBP and productive initiation [193].

The precise positioning of nucleosomes at promoters may also be important for establishing regulated gene expression, as illustrated by the actions of the RSC complex at the *CHA1* promoter in budding yeast. In uninduced conditions, RSC represses *CHA1* expression by placing a nucleosome over the TATA box, resulting in a decreased level of TBP binding [177, 194]. Crucially, in the absence of two key RSC components (Swh3 and Sth1), the expression levels of *CHA1* in uninduced cells are approximately equal to those observed in fully induced cells. Thus, the presence of an inhibitory nucleosome over binding motifs recognized by the basal transcription machinery is vital for maintaining activator-regulated expression of *CHA1*. Similar regulation mechanisms are likely far more widespread, given the aforementioned observation that yeast genes with variable expression levels tend to have increased nucleosome occupancy within their promoter regions, often overlapping TATA boxes [142]. Taken together, these various observations show that the complex interplay between chromatin, CMs and TFs affects all aspects of transcription regulation.

11.6 TFs and Higher Order Chromatin Organization

In addition to the localized organization at the level of individual regulatory regions, chromatin is also arranged into higher-order structures that can span broad regions and affect multiple genes. These domains typically share a common chromatin environment that is characterized by a specific signature of histone marks and associated proteins. Classic examples of such domains include the condensed heterochromatin regions found at telomeres and in the pericentric regions surrounding centromeres in most organisms, as well as the mating-type loci in yeasts [195]. The heterochromatin in these regions is characterized by the presence of heterochromatin protein 1 (HP1) [196], histone hypoacetylation and H3K9 methylation (H3K9me) [197]. The co-occurrence of these marks is no coincidence, as H3K9me serves as an anchor point for the chromodomain that is present in HP1 [75]. Homologues of HP1 have been identified in *Drosophila*, vertebrates and fission yeast and its loss invariably leads to defects in telomere and centromere function. Additional domains marked by HP1 and H3K9me have also been associated with silencing of a number of genes dispersed throughout the genome [198–200].

A second important type of chromatin domain involved in gene silencing is established by Polycomb group (PcG) proteins. PcG proteins were initially identified as key developmental regulators of the Hox gene cluster in *Drosophila* (Reviewed in [201]), and two main PcG protein complexes have since been characterized with distinct roles in silencing in plants, vertebrates and flies. Polycomb repressive complex 2 (PRC2) has histone modifier activity and trimethylates H3K27, a characteristic signature of PcG chromatin domains, which can span up to 100 kb [202–204]. This methylation mark can be read by PRC1, which possesses ubiquitination activity. The specific mechanisms underlying HP1 and PcG silencing have been discussed in great detail elsewhere [195, 205–207]. Here, I will use these two domain types to illustrate the role of TFs in establishing higher order chromatin structure.

Heterochromatin typically originates at specific nucleation sites from which chromatin condensation spreads along the chromatin fibre. At telomeres, pericentric regions and yeast mating type loci, these nucleation sites often consist of highly repetitive DNA elements [208–210]. Studies in fission yeast have shown that repeat-based silencing depends on transcription of the repetitive regions and RNAi pathways [211, 212], and similar mechanisms have since been found to operate in fly, plants and vertebrates (Reviewed in [213]). There are also many examples where silencing is nucleated by TF binding, however. In fission yeast, the Pcr1 and Atf1 TFs can bind a heptamer sequence in the REIII element at the mating-type locus [214] and recruit the Clr4 histone methylase, the HP1 homolog Swi6, and the histone deacetylase Clr3 silencing factors [215, 216]. Budding yeast lacks HP1 homologs, but possesses silent information regulator (SIR) proteins that perform similar functions and which can be recruited to telomeres and mating-type loci by the synergistic actions of Rap1, Abf1 and Orc1 [217]. In tetrapods (four-limbed vertebrates), a large family of kruppel-associated box domain zinc finger TFs (KRAB-ZF) has also been implicated in silencing. The KRAB domain that characterizes this family interacts with KRAB associated protein 1 (KAP1) [218, 219], which acts as a scaffold for several heterochromatin-associated proteins, including HP1 [220–222]. Synthetic TF constructs with KRAB domains have been shown to induce heterochromatin silencing over broad regions, up to 12 kb away from their binding site [223, 224]. Natural KRAB-ZF proteins have been linked to the autoregulation of large clusters of KRAB-ZF genes [199, 200], but given that KRAB domains are present in more than 200 human TFs, they likely play a much wider role in chromatin metabolism. The KRAB domain is also discussed in Chapters 4 and 12 of this volume.

In contrast to HP1-associated heterochromatin, the origins of Polycomb domains are less well understood. In *Drosophila*, silencing by PcG proteins is driven by Polycomb response elements (PREs), which contain binding sites for the

Pleiohomeotic (PHO) and PHO-like zinc finger TFs [225, 226], the only PcG proteins identified to date with DNA sequence specificity. The importance of PHO and PHO-like for PRE function is firmly established, as their disruption results in silencing defects at Hox genes [225, 227, 228] and a loss of PRC1 and PRC2 components [228]; however, PHO binding sites alone are insufficient to confer PRE-mediated silencing [225, 226, 229]. Many other TFs have been shown to bind PREs in *Drosophila*, including Pipsqueak, Zeste and GAGA factor (GAF) (Reviewed in [72]), but their role in silencing is unclear, given that null mutants for many of these genes do not show obvious PcG phenotypes. One possible explanation is that these TFs act synergistically at PREs, which is consistent with computational analyses that show that clusters of TF binding motifs – but not individual sites – can distinguish PRE from non-PRE sequences [230]. Redundancy between factors may explain why some null mutants do not show phenotypes.

Even less is known about PRC recruitment in vertebrates, where it has proved challenging to identify PREs because PcG proteins are often distributed over broad regions [202, 204, 231, 232]. A 3kb DNA fragment in the MafB gene region that possesses activities consistent with a PRE was recently identified in mouse [233]. This fragment, named PRE-*kr*, was shown to bind PcG proteins and contains conserved binding sites for the mammalian PHO homolog YY1, as well as GAGAG motifs that are known to be bound by GAF and Pipsqueak in *Drosophila*. Another PRE with conserved YY1 binding sites has since been characterized in the human *HOXD* cluster, and disruption of these sites negatively affected binding of the PRC1 component BMI1 [234]. The role of YY1 in PcG silencing is consistent with earlier observations that YY1 knockdown results in loss of recruitment of the PRC2 component Ezh2 and H3K27me [235], as well as with other studies that have shown that YY1 interacts with PcG components [236–238]. Taken together, these data suggest that at least some of the PcG-targeting mechanisms are conserved between flies and mammals. Nonetheless, other TFs such as the embryonic stem cell regulators OCT4 and NANOG may also be involved in targeting PcG proteins in mammals, based on their high degree of overlap with PcG proteins in ChIP studies [202, 231, 239]. Moreover, the discovery of the HOTAIR transcript, which targets PRC2 to the human *HOXD* locus, indicates that ncRNAs also play a role in directing Polycomb silencing [240]. Future studies will undoubtedly reveal whether this latter mechanism is more widespread.

Several mechanisms are believed to operate to expand chromatin domains beyond their initial nucleation sites (Reviewed in [241]). One model of spreading described for HP1 family members depends on a self-sustaining wave of silencing complex assembly, which is based in the ability of HP1 to bind both H3K9 methylated histones as well as the methyltransferase responsible for this modification (Fig. 11.4a) [75, 77, 242]. Starting at the nucleation site, H3K9 methylation of neighboring nucleosomes by HP1-recruited methyltransferases creates new HP1 binding sites, resulting in more HP1 binding and further propagation of the signal. A similar mechanism involving repeated cycles of deacetylation has also been

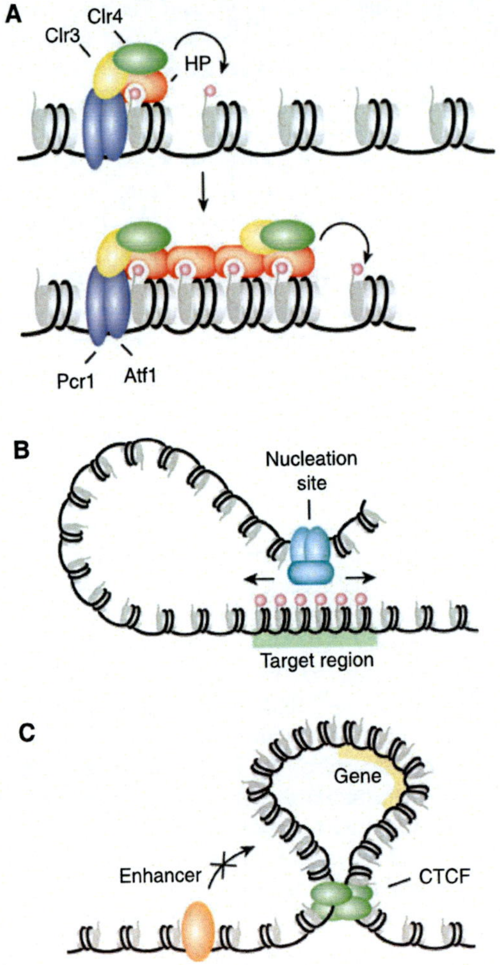

Fig. 11.4 Formation of chromatin domains. **a** Mechanism of spreading for HP1 heterochromatin at the *S. pombe* mating type locus from TF nucleation sites (Modified from [214]). Atf1 and Pcr1 binding results in the recruitment of the Clr3 histone deacetylase, which subsequently cooperates with heterochromatin proteins (HP) such as the HP1 homolog Swi6 to promote H3K9me of neighbouring nucleosomes. This creates additional HP1 binding sites, which form the basis for the spreading process. **b** Schematic representation of spreading of chromatin domains by looping interactions between the nucleation site and the surrounding DNA. **c** Model for the enhancer-blocker function of CTCF. Interactions between distant CTCF binding sites can form looped domains, thereby isolating genes from the actions of upstream enhancers

described for SIR proteins in budding yeast [243, 244]. Recurrent assembly cannot completely account for all observations of spreading from a nucleation site, however, as indicated by the following examples. In budding yeast, individual Rap1 and Abf1 binding sites that are unable to direct silencing independently can enhance the actions of a silencer that is 4 kb away [245], suggesting long–range interactions between these sites. Another signal spreading from a subtelomeric silencer was shown to "skip over" an active reporter gene flanked by subtelomeric antisilencing regions (STARS), but still affected a second distal reporter gene [246]. Finally,

ChIP studies of PcG proteins in *Drosophila* have revealed distribution patterns that seem inconsistent with a progressive spreading of Polycomb complexes. For example, while the H3K27me3 mark is consistently found in large domains [203, 247–250], the PRC1 components Ph and Psc and the PRC2 methyltransferase E(z) are concentrated in much smaller peaks [203, 247]. Currently, the most favoured model to explain these various observations involves folding of the DNA in a manner that allows nucleation sites to contact and modify the surrounding chromatin (Fig. 11.4b), and has been proposed to explain the difference in distribution patterns of PcG components and H3K27me3 [251]. Several cases of long–range interactions between PREs and distant regulatory sites have also been described, forming higher order chromatin loop configurations that may facilitate gene silencing across broad domains [252, 253]. The relationship of TFs to higher-order chromatin structure is described in more detail in Chapter 13.

Given that silencing can propagate autonomously along the chromatin fibre, and that distal regulatory elements such as PREs and enhancers can operate over large distances, how are their effects on one region of the genome kept from spilling over to nearby genes? The answer to this question lies in yet another group of regulatory elements called insulators [254–256], which possess one of two distinct characteristics: (1) they can block enhancers from activating genes when placed between the enhancer and the gene or (2) they can act as boundary elements to prevent the spread of the silencing effects of heterochromatin. These two activities are separate and measured in different assays, though many insulators can perform both functions in vivo, such as the 5′HS4 insulator in the chicken β-globin locus [257, 258]. Once again, TFs play a central role in establishing insulator regions, and at least five different insulator-binding TFs have been identified in *Drosophila* to date: ZW5, Su(Hw), dCTCF, BEAF, and GAGA (Reviewed in [259]). In contrast, most vertebrate insulators appear to depend on only a single TF, the CCCTC-binding factor (CTCF) [257]. CTCF is considered to mainly function as an enhancer blocker rather than as a boundary protein, as evidenced by the fact that it is dispensable for blocking the spread of heterochromatin at the chicken β-globin locus [260]. Instead, this latter function depends on the USF1 TF, which binds boundary elements in the 5′HS4 insulator as a heterodimer with USF2 [258, 261]. The USF1/USF2 heterodimer recruits HATs and the SET 7/9 methyltransferase, which establish a region of open chromatin that is thought to prevent the progression of silencing analogous to the manner in which firewalls prevent forest fires from spreading. In contrast, enhancer-blocking insulators such as those bound by Su(Hw) in *Drosophila* (Reviewed in [262]) or CTCF in vertebrates (Reviewed in [263]) have been suggested to operate by organizing chromatin into looped domains, isolating the genes contained inside from their distant regulatory elements (Fig. 11.4c). In addition, CTCF has also been implicated in anchoring DNA to the nuclear periphery, an area that is typically associated with a repressive chromatin environment, as it was found to be enriched at the boundaries of domains that are linked to the nuclear lamina [264].

11.7 Concluding Remarks

The complexity of chromatin–TF interactions is reflected in the considerable variability in initiation mechanisms for the few genes studied in great detail [83] suggesting that there are many routes leading to productive transcription. Indeed, considering that the requirement for coregulators at a single gene can vary depending on external conditions, and that promoters are typically unique in a genome, the number of transcriptional activation mechanisms may yet prove to be larger than the number of genes. Nonetheless, the number of possibilities is clearly not unlimited, since at any given regulatory region only a subset of TFs and their coregulators play a dominant role. Thus, it should be possible to build a catalogue of the proteins most commonly bound to these elements in specific cell types, and eventually decode the mechanisms that control gene expression. ChIP in combination with either microarrays or next-generation sequencing is currently the most widely used method for the identification of the proteins and histone modifications associated with DNA [265, 266]; however, this technique has several drawbacks. First, it can only identify the location of a handful of proteins at the same time, and second, it requires advance knowledge of the factor(s) to study. An alternative approach called proteomics of isolated chromatin segments (PICh) was recently developed that does not suffer from these limitations, and uses mass-spectrometry to detect proteins associated with a chromatin segment [267]. If this approach were to be applied to the large collections of regulatory regions that are now being identified in genome-wide nuclease hypersensitivity assays such as those undertaken by the ENCODE and modENCODE consortia [268], it might greatly expand our knowledge of the interplay between TFs and chromatin at these locations.

Simply knowing which proteins are associated with a given genomic region will not be enough to understand how these proteins operate to regulate transcription, since they generally do not work in isolation. Protein–protein interaction maps should also greatly facilitate mapping gene regulatory mechanisms, since they reveal interactions between and among TFs and CMs [269]. Moreover, maps of long range interactions between regulatory regions are needed to understand the interplay between promoters, enhancers, silencers and insulators. The advent of new technologies such as the numerous derivatives of chromosome conformation capture (3C) [270, 271] now make such approaches possible at a genome-wide level (see Chapter 13). Finally, detailed knowledge of the affinities of TFs and their coregulators for DNA, as well as for their protein binding partners will also be essential. This will require the application of techniques that can assess both the intrinsic DNA sequence specificities of TFs (see Chapter 8) and the binding kinetics of proteins, in a high-throughput and quantitative fashion. Potential strategies for the latter have been outlined by Segal and Widom [272]. Together, these various types of data will provide valuable insight into the ground rules that govern the interactions between DNA, chromatin and the transcription machinery. These rules can then form the basis for in silico modeling of these processes, which will be essential if we are to fully understand the intricate relationships between TFs and chromatin.

Glossary

Chromatin The combination of DNA and accessory proteins, such as histones, that together constitute chromosomes.

Transcriptional coregulator An accessory factor recruited by transcription factors to modulate gene expression. Cofactors typically lack intrinsic DNA binding specificity and rely on transcription factors for targeting. Most cofactors excert their effects by locally modifying chromatin structure.

Transcriptional coactivator A coregulator that positively affects gene expression.

Transcriptional corepressor A coregulator that negatively affects gene expression.

Chromatin modifiers Proteins or protein complexes that can effect changes in chromatin structure by covalently modifying histones or moving nucleosomes. In this chapter the term chromatin modifier is used generally to refer to histone modifiers and ATPase nucleosome remodelers.

Histone modifiers The enzymes responsible for adding or removing covalent modifications on histones, the majority of which are are found on the flexible histone tails. Some histone modifiers, such as HDACs and HATs can also have non-histone targets.

ATPase nucleosome remodelers Protein complexes that use the energy generated by ATP hydrolysis to alter nucleosome-DNA interactions and displace nucleosomes.

Heterochromatin A tightly packed form of chromatin where DNA is typically rendered inaccessible to the transcriptional machinery. Different types of heterochromatin are associated with distinct chromatin marks, such as HP1 heterochromatin (HP1 binding and H3K9me) or Polycomb domains (H3K27me).

Euchromatin An open chromatin conformation in which DNA is easily accessible. This type of chromatin is often, but not exclusively, associated with active transcription.

Histone code Distinct patterns of histone modifications are believed to constitute a code that is used to direct specific activities on DNA, such as during transcriptional silencing or during the various stages of the transcriptional cycle. For example, the initiation, elongation and termination of transcription are each associated with different patterns of histone modifications that are believed to contribute to the recruitment and regulation of the proteins required in each stage.

Epigenetics Inherited changes in phenotypes or expression profiles that are not due to changes in the underlying DNA sequence. Examples of epigenetic modifications include DNA methylation and covalent histone modifications, which play an important role in a variety of processes, including cell differentiation, X chromosome inactivation and imprinting.

Polycomb-group proteins A family of proteins, initially discovered in *Drosophila*, that are involved in epigenetic silencing of genes by inducing a repressive chromatin structure. Polycomb group proteins are predominantly found as part of two main protein complexes: Polycomb-group Repressive Complex 1 and 2 (PRC1 and PRC2).

Nucleosome The basic building block of chromatin, consisting of ~147 bp of DNA wrapped around an octamer of two of each of the histones H2A, H2B, H3 and H4.

Effector domains The domains in transcription factors that are responsible for mediating their effects on gene expression. These effects can be activating or inhibitory and involve a variety of mechanisms, including recruitment of chromatin modifiers, or interactions with components of the basal transcriptional machinery and other transcription factors.

DNA binding domain A protein domain with DNA binding activity. In the case of transcription factors, these domains typically possess specificity affinity for a limited number of DNA sequences.

Enhancer A DNA element bound by transcription factors that can operate over long distances (up to thousands of basepairs) to stimulate transcription of its target gene(s). Enhancers are thought to operate through looping interactions with promoter regions. In addition to their distance to genes, enhancers can also be distinguished from promoters by a unique chromatin profile. Though most enhancers act in cis, they can also be located on different chromosomes.

Silencer Like enhancers, silencers are DNA elements that can be located far away from the genes they control, but their effect on gene expression is negative. Silencers can also act as nucleation sites for repressive chromatin domains.

Insulator A DNA element that either prevents an enhancer from activating target genes, or acts as a boundary element to delineate different chromatin domains. Insulators are distinct from from silencer regions in that an insulator needs to be located between an enhancer and a gene to affect expression, while silencers can typically operate in any orientation relative to a gene.

Chromatin domain A relatively uniform region of chromatin characterized by distinct histone and/or DNA modifications. Examples include Polycomb domains as well as telomeric- and pericentromeric heterochromatin.

Preinitiation complex Large complex of proteins required for successful transcription initiation by RNA Polymerase II. Major components include the basal transcription factors TFIIA, TFIIB, TFIID, TFIIE, TFIIF, and TFIIH. The preinitiation complex plays a role in positioning polymerase and melting the DNA so that it is properly configured to fit in the active site. Positioning is aided by motifs recognized by the general transcription factors.

CpG island Sequence elements rich in CG dinucleotides that are found at a large number of mammalian promoters.

General transcription factors Transcription factors that are universally required for RNA polymerase II transcription. Most GTFs are part of the preinitiation complex.

References

1. Lee TI, Young RA (2000) Transcription of eukaryotic protein-coding genes. Annu Rev Genet 34:77–137
2. van Hijum SA, Medema MH, Kuipers OP (2009) Mechanisms and evolution of control logic in prokaryotic transcriptional regulation. Microbiol Mol Biol Rev 73 (3):481–509
3. Kornberg RD, Thomas JO (1974) Chromatin structure; oligomers of the histones. Science 184 (139):865–868
4. Richmond TJ, Finch JT, Rushton B, Rhodes D, Klug A (1984) Structure of the nucleosome core particle at 7 A resolution. Nature 311 (5986):532–537
5. Knezetic JA, Luse DS (1986) The presence of nucleosomes on a DNA template prevents initiation by RNA polymerase II in vitro. Cell 45 (1):95–104
6. Lorch Y, LaPointe JW, Kornberg RD (1987) Nucleosomes inhibit the initiation of transcription but allow chain elongation with the displacement of histones. Cell 49 (2):203–210
7. Han M, Grunstein M (1988) Nucleosome loss activates yeast downstream promoters in vivo. Cell 55 (6):1137–1145
8. Hager GL, McNally JG, Misteli T (2009) Transcription dynamics. Mol Cell 35 (6): 741–753
9. Segal E, Widom J (2009a) What controls nucleosome positions? Trends Genet 25 (8): 335–343
10. Wu C, Bingham PM, Livak KJ, Holmgren R, Elgin SC (1979) The chromatin structure of specific genes: I. Evidence for higher order domains of defined DNA sequence. Cell 16 (4):797–806
11. Giresi PG, Kim J, McDaniell RM, Iyer VR, Lieb JD (2007) FAIRE (Formaldehyde-Assisted Isolation of Regulatory Elements) isolates active regulatory elements from human chromatin. Genome Res 17 (6):877–885
12. Lee W, Tillo D, Bray N, Morse RH, Davis RW, Hughes TR, Nislow C (2007) A high-resolution atlas of nucleosome occupancy in yeast. Nat Genet 39 (10):1235–1244
13. Mavrich TN, Jiang C, Ioshikhes IP, Li X, Venters BJ, Zanton SJ, Tomsho LP, Qi J, Glaser RL, Schuster SC, Gilmour DS, Albert I, Pugh BF (2008a) Nucleosome organization in the *Drosophila* genome. Nature 453 (7193):358–362
14. Schones DE, Cui K, Cuddapah S, Roh TY, Barski A, Wang Z, Wei G, Zhao K (2008) Dynamic regulation of nucleosome positioning in the human genome. Cell 132 (5):887–898
15. Yuan GC, Liu YJ, Dion MF, Slack MD, Wu LF, Altschuler SJ, Rando OJ (2005) Genome-scale identification of nucleosome positions in *S. cerevisiae*. Science 309 (5734):626–630
16. Boyle AP, Davis S, Shulha HP, Meltzer P, Margulies EH, Weng Z, Furey TS, Crawford GE (2008) High-resolution mapping and characterization of open chromatin across the genome. Cell 132 (2):311–322
17. Hesselberth JR, Chen X, Zhang Z, Sabo PJ, Sandstrom R, Reynolds AP, Thurman RE, Neph S, Kuehn MS, Noble WS, Fields S, Stamatoyannopoulos JA (2009) Global mapping of protein-DNA interactions in vivo by digital genomic footprinting. Nat Methods 6 (4):283–289

18. Barski A, Cuddapah S, Cui K, Roh TY, Schones DE, Wang Z, Wei G, Chepelev I, Zhao K (2007) High-resolution profiling of histone methylations in the human genome. Cell 129 (4):823–837
19. Harbison CT, Gordon DB, Lee TI, Rinaldi NJ, Macisaac KD, Danford TW, Hannett NM, Tagne JB, Reynolds DB, Yoo J, Jennings EG, Zeitlinger J, Pokholok DK, Kellis M, Rolfe PA, Takusagawa KT, Lander ES, Gifford DK, Fraenkel E, Young RA (2004) Transcriptional regulatory code of a eukaryotic genome. Nature 431 (7004):99–104
20. Heintzman ND, Stuart RK, Hon G, Fu Y, Ching CW, Hawkins RD, Barrera LO, Van Calcar S, Qu C, Ching KA, Wang W, Weng Z, Green RD, Crawford GE, Ren B (2007) Distinct and predictive chromatin signatures of transcriptional promoters and enhancers in the human genome. Nat Genet 39 (3):311–318
21. Kim TH, Barrera LO, Zheng M, Qu C, Singer MA, Richmond TA, Wu Y, Green RD, Ren B (2005) A high-resolution map of active promoters in the human genome. Nature 436 (7052):876–880
22. Pokholok DK, Harbison CT, Levine S, Cole M, Hannett NM, Lee TI, Bell GW, Walker K, Rolfe PA, Herbolsheimer E, Zeitlinger J, Lewitter F, Gifford DK, Young RA (2005) Genome-wide map of nucleosome acetylation and methylation in yeast. Cell 122 (4): 517–527
23. Wang Z, Zang C, Rosenfeld JA, Schones DE, Barski A, Cuddapah S, Cui K, Roh TY, Peng W, Zhang MQ, Zhao K (2008) Combinatorial patterns of histone acetylations and methylations in the human genome. Nat Genet 40 (7):897–903
24. Dion MF, Kaplan T, Kim M, Buratowski S, Friedman N, Rando OJ (2007) Dynamics of replication-independent histone turnover in budding yeast. Science 315 (5817):1405–1408
25. Linger J, Tyler JK (2006) Global replication-independent histone H4 exchange in budding yeast. Eukaryot Cell 5 (10):1780–1787
26. Fletcher TM, Xiao N, Mautino G, Baumann CT, Wolford R, Warren BS, Hager GL (2002) ATP-dependent mobilization of the glucocorticoid receptor during chromatin remodeling. Mol Cell Biol 22 (10):3255–3263
27. Karpova TS, Chen TY, Sprague BL, McNally JG (2004) Dynamic interactions of a transcription factor with DNA are accelerated by a chromatin remodeller. EMBO Rep 5 (11):1064–1070
28. McNally JG, Muller WG, Walker D, Wolford R, Hager GL (2000) The glucocorticoid receptor: rapid exchange with regulatory sites in living cells. Science 287 (5456):1262–1265
29. Sharp ZD, Mancini MG, Hinojos CA, Dai F, Berno V, Szafran AT, Smith KP, Lele TP, Ingber DE, Mancini MA (2006) Estrogen-receptor-alpha exchange and chromatin dynamics are ligand- and domain-dependent. J Cell Sci 119 (Pt 19):4101–4116
30. Farnham PJ (2009) Insights from genomic profiling of transcription factors. Nat Rev Genet 10 (9):605–616
31. Li B, Carey M, Workman JL (2007) The role of chromatin during transcription. Cell 128 (4):707–719
32. Venters BJ, Pugh BF (2009) How eukaryotic genes are transcribed. Crit Rev Biochem Mol Biol 44 (2-3):117–141
33. Weake VM, Workman JL (2010) Inducible gene expression: diverse regulatory mechanisms. Nat Rev Genet 11 (6):426–437
34. Kouzarides T (2007) Chromatin modifications and their function. Cell 128 (4):693–705
35. Reid G, Gallais R, Metivier R (2009) Marking time: the dynamic role of chromatin and covalent modification in transcription. Int J Biochem Cell Biol 41 (1):155–163
36. Strahl BD, Allis CD (2000) The language of covalent histone modifications. Nature 403 (6765):41–45
37. Rice JC, Allis CD (2001) Histone methylation versus histone acetylation: new insights into epigenetic regulation. Curr Opin Cell Biol 13 (3):263–273
38. Kurdistani SK, Tavazoie S, Grunstein M (2004) Mapping global histone acetylation patterns to gene expression. Cell 117 (6):721–733

39. Liu CL, Kaplan T, Kim M, Buratowski S, Schreiber SL, Friedman N, Rando OJ (2005) Single-nucleosome mapping of histone modifications in *S. cerevisiae*. PLoS Biol 3 (10):e328
40. Roh TY, Ngau WC, Cui K, Landsman D, Zhao K (2004) High-resolution genome-wide mapping of histone modifications. Nat Biotechnol 22 (8):1013–1016
41. Sinha I, Wiren M, Ekwall K (2006) Genome-wide patterns of histone modifications in fission yeast. Chromosome Res 14 (1):95–105
42. Bernstein BE, Kamal M, Lindblad-Toh K, Bekiranov S, Bailey DK, Huebert DJ, McMahon S, Karlsson EK, Kulbokas EJ, 3rd, Gingeras TR, Schreiber SL, Lander ES (2005) Genomic maps and comparative analysis of histone modifications in human and mouse. Cell 120 (2):169–181
43. Liang G, Lin JC, Wei V, Yoo C, Cheng JC, Nguyen CT, Weisenberger DJ, Egger G, Takai D, Gonzales FA, Jones PA (2004) Distinct localization of histone H3 acetylation and H3-K4 methylation to the transcription start sites in the human genome. Proc Natl Acad Sci U S A 101 (19):7357–7362
44. Roh TY, Cuddapah S, Zhao K (2005) Active chromatin domains are defined by acetylation islands revealed by genome-wide mapping. Genes Dev 19 (5):542–552
45. Schubeler D, MacAlpine DM, Scalzo D, Wirbelauer C, Kooperberg C, van Leeuwen F, Gottschling DE, O'Neill LP, Turner BM, Delrow J, Bell SP, Groudine M (2004) The histone modification pattern of active genes revealed through genome-wide chromatin analysis of a higher eukaryote. Genes Dev 18 (11):1263–1271
46. Shahbazian MD, Grunstein M (2007) Functions of site-specific histone acetylation and deacetylation. Annu Rev Biochem 76:75–100
47. Wiren M, Silverstein RA, Sinha I, Walfridsson J, Lee HM, Laurenson P, Pillus L, Robyr D, Grunstein M, Ekwall K (2005) Genomewide analysis of nucleosome density histone acetylation and HDAC function in fission yeast. Embo J 24 (16):2906–2918
48. Bannister AJ, Kouzarides T (2005) Reversing histone methylation. Nature 436 (7054):1103–1106
49. Kouzarides T (2000) Acetylation: a regulatory modification to rival phosphorylation? Embo J 19 (6):1176–1179
50. Spange S, Wagner T, Heinzel T, Kramer OH (2009) Acetylation of non-histone proteins modulates cellular signalling at multiple levels. Int J Biochem Cell Biol 41 (1):185–198
51. Yang XJ (2004) The diverse superfamily of lysine acetyltransferases and their roles in leukemia and other diseases. Nucleic Acids Res 32 (3):959–976
52. Ge Z, Liu C, Bjorkholm M, Gruber A, Xu D (2006) Mitogen-activated protein kinase cascade-mediated histone H3 phosphorylation is critical for telomerase reverse transcriptase expression/telomerase activation induced by proliferation. Mol Cell Biol 26 (1):230–237
53. Mahadevan LC, Willis AC, Barratt MJ (1991) Rapid histone H3 phosphorylation in response to growth factors, phorbol esters, okadaic acid, and protein synthesis inhibitors. Cell 65 (5):775–783
54. Nathan D, Ingvarsdottir K, Sterner DE, Bylebyl GR, Dokmanovic M, Dorsey JA, Whelan KA, Krsmanovic M, Lane WS, Meluh PB, Johnson ES, Berger SL (2006) Histone sumoylation is a negative regulator in *Saccharomyces cerevisiae* and shows dynamic interplay with positive-acting histone modifications. Genes Dev 20 (8):966–976
55. Van Rechem C, Boulay G, Pinte S, Stankovic-Valentin N, Guerardel C, Leprince D (2010) Differential regulation of HIC1 target genes by CtBP and NuRD, via an acetylation/SUMOylation switch, in quiescent versus proliferating cells. Mol Cell Biol 30 (16):4045–4059
56. Chandrasekharan MB, Huang F, Sun ZW (2010) Histone H2B ubiquitination and beyond: Regulation of nucleosome stability, chromatin dynamics and the trans-histone H3 methylation. Epigenetics 5 (6)
57. Wang H, Zhai L, Xu J, Joo HY, Jackson S, Erdjument-Bromage H, Tempst P, Xiong Y, Zhang Y (2006) Histone H3 and H4 ubiquitylation by the CUL4-DDB-ROC1 ubiquitin ligase facilitates cellular response to DNA damage. Mol Cell 22 (3):383–394

58. Zhu B, Zheng Y, Pham AD, Mandal SS, Erdjument-Bromage H, Tempst P, Reinberg D (2005) Monoubiquitination of human histone H2B: the factors involved and their roles in HOX gene regulation. Mol Cell 20 (4):601–611
59. Cheung P, Tanner KG, Cheung WL, Sassone-Corsi P, Denu JM, Allis CD (2000) Synergistic coupling of histone H3 phosphorylation and acetylation in response to epidermal growth factor stimulation. Mol Cell 5 (6):905–915
60. Lo WS, Trievel RC, Rojas JR, Duggan L, Hsu JY, Allis CD, Marmorstein R, Berger SL (2000) Phosphorylation of serine 10 in histone H3 is functionally linked in vitro and in vivo to Gcn5-mediated acetylation at lysine 14. Mol Cell 5 (6):917–926
61. Rea S, Eisenhaber F, O'Carroll D, Strahl BD, Sun ZW, Schmid M, Opravil S, Mechtler K, Ponting CP, Allis CD, Jenuwein T (2000) Regulation of chromatin structure by site-specific histone H3 methyltransferases. Nature 406 (6796):593–599
62. Clapier CR, Cairns BR (2009) The biology of chromatin remodeling complexes. Annu Rev Biochem 78:273–304
63. Floer M, Wang X, Prabhu V, Berrozpe G, Narayan S, Spagna D, Alvarez D, Kendall J, Krasnitz A, Stepansky A, Hicks J, Bryant GO, Ptashne M (2010) A RSC/nucleosome complex determines chromatin architecture and facilitates activator binding. Cell 141 (3):407–418
64. Narlikar GJ, Phelan ML, Kingston RE (2001) Generation and interconversion of multiple distinct nucleosomal states as a mechanism for catalyzing chromatin fluidity. Mol Cell 8 (6):1219–1230
65. Dechassa ML, Sabri A, Pondugula S, Kassabov SR, Chatterjee N, Kladde MP, Bartholomew B (2010) SWI/SNF has intrinsic nucleosome disassembly activity that is dependent on adjacent nucleosomes. Mol Cell 38 (4):590–602
66. Lorch Y, Maier-Davis B, Kornberg RD (2006) Chromatin remodeling by nucleosome disassembly in vitro. Proc Natl Acad Sci U S A 103 (9):3090–3093
67. Vicent GP, Nacht AS, Smith CL, Peterson CL, Dimitrov S, Beato M (2004) DNA instructed displacement of histones H2A and H2B at an inducible promoter. Mol Cell 16 (3): 439–452
68. Bruno M, Flaus A, Stockdale C, Rencurel C, Ferreira H, Owen-Hughes T (2003) Histone H2A/H2B dimer exchange by ATP-dependent chromatin remodeling activities. Mol Cell 12 (6):1599–1606
69. Mizuguchi G, Shen X, Landry J, Wu WH, Sen S, Wu C (2004) ATP-driven exchange of histone H2AZ variant catalyzed by SWR1 chromatin remodeling complex. Science 303 (5656):343–348
70. Bao Y, Shen X (2007) SnapShot: chromatin remodeling complexes. Cell 129 (3):632
71. Eisen JA, Sweder KS, Hanawalt PC (1995) Evolution of the SNF2 family of proteins: subfamilies with distinct sequences and functions. Nucleic Acids Res 23 (14):2715–2723
72. Bottomley MJ (2004) Structures of protein domains that create or recognize histone modifications. EMBO Rep 5 (5):464–469
73. de la Cruz X, Lois S, Sanchez-Molina S, Martinez-Balbas MA (2005) Do protein motifs read the histone code? Bioessays 27 (2):164–175
74. Hassan AH, Prochasson P, Neely KE, Galasinski SC, Chandy M, Carrozza MJ, Workman JL (2002) Function and selectivity of bromodomains in anchoring chromatin-modifying complexes to promoter nucleosomes. Cell 111 (3):369–379
75. Bannister AJ, Zegerman P, Partridge JF, Miska EA, Thomas JO, Allshire RC, Kouzarides T (2001) Selective recognition of methylated lysine 9 on histone H3 by the HP1 chromo domain. Nature 410 (6824):120–124
76. Flanagan JF, Mi LZ, Chruszcz M, Cymborowski M, Clines KL, Kim Y, Minor W, Rastinejad F, Khorasanizadeh S (2005) Double chromodomains cooperate to recognize the methylated histone H3 tail. Nature 438 (7071):1181–1185
77. Lachner M, O'Carroll D, Rea S, Mechtler K, Jenuwein T (2001) Methylation of histone H3 lysine 9 creates a binding site for HP1 proteins. Nature 410 (6824):116–120

78. Sims RJ, 3rd, Chen CF, Santos-Rosa H, Kouzarides T, Patel SS, Reinberg D (2005) Human but not yeast CHD1 binds directly and selectively to histone H3 methylated at lysine 4 via its tandem chromodomains. J Biol Chem 280 (51):41789–41792
79. Boyer LA, Latek RR, Peterson CL (2004) The SANT domain: a unique histone-tail-binding module? Nat Rev Mol Cell Biol 5 (2):158–163
80. Ho L, Crabtree GR (2010) Chromatin remodelling during development. Nature 463 (7280):474–484
81. Bowen NJ, Fujita N, Kajita M, Wade PA (2004) Mi-2/NuRD: multiple complexes for many purposes. Biochim Biophys Acta 1677 (1-3):52–57
82. Dirscherl SS, Krebs JE (2004) Functional diversity of ISWI complexes. Biochem Cell Biol 82 (4):482–489
83. Biddick R, Young ET (2009) The disorderly study of ordered recruitment. Yeast 26 (4):205–220
84. Dyson HJ, Wright PE (2005) Intrinsically unstructured proteins and their functions. Nat Rev Mol Cell Biol 6 (3):197–208
85. Liu J, Perumal NB, Oldfield CJ, Su EW, Uversky VN, Dunker AK (2006a) Intrinsic disorder in transcription factors. Biochemistry 45 (22):6873–6888
86. Dunker AK, Lawson JD, Brown CJ, Williams RM, Romero P, Oh JS, Oldfield CJ, Campen AM, Ratliff CM, Hipps KW, Ausio J, Nissen MS, Reeves R, Kang C, Kissinger CR, Bailey RW, Griswold MD, Chiu W, Garner EC, Obradovic Z (2001) Intrinsically disordered protein. J Mol Graph Model 19 (1):26–59
87. Hassan AH, Neely KE, Vignali M, Reese JC, Workman JL (2001) Promoter targeting of chromatin-modifying complexes. Front Biosci 6:D1054–1064
88. Narlikar GJ, Fan HY, Kingston RE (2002) Cooperation between complexes that regulate chromatin structure and transcription. Cell 108 (4):475–487
89. Peterson CL, Workman JL (2000) Promoter targeting and chromatin remodeling by the SWI/SNF complex. Curr Opin Genet Dev 10 (2):187–192
90. Cairns BR, Lorch Y, Li Y, Zhang M, Lacomis L, Erdjument-Bromage H, Tempst P, Du J, Laurent B, Kornberg RD (1996) RSC, an essential, abundant chromatin-remodeling complex. Cell 87 (7):1249–1260
91. Badis G, Chan ET, van Bakel H, Pena-Castillo L, Tillo D, Tsui K, Carlson CD, Gossett AJ, Hasinoff MJ, Warren CL, Gebbia M, Talukder S, Yang A, Mnaimneh S, Terterov D, Coburn D, Li Yeo A, Yeo ZX, Clarke ND, Lieb JD, Ansari AZ, Nislow C, Hughes TR (2008) A library of yeast transcription factor motifs reveals a widespread function for Rsc3 in targeting nucleosome exclusion at promoters. Mol Cell 32 (6):878–887
92. Hogan CJ, Aligianni S, Durand-Dubief M, Persson J, Will WR, Webster J, Wheeler L, Mathews CK, Elderkin S, Oxley D, Ekwall K, Varga-Weisz PD (2009) Fission yeast Iec1-ino80-mediated nucleosome eviction regulates nucleotide and phosphate metabolism. Mol Cell Biol 30 (3):657–674
93. Mohrmann L, Verrijzer CP (2005) Composition and functional specificity of SWI2/SNF2 class chromatin remodeling complexes. Biochim Biophys Acta 1681 (2-3):59–73
94. Patsialou A, Wilsker D, Moran E (2005) DNA-binding properties of ARID family proteins. Nucleic Acids Res 33 (1):66–80
95. Thomas JO, Travers AA (2001) HMG1 and 2, and related 'architectural' DNA-binding proteins. Trends Biochem Sci 26 (3):167–174
96. Wilson B, Erdjument-Bromage H, Tempst P, Cairns BR (2006) The RSC chromatin remodeling complex bears an essential fungal-specific protein module with broad functional roles. Genetics 172 (2):795–809
97. Trotter KW, Archer TK (2008) The BRG1 transcriptional coregulator. Nucl Recept Signal 6:e004
98. Kitagawa H, Fujiki R, Yoshimura K, Mezaki Y, Uematsu Y, Matsui D, Ogawa S, Unno K, Okubo M, Tokita A, Nakagawa T, Ito T, Ishimi Y, Nagasawa H, Matsumoto T, Yanagisawa J, Kato S (2003) The chromatin-remodeling complex WINAC targets a nuclear receptor to promoters and is impaired in Williams syndrome. Cell 113 (7):905–917

99. Xu W, Cho H, Kadam S, Banayo EM, Anderson S, Yates JR, 3rd, Emerson BM, Evans RM (2004) A methylation-mediator complex in hormone signaling. Genes Dev 18 (2):144–156
100. Lessard J, Wu JI, Ranish JA, Wan M, Winslow MM, Staahl BT, Wu H, Aebersold R, Graef IA, Crabtree GR (2007) An essential switch in subunit composition of a chromatin remodeling complex during neural development. Neuron 55 (2):201–215
101. Wu JI, Lessard J, Olave IA, Qiu Z, Ghosh A, Graef IA, Crabtree GR (2007) Regulation of dendritic development by neuron-specific chromatin remodeling complexes. Neuron 56 (1):94–108
102. Takeuchi JK, Bruneau BG (2009) Directed transdifferentiation of mouse mesoderm to heart tissue by defined factors. Nature 459 (7247):708–711
103. Bhaumik SR, Green MR (2001) SAGA is an essential in vivo target of the yeast acidic activator Gal4p. Genes Dev 15 (15):1935–1945
104. Bhaumik SR, Green MR (2002) Differential requirement of SAGA components for recruitment of TATA-box-binding protein to promoters in vivo. Mol Cell Biol 22 (21):7365–7371
105. Larschan E, Winston F (2001) The *S. cerevisiae* SAGA complex functions in vivo as a coactivator for transcriptional activation by Gal4. Genes Dev 15 (15):1946–1956
106. Chan HM, La Thangue NB (2001) p300/CBP proteins: HATs for transcriptional bridges and scaffolds. J Cell Sci 114 (Pt 13):2363–2373
107. Kalkhoven E (2004) CBP and p300: HATs for different occasions. Biochem Pharmacol 68 (6):1145–1155
108. Panne D, Maniatis T, Harrison SC (2007) An atomic model of the interferon-beta enhanceosome. Cell 129 (6):1111–1123
109. Merika M, Williams AJ, Chen G, Collins T, Thanos D (1998) Recruitment of CBP/p300 by the IFN beta enhanceosome is required for synergistic activation of transcription. Mol Cell 1 (2):277–287
110. Birney E, Stamatoyannopoulos JA, Dutta A, Guigo R, Gingeras TR, Margulies EH, Weng Z, Snyder M, Dermitzakis ET, Thurman RE, Kuehn MS, Taylor CM, Neph S, Koch CM, Asthana S, Malhotra A, Adzhubei I, Greenbaum JA, Andrews RM, Flicek P, Boyle PJ, Cao H, Carter NP, Clelland GK, Davis S, Day N, Dhami P, Dillon SC, Dorschner MO, Fiegler H, Giresi PG, Goldy J, Hawrylycz M, Haydock A, Humbert R, James KD, Johnson BE, Johnson EM, Frum TT, Rosenzweig ER, Karnani N, Lee K, Lefebvre GC, Navas PA, Neri F, Parker SC, Sabo PJ, Sandstrom R, Shafer A, Vetrie D, Weaver M, Wilcox S, Yu M, Collins FS, Dekker J, Lieb JD, Tullius TD, Crawford GE, Sunyaev S, Noble WS, Dunham I, Denoeud F, Reymond A, Kapranov P, Rozowsky J, Zheng D, Castelo R, Frankish A, Harrow J, Ghosh S, Sandelin A, Hofacker IL, Baertsch R, Keefe D, Dike S, Cheng J, Hirsch HA, Sekinger EA, Lagarde J, Abril JF, Shahab A, Flamm C, Fried C, Hackermuller J, Hertel J, Lindemeyer M, Missal K, Tanzer A, Washietl S, Korbel J, Emanuelsson O, Pedersen JS, Holroyd N, Taylor R, Swarbreck D, Matthews N, Dickson MC, Thomas DJ, Weirauch MT, Gilbert J, Drenkow J, Bell I, Zhao X, Srinivasan KG, Sung WK, Ooi HS, Chiu KP, Foissac S, Alioto T, Brent M, Pachter L, Tress ML, Valencia A, Choo SW, Choo CY, Ucla C, Manzano C, Wyss C, Cheung E, Clark TG, Brown JB, Ganesh M, Patel S, Tammana H, Chrast J, Henrichsen CN, Kai C, Kawai J, Nagalakshmi U, Wu J, Lian Z, Lian J, Newburger P, Zhang X, Bickel P, Mattick JS, Carninci P, Hayashizaki Y, Weissman S, Hubbard T, Myers RM, Rogers J, Stadler PF, Lowe TM, Wei CL, Ruan Y, Struhl K, Gerstein M, Antonarakis SE, Fu Y, Green ED, Karaoz U, Siepel A, Taylor J, Liefer LA, Wetterstrand KA, Good PJ, Feingold EA, Guyer MS, Cooper GM, Asimenos G, Dewey CN, Hou M, Nikolaev S, Montoya-Burgos JI, Loytynoja A, Whelan S, Pardi F, Massingham T, Huang H, Zhang NR, Holmes I, Mullikin JC, Ureta-Vidal A, Paten B, Seringhaus M, Church D, Rosenbloom K, Kent WJ, Stone EA, Batzoglou S, Goldman N, Hardison RC, Haussler D, Miller W, Sidow A, Trinklein ND, Zhang ZD, Barrera L, Stuart R, King DC, Ameur A, Enroth S, Bieda MC, Kim J, Bhinge AA, Jiang N, Liu J, Yao F, Vega VB, Lee CW, Ng P, Shahab A, Yang A, Moqtaderi Z, Zhu Z, Xu X, Squazzo S, Oberley MJ, Inman D, Singer MA, Richmond TA, Munn KJ, Rada-Iglesias A, Wallerman O, Komorowski J, Fowler JC, Couttet P, Bruce AW, Dovey OM,

Ellis PD, Langford CF, Nix DA, Euskirchen G, Hartman S, Urban AE, Kraus P, Van Calcar S, Heintzman N, Kim TH, Wang K, Qu C, Hon G, Luna R, Glass CK, Rosenfeld MG, Aldred SF, Cooper SJ, Halees A, Lin JM, Shulha HP, Zhang X, Xu M, Haidar JN, Yu Y, Ruan Y, Iyer VR, Green RD, Wadelius C, Farnham PJ, Ren B, Harte RA, Hinrichs AS, Trumbower H, Clawson H, Hillman-Jackson J, Zweig AS, Smith K, Thakkapallayil A, Barber G, Kuhn RM, Karolchik D, Armengol L, Bird CP, de Bakker PI, Kern AD, Lopez-Bigas N, Martin JD, Stranger BE, Woodroffe A, Davydov E, Dimas A, Eyras E, Hallgrimsdottir IB, Huppert J, Zody MC, Abecasis GR, Estivill X, Bouffard GG, Guan X, Hansen NF, Idol JR, Maduro VV, Maskeri B, McDowell JC, Park M, Thomas PJ, Young AC, Blakesley RW, Muzny DM, Sodergren E, Wheeler DA, Worley KC, Jiang H, Weinstock GM, Gibbs RA, Graves T, Fulton R, Mardis ER, Wilson RK, Clamp M, Cuff J, Gnerre S, Jaffe DB, Chang JL, Lindblad-Toh K, Lander ES, Koriabine M, Nefedov M, Osoegawa K, Yoshinaga Y, Zhu B, de Jong PJ (2007) Identification and analysis of functional elements in 1% of the human genome by the ENCODE pilot project. Nature 447 (7146):799–816

111. Heintzman ND, Hon GC, Hawkins RD, Kheradpour P, Stark A, Harp LF, Ye Z, Lee LK, Stuart RK, Ching CW, Ching KA, Antosiewicz-Bourget JE, Liu H, Zhang X, Green RD, Lobanenkov VV, Stewart R, Thomson JA, Crawford GE, Kellis M, Ren B (2009) Histone modifications at human enhancers reflect global cell-type-specific gene expression. Nature 459 (7243):108–112
112. Kim TK, Hemberg M, Gray JM, Costa AM, Bear DM, Wu J, Harmin DA, Laptewicz M, Barbara-Haley K, Kuersten S, Markenscoff-Papadimitriou E, Kuhl D, Bito H, Worley PF, Kreiman G, Greenberg ME (2010) Widespread transcription at neuronal activity-regulated enhancers. Nature 465 (7295):182–187
113. Koch CM, Andrews RM, Flicek P, Dillon SC, Karaoz U, Clelland GK, Wilcox S, Beare DM, Fowler JC, Couttet P, James KD, Lefebvre GC, Bruce AW, Dovey OM, Ellis PD, Dhami P, Langford CF, Weng Z, Birney E, Carter NP, Vetrie D, Dunham I (2007) The landscape of histone modifications across 1% of the human genome in five human cell lines. Genome Res 17 (6):691–707
114. Visel A, Blow MJ, Li Z, Zhang T, Akiyama JA, Holt A, Plajzer-Frick I, Shoukry M, Wright C, Chen F, Afzal V, Ren B, Rubin EM, Pennacchio LA (2009) ChIP-seq accurately predicts tissue-specific activity of enhancers. Nature 457 (7231):854–858
115. Ramos YF, Hestand MS, Verlaan M, Krabbendam E, Ariyurek Y, van Galen M, van Dam H, van Ommen GJ, den Dunnen JT, Zantema A, t Hoen PA (2010) Genome-wide assessment of differential roles for p300 and CBP in transcription regulation. Nucleic Acids Res 36 (16):5396–5408
116. Badis G, Berger MF, Philippakis AA, Talukder S, Gehrke AR, Jaeger SA, Chan ET, Metzler G, Vedenko A, Chen X, Kuznetsov H, Wang CF, Coburn D, Newburger DE, Morris Q, Hughes TR, Bulyk ML (2009) Diversity and complexity in DNA recognition by transcription factors. Science 324 (5935):1720–1723
117. Liu X, Lee CK, Granek JA, Clarke ND, Lieb JD (2006b) Whole-genome comparison of Leu3 binding in vitro and in vivo reveals the importance of nucleosome occupancy in target site selection. Genome Res 16 (12):1517–1528
118. Struhl K (1999) Fundamentally different logic of gene regulation in eukaryotes and prokaryotes. Cell 98 (1):1–4
119. Cheung V, Chua G, Batada NN, Landry CR, Michnick SW, Hughes TR, Winston F (2008) Chromatin- and transcription-related factors repress transcription from within coding regions throughout the *Saccharomyces cerevisiae* genome. PLoS Biol 6 (11): e277
120. Kaplan CD, Laprade L, Winston F (2003) Transcription elongation factors repress transcription initiation from cryptic sites. Science 301 (5636):1096–1099
121. Berman BP, Nibu Y, Pfeiffer BD, Tomancak P, Celniker SE, Levine M, Rubin GM, Eisen MB (2002) Exploiting transcription factor binding site clustering to identify cis-regulatory modules involved in pattern formation in the *Drosophila* genome. Proc Natl Acad Sci U S A 99 (2):757–762

122. Georges AB, Benayoun BA, Caburet S, Veitia RA (2010) Generic binding sites, generic DNA-binding domains: where does specific promoter recognition come from? Faseb J 24 (2):346–356
123. Papatsenko DA, Makeev VJ, Lifanov AP, Regnier M, Nazina AG, Desplan C (2002) Extraction of functional binding sites from unique regulatory regions: the *Drosophila* early developmental enhancers. Genome Res 12 (3):470–481
124. Valouev A, Ichikawa J, Tonthat T, Stuart J, Ranade S, Peckham H, Zeng K, Malek JA, Costa G, McKernan K, Sidow A, Fire A, Johnson SM (2008) A high-resolution, nucleosome position map of *C. elegans* reveals a lack of universal sequence-dictated positioning. Genome Res 18 (7):1051–1063
125. Kaplan N, Moore IK, Fondufe-Mittendorf Y, Gossett AJ, Tillo D, Field Y, LeProust EM, Hughes TR, Lieb JD, Widom J, Segal E (2009) The DNA-encoded nucleosome organization of a eukaryotic genome. Nature 458 (7236):362–366
126. Sekinger EA, Moqtaderi Z, Struhl K (2005) Intrinsic histone-DNA interactions and low nucleosome density are important for preferential accessibility of promoter regions in yeast. Mol Cell 18 (6):735–748
127. Anderson JD, Widom J (2001) Poly(dA-dT) promoter elements increase the equilibrium accessibility of nucleosomal DNA target sites. Mol Cell Biol 21 (11):3830–3839
128. Bao Y, White CL, Luger K (2006) Nucleosome core particles containing a poly(dA.dT) sequence element exhibit a locally distorted DNA structure. J Mol Biol 361 (4): 617–624
129. Iyer V, Struhl K (1995) Poly(dA:dT), a ubiquitous promoter element that stimulates transcription via its intrinsic DNA structure. Embo J 14 (11):2570–2579
130. Suter B, Schnappauf G, Thoma F (2000) Poly(dA.dT) sequences exist as rigid DNA structures in nucleosome-free yeast promoters in vivo. Nucleic Acids Res 28 (21):4083–4089
131. Shimizu M, Mori T, Sakurai T, Shindo H (2000) Destabilization of nucleosomes by an unusual DNA conformation adopted by poly(dA) small middle dotpoly(dT) tracts in vivo. Embo J 19 (13):3358–3365
132. White CL, Luger K (2004) Defined structural changes occur in a nucleosome upon Amt1 transcription factor binding. J Mol Biol 342 (5):1391–1402
133. Zhu Z, Thiele DJ (1996) A specialized nucleosome modulates transcription factor access to a C. glabrata metal responsive promoter. Cell 87 (3):459–470
134. Field Y, Kaplan N, Fondufe-Mittendorf Y, Moore IK, Sharon E, Lubling Y, Widom J, Segal E (2008) Distinct modes of regulation by chromatin encoded through nucleosome positioning signals. PLoS Comput Biol 4 (11):e1000216
135. Peckham HE, Thurman RE, Fu Y, Stamatoyannopoulos JA, Noble WS, Struhl K, Weng Z (2007) Nucleosome positioning signals in genomic DNA. Genome Res 17 (8):1170–1177
136. Segal E, Fondufe-Mittendorf Y, Chen L, Thastrom A, Field Y, Moore IK, Wang JP, Widom J (2006) A genomic code for nucleosome positioning. Nature 442 (7104):772–778
137. Kaplan N, Moore I, Fondufe-Mittendorf Y, Gossett AJ, Tillo D, Field Y, Hughes TR, Lieb JD, Widom J, Segal E (2010) Nucleosome sequence preferences influence in vivo nucleosome organization. Nat Struct Mol Biol 17 (8):918–920
138. Mavrich TN, Ioshikhes IP, Venters BJ, Jiang C, Tomsho LP, Qi J, Schuster SC, Albert I, Pugh BF (2008b) A barrier nucleosome model for statistical positioning of nucleosomes throughout the yeast genome. Genome Res 18 (7):1073–1083
139. Zhang Y, Moqtaderi Z, Rattner BP, Euskirchen G, Snyder M, Kadonaga JT, Liu XS, Struhl K (2009) Intrinsic histone-DNA interactions are not the major determinant of nucleosome positions in vivo. Nat Struct Mol Biol 16 (8):847–852
140. Zhang Y, Moqtaderi Z, Rattner BP, Euskirchen G, Snyder M, Kadonaga JT, Liu XS, Struhl K (2010) Reply to "Evidence against a genomic code for nucleosome positioning". Nat Struct Mol Biol 17 (8):920–923
141. Hartley PD, Madhani HD (2009) Mechanisms that specify promoter nucleosome location and identity. Cell 137 (3):445–458

142. Choi JK, Kim YJ (2009) Intrinsic variability of gene expression encoded in nucleosome positioning sequences. Nat Genet 41 (4):498–503
143. Tillo D, Kaplan N, Moore IK, Fondufe-Mittendorf Y, Gossett AJ, Field Y, Lieb JD, Widom J, Segal E, Hughes TR (2010) High nucleosome occupancy is encoded at human regulatory sequences. PLoS One 5 (2):e9129
144. Fu Y, Sinha M, Peterson CL, Weng Z (2008) The insulator binding protein CTCF positions 20 nucleosomes around its binding sites across the human genome. PLoS Genet 4 (7):e1000138
145. Lidor Nili E, Field Y, Lubling Y, Widom J, Oren M, Segal E (2010) p53 binds preferentially to genomic regions with high DNA-encoded nucleosome occupancy. Genome Res 20 (10):1361–1368
146. Li G, Levitus M, Bustamante C, Widom J (2005) Rapid spontaneous accessibility of nucleosomal DNA. Nat Struct Mol Biol 12 (1):46–53
147. Li G, Widom J (2004) Nucleosomes facilitate their own invasion. Nat Struct Mol Biol 11 (8):763–769
148. Zlatanova J, Seebart C, Tomschik M (2008) The linker-protein network: control of nucleosomal DNA accessibility. Trends Biochem Sci 33 (6):247–253
149. Tomschik M, Zheng H, van Holde K, Zlatanova J, Leuba SH (2005) Fast, long-range, reversible conformational fluctuations in nucleosomes revealed by single-pair fluorescence resonance energy transfer. Proc Natl Acad Sci U S A 102 (9):3278–3283
150. Albert I, Mavrich TN, Tomsho LP, Qi J, Zanton SJ, Schuster SC, Pugh BF (2007) Translational and rotational settings of H2A.Z nucleosomes across the *Saccharomyces cerevisiae* genome. Nature 446 (7135):572–576
151. Jiang C, Pugh BF (2009) Nucleosome positioning and gene regulation: advances through genomics. Nat Rev Genet 10 (3):161–172
152. Koerber RT, Rhee HS, Jiang C, Pugh BF (2009) Interaction of transcriptional regulators with specific nucleosomes across the Saccharomyces genome. Mol Cell 35 (6): 889–902
153. Polach KJ, Widom J (1995) Mechanism of protein access to specific DNA sequences in chromatin: a dynamic equilibrium model for gene regulation. J Mol Biol 254 (2): 130–149
154. Adams CC, Workman JL (1995) Binding of disparate transcriptional activators to nucleosomal DNA is inherently cooperative. Mol Cell Biol 15 (3):1405–1421
155. Miller JA, Widom J (2003) Collaborative competition mechanism for gene activation in vivo. Mol Cell Biol 23 (5):1623–1632
156. Pettersson M, Schaffner W (1990) Synergistic activation of transcription by multiple binding sites for NF-kappa B even in absence of co-operative factor binding to DNA. J Mol Biol 214 (2):373–380
157. Vashee S, Melcher K, Ding WV, Johnston SA, Kodadek T (1998) Evidence for two modes of cooperative DNA binding in vivo that do not involve direct protein-protein interactions. Curr Biol 8 (8):452–458
158. Cirillo LA, Zaret KS (1999) An early developmental transcription factor complex that is more stable on nucleosome core particles than on free DNA. Mol Cell 4 (6):961–969
159. Cirillo LA, Lin FR, Cuesta I, Friedman D, Jarnik M, Zaret KS (2002) Opening of compacted chromatin by early developmental transcription factors HNF3 (FoxA) and GATA-4. Mol Cell 9 (2):279–289
160. Pennings S, Meersseman G, Bradbury EM (1994) Linker histones H1 and H5 prevent the mobility of positioned nucleosomes. Proc Natl Acad Sci U S A 91 (22):10275–10279
161. Ura K, Hayes JJ, Wolffe AP (1995) A positive role for nucleosome mobility in the transcriptional activity of chromatin templates: restriction by linker histones. Embo J 14 (15):3752–3765
162. Holmqvist PH, Belikov S, Zaret KS, Wrange O (2005) FoxA1 binding to the MMTV LTR modulates chromatin structure and transcription. Exp Cell Res 304 (2):593–603

163. Li G, Margueron R, Hu G, Stokes D, Wang YH, Reinberg D (2010) Highly compacted chromatin formed in vitro reflects the dynamics of transcription activation in vivo. Mol Cell 38 (1):41–53
164. Lupien M, Eeckhoute J, Meyer CA, Wang Q, Zhang Y, Li W, Carroll JS, Liu XS, Brown M (2008) FoxA1 translates epigenetic signatures into enhancer-driven lineage-specific transcription. Cell 132 (6):958–970
165. Plachetka A, Chayka O, Wilczek C, Melnik S, Bonifer C, Klempnauer KH (2008) C/EBPbeta induces chromatin opening at a cell-type-specific enhancer. Mol Cell Biol 28 (6):2102–2112
166. Dobi KC, Winston F (2007) Analysis of transcriptional activation at a distance in *Saccharomyces cerevisiae*. Mol Cell Biol 27 (15):5575–5586
167. Lee KK, Workman JL (2007) Histone acetyltransferase complexes: one size doesn't fit all. Nat Rev Mol Cell Biol 8 (4):284–295
168. Perissi V, Jepsen K, Glass CK, Rosenfeld MG (2010) Deconstructing repression: evolving models of co-repressor action. Nat Rev Genet 11 (2):109–123
169. Glozak MA, Sengupta N, Zhang X, Seto E (2005) Acetylation and deacetylation of non-histone proteins. Gene 363:15–23
170. Underhill C, Qutob MS, Yee SP, Torchia J (2000) A novel nuclear receptor corepressor complex, N-CoR, contains components of the mammalian SWI/SNF complex and the corepressor KAP-1. J Biol Chem 275 (51):40463–40470
171. Dokmanovic M, Clarke C, Marks PA (2007) Histone deacetylase inhibitors: overview and perspectives. Mol Cancer Res 5 (10):981–989
172. Goldmark JP, Fazzio TG, Estep PW, Church GM, Tsukiyama T (2000) The Isw2 chromatin remodeling complex represses early meiotic genes upon recruitment by Ume6p. Cell 103 (3):423–433
173. Inayoshi Y, Kaneoka H, Machida Y, Terajima M, Dohda T, Miyake K, Iijima S (2005) Repression of GR-mediated expression of the tryptophan oxygenase gene by the SWI/SNF complex during liver development. J Biochem 138 (4):457–465
174. Murphy DJ, Hardy S, Engel DA (1999) Human SWI-SNF component BRG1 represses transcription of the c-fos gene. Mol Cell Biol 19 (4):2724–2733
175. Ooi L, Belyaev ND, Miyake K, Wood IC, Buckley NJ (2006) BRG1 chromatin remodeling activity is required for efficient chromatin binding by repressor element 1-silencing transcription factor (REST) and facilitates REST-mediated repression. J Biol Chem 281 (51):38974–38980
176. Sif S, Saurin AJ, Imbalzano AN, Kingston RE (2001) Purification and characterization of mSin3A-containing Brg1 and hBrm chromatin remodeling complexes. Genes Dev 15 (5):603–618
177. Moreira JM, Holmberg S (1999) Transcriptional repression of the yeast CHA1 gene requires the chromatin-remodeling complex RSC. Embo J 18 (10):2836–2844
178. Buck MJ, Lieb JD (2006) A chromatin-mediated mechanism for specification of conditional transcription factor targets. Nat Genet 38 (12):1446–1451
179. Keleher CA, Redd MJ, Schultz J, Carlson M, Johnson AD (1992) Ssn6-Tup1 is a general repressor of transcription in yeast. Cell 68 (4):709–719
180. De Vit MJ, Waddle JA, Johnston M (1997) Regulated nuclear translocation of the Mig1 glucose repressor. Mol Biol Cell 8 (8):1603–1618
181. Park SH, Koh SS, Chun JH, Hwang HJ, Kang HS (1999) Nrg1 is a transcriptional repressor for glucose repression of STA1 gene expression in *Saccharomyces cerevisiae*. Mol Cell Biol 19 (3):2044–2050
182. Proft M, Serrano R (1999) Repressors and upstream repressing sequences of the stress-regulated ENA1 gene in *Saccharomyces cerevisiae*: bZIP protein Sko1p confers HOG-dependent osmotic regulation. Mol Cell Biol 19 (1):537–546
183. Tzamarias D, Struhl K (1994) Functional dissection of the yeast Cyc8-Tup1 transcriptional co-repressor complex. Nature 369 (6483):758–761

184. Davie JK, Edmondson DG, Coco CB, Dent SY (2003) Tup1-Ssn6 interacts with multiple class I histone deacetylases in vivo. J Biol Chem 278 (50):50158–50162
185. Watson AD, Edmondson DG, Bone JR, Mukai Y, Yu Y, Du W, Stillman DJ, Roth SY (2000) Ssn6-Tup1 interacts with class I histone deacetylases required for repression. Genes Dev 14 (21):2737–2744
186. Lieb JD, Liu X, Botstein D, Brown PO (2001) Promoter-specific binding of Rap1 revealed by genome-wide maps of protein-DNA association. Nat Genet 28 (4):327–334
187. Shore D (1994) RAP1: a protean regulator in yeast. Trends Genet 10 (11):408–412
188. Kurdistani SK, Robyr D, Tavazoie S, Grunstein M (2002) Genome-wide binding map of the histone deacetylase Rpd3 in yeast. Nat Genet 31 (3):248–254
189. Wang Z, Zang C, Cui K, Schones DE, Barski A, Peng W, Zhao K (2009) Genome-wide mapping of HATs and HDACs reveals distinct functions in active and inactive genes. Cell 138 (5):1019–1031
190. Wang A, Kurdistani SK, Grunstein M (2002) Requirement of Hos2 histone deacetylase for gene activity in yeast. Science 298 (5597):1412–1414
191. Nagaich AK, Walker DA, Wolford R, Hager GL (2004) Rapid periodic binding and displacement of the glucocorticoid receptor during chromatin remodeling. Mol Cell 14 (2):163–174
192. Auble DT, Hansen KE, Mueller CG, Lane WS, Thorner J, Hahn S (1994) Mot1, a global repressor of RNA polymerase II transcription, inhibits TBP binding to DNA by an ATP-dependent mechanism. Genes Dev 8 (16):1920–1934
193. van Werven FJ, van Bakel H, van Teeffelen HA, Altelaar AF, Koerkamp MG, Heck AJ, Holstege FC, Timmers HT (2008) Cooperative action of NC2 and Mot1p to regulate TATA-binding protein function across the genome. Genes Dev 22 (17):2359–2369
194. Li G, Chandler SP, Wolffe AP, Hall TC (1998) Architectural specificity in chromatin structure at the TATA box in vivo: nucleosome displacement upon beta-phaseolin gene activation. Proc Natl Acad Sci U S A 95 (8):4772–4777
195. Grewal SI, Jia S (2007) Heterochromatin revisited. Nat Rev Genet 8 (1):35–46
196. James TC, Elgin SC (1986) Identification of a nonhistone chromosomal protein associated with heterochromatin in *Drosophila melanogaster* and its gene. Mol Cell Biol 6 (11):3862–3872
197. Ebert A, Lein S, Schotta G, Reuter G (2006) Histone modification and the control of heterochromatic gene silencing in *Drosophila*. Chromosome Res 14 (4):377–392
198. Nielsen SJ, Schneider R, Bauer UM, Bannister AJ, Morrison A, O'Carroll D, Firestein R, Cleary M, Jenuwein T, Herrera RE, Kouzarides T (2001) Rb targets histone H3 methylation and HP1 to promoters. Nature 412 (6846):561–565
199. O'Geen H, Squazzo SL, Iyengar S, Blahnik K, Rinn JL, Chang HY, Green R, Farnham PJ (2007) Genome-wide analysis of KAP1 binding suggests autoregulation of KRAB-ZNFs. PLoS Genet 3 (6):e89
200. Vogel MJ, Guelen L, de Wit E, Peric-Hupkes D, Loden M, Talhout W, Feenstra M, Abbas B, Classen AK, van Steensel B (2006) Human heterochromatin proteins form large domains containing KRAB-ZNF genes. Genome Res 16 (12):1493–1504
201. Schwartz YB, Pirrotta V (2007) Polycomb silencing mechanisms and the management of genomic programmes. Nat Rev Genet 8 (1):9–22
202. Lee TI, Jenner RG, Boyer LA, Guenther MG, Levine SS, Kumar RM, Chevalier B, Johnstone SE, Cole MF, Isono K, Koseki H, Fuchikami T, Abe K, Murray HL, Zucker JP, Yuan B, Bell GW, Herbolsheimer E, Hannett NM, Sun K, Odom DT, Otte AP, Volkert TL, Bartel DP, Melton DA, Gifford DK, Jaenisch R, Young RA (2006) Control of developmental regulators by Polycomb in human embryonic stem cells. Cell 125 (2):301–313
203. Schwartz YB, Kahn TG, Nix DA, Li XY, Bourgon R, Biggin M, Pirrotta V (2006) Genome-wide analysis of Polycomb targets in *Drosophila melanogaster*. Nat Genet 38 (6):700–705

204. Squazzo SL, O'Geen H, Komashko VM, Krig SR, Jin VX, Jang SW, Margueron R, Reinberg D, Green R, Farnham PJ (2006) Suz12 binds to silenced regions of the genome in a cell-type-specific manner. Genome Res 16 (7):890–900
205. Fanti L, Pimpinelli S (2008) HP1: a functionally multifaceted protein. Curr Opin Genet Dev 18 (2):169–174
206. Muller J, Verrijzer P (2009) Biochemical mechanisms of gene regulation by polycomb group protein complexes. Curr Opin Genet Dev 19 (2):150–158
207. Simon JA, Kingston RE (2009) Mechanisms of polycomb gene silencing: knowns and unknowns. Nat Rev Mol Cell Biol 10 (10):697–708
208. Dorer DR, Henikoff S (1994) Expansions of transgene repeats cause heterochromatin formation and gene silencing in Drosophila. Cell 77 (7):993–1002
209. Luff B, Pawlowski L, Bender J (1999) An inverted repeat triggers cytosine methylation of identical sequences in *Arabidopsis*. Mol Cell 3 (4):505–511
210. Selker EU (2002) Repeat-induced gene silencing in fungi. Adv Genet 46:439–450
211. Hall IM, Shankaranarayana GD, Noma K, Ayoub N, Cohen A, Grewal SI (2002) Establishment and maintenance of a heterochromatin domain. Science 297 (5590):2232–2237
212. Volpe TA, Kidner C, Hall IM, Teng G, Grewal SI, Martienssen RA (2002) Regulation of heterochromatic silencing and histone H3 lysine-9 methylation by RNAi. Science 297 (5588):1833–1837
213. Grewal SI (2010) RNAi-dependent formation of heterochromatin and its diverse functions. Curr Opin Genet Dev 20 (2):134–141
214. Yamada T, Fischle W, Sugiyama T, Allis CD, Grewal SI (2005) The nucleation and maintenance of heterochromatin by a histone deacetylase in fission yeast. Mol Cell 20 (2):173–185
215. Jia S, Noma K, Grewal SI (2004) RNAi-independent heterochromatin nucleation by the stress-activated ATF/CREB family proteins. Science 304 (5679):1971–1976
216. Kim HS, Choi ES, Shin JA, Jang YK, Park SD (2004) Regulation of Swi6/HP1-dependent heterochromatin assembly by cooperation of components of the mitogen-activated protein kinase pathway and a histone deacetylase Clr6. J Biol Chem 279 (41): 42850–42859
217. Rusche LN, Kirchmaier AL, Rine J (2003) The establishment, inheritance, and function of silenced chromatin in *Saccharomyces cerevisiae*. Annu Rev Biochem 72:481–516
218. Abrink M, Ortiz JA, Mark C, Sanchez C, Looman C, Hellman L, Chambon P, Losson R (2001) Conserved interaction between distinct Kruppel-associated box domains and the transcriptional intermediary factor 1 beta. Proc Natl Acad Sci U S A 98 (4):1422–1426
219. Peng H, Begg GE, Schultz DC, Friedman JR, Jensen DE, Speicher DW, Rauscher FJ, 3rd (2000) Reconstitution of the KRAB-KAP-1 repressor complex: a model system for defining the molecular anatomy of RING-B box-coiled-coil domain-mediated protein-protein interactions. J Mol Biol 295 (5):1139–1162
220. Lechner MS, Begg GE, Speicher DW, Rauscher FJ, 3rd (2000) Molecular determinants for targeting heterochromatin protein 1-mediated gene silencing: direct chromoshadow domain-KAP-1 corepressor interaction is essential. Mol Cell Biol 20 (17):6449–6465
221. Schultz DC, Ayyanathan K, Negorev D, Maul GG, Rauscher FJ, 3rd (2002) SETDB1: a novel KAP-1-associated histone H3, lysine 9-specific methyltransferase that contributes to HP1-mediated silencing of euchromatic genes by KRAB zinc-finger proteins. Genes Dev 16 (8):919–932
222. Schultz DC, Friedman JR, Rauscher FJ, 3rd (2001) Targeting histone deacetylase complexes via KRAB-zinc finger proteins: the PHD and bromodomains of KAP-1 form a cooperative unit that recruits a novel isoform of the Mi-2alpha subunit of NuRD. Genes Dev 15 (4):428–443
223. Ayyanathan K, Lechner MS, Bell P, Maul GG, Schultz DC, Yamada Y, Tanaka K, Torigoe K, Rauscher FJ, 3rd (2003) Regulated recruitment of HP1 to a euchromatic gene induces

mitotically heritable, epigenetic gene silencing: a mammalian cell culture model of gene variegation. Genes Dev 17 (15):1855–1869
224. Groner AC, Meylan S, Ciuffi A, Zangger N, Ambrosini G, Denervaud N, Bucher P, Trono D (2010) KRAB-zinc finger proteins and KAP1 can mediate long-range transcriptional repression through heterochromatin spreading. PLoS Genet 6 (3):e1000869
225. Brown JL, Fritsch C, Mueller J, Kassis JA (2003) The *Drosophila pho-like* gene encodes a YY1-related DNA binding protein that is redundant with *pleiohomeotic* in homeotic gene silencing. Development 130 (2):285–294
226. Fritsch C, Brown JL, Kassis JA, Muller J (1999) The DNA-binding polycomb group protein pleiohomeotic mediates silencing of a *Drosophila* homeotic gene. Development 126 (17):3905–3913
227. Klymenko T, Papp B, Fischle W, Kocher T, Schelder M, Fritsch C, Wild B, Wilm M, Muller J (2006) A Polycomb group protein complex with sequence-specific DNA-binding and selective methyl-lysine-binding activities. Genes Dev 20 (9):1110–1122
228. Wang L, Brown JL, Cao R, Zhang Y, Kassis JA, Jones RS (2004) Hierarchical recruitment of polycomb group silencing complexes. Mol Cell 14 (5):637–646
229. Dejardin J, Rappailles A, Cuvier O, Grimaud C, Decoville M, Locker D, Cavalli G (2005) Recruitment of *Drosophila* Polycomb group proteins to chromatin by DSP1. Nature 434 (7032):533–538
230. Ringrose L, Rehmsmeier M, Dura JM, Paro R (2003) Genome-wide prediction of Polycomb/Trithorax response elements in *Drosophila melanogaster*. Dev Cell 5 (5):759–771
231. Boyer LA, Plath K, Zeitlinger J, Brambrink T, Medeiros LA, Lee TI, Levine SS, Wernig M, Tajonar A, Ray MK, Bell GW, Otte AP, Vidal M, Gifford DK, Young RA, Jaenisch R (2006) Polycomb complexes repress developmental regulators in murine embryonic stem cells. Nature 441 (7091):349–353
232. Ku M, Koche RP, Rheinbay E, Mendenhall EM, Endoh M, Mikkelsen TS, Presser A, Nusbaum C, Xie X, Chi AS, Adli M, Kasif S, Ptaszek LM, Cowan CA, Lander ES, Koseki H, Bernstein BE (2008) Genomewide analysis of PRC1 and PRC2 occupancy identifies two classes of bivalent domains. PLoS Genet 4 (10):e1000242
233. Sing A, Pannell D, Karaiskakis A, Sturgeon K, Djabali M, Ellis J, Lipshitz HD, Cordes SP (2009) A vertebrate Polycomb response element governs segmentation of the posterior hindbrain. Cell 138 (5):885–897
234. Woo CJ, Kharchenko PV, Daheron L, Park PJ, Kingston RE (2010) A region of the human HOXD cluster that confers polycomb-group responsiveness. Cell 140 (1):99–110
235. Caretti G, Di Padova M, Micales B, Lyons GE, Sartorelli V (2004) The Polycomb Ezh2 methyltransferase regulates muscle gene expression and skeletal muscle differentiation. Genes Dev 18 (21):2627–2638
236. Garcia E, Marcos-Gutierrez C, del Mar Lorente M, Moreno JC, Vidal M (1999) RYBP, a new repressor protein that interacts with components of the mammalian Polycomb complex, and with the transcription factor YY1. Embo J 18 (12):3404–3418
237. Kim SY, Paylor SW, Magnuson T, Schumacher A (2006) Juxtaposed Polycomb complexes co-regulate vertebral identity. Development 133 (24):4957–4968
238. Satijn DP, Hamer KM, den Blaauwen J, Otte AP (2001) The polycomb group protein EED interacts with YY1, and both proteins induce neural tissue in *Xenopus* embryos. Mol Cell Biol 21 (4):1360–1369
239. Endoh M, Endo TA, Endoh T, Fujimura Y, Ohara O, Toyoda T, Otte AP, Okano M, Brockdorff N, Vidal M, Koseki H (2008) Polycomb group proteins Ring1A/B are functionally linked to the core transcriptional regulatory circuitry to maintain ES cell identity. Development 135 (8):1513–1524
240. Rinn JL, Kertesz M, Wang JK, Squazzo SL, Xu X, Brugmann SA, Goodnough LH, Helms JA, Farnham PJ, Segal E, Chang HY (2007) Functional demarcation of active and silent chromatin domains in human HOX loci by noncoding RNAs. Cell 129 (7):1311–1323
241. Talbert PB, Henikoff S (2006) Spreading of silent chromatin: inaction at a distance. Nat Rev Genet 7 (10):793–803

242. Nakayama J, Rice JC, Strahl BD, Allis CD, Grewal SI (2001) Role of histone H3 lysine 9 methylation in epigenetic control of heterochromatin assembly. Science 292 (5514):110–113
243. Hoppe GJ, Tanny JC, Rudner AD, Gerber SA, Danaie S, Gygi SP, Moazed D (2002) Steps in assembly of silent chromatin in yeast: Sir3-independent binding of a Sir2/Sir4 complex to silencers and role for Sir2-dependent deacetylation. Mol Cell Biol 22 (12): 4167–4180
244. Rusche LN, Kirchmaier AL, Rine J (2002) Ordered nucleation and spreading of silenced chromatin in *Saccharomyces cerevisiae*. Mol Biol Cell 13 (7):2207–2222
245. Boscheron C, Maillet L, Marcand S, Tsai-Pflugfelder M, Gasser SM, Gilson E (1996) Cooperation at a distance between silencers and proto-silencers at the yeast HML locus. Embo J 15 (9):2184–2195
246. Fourel G, Revardel E, Koering CE, Gilson E (1999) Cohabitation of insulators and silencing elements in yeast subtelomeric regions. Embo J 18 (9):2522–2537
247. Beisel C, Buness A, Roustan-Espinosa IM, Koch B, Schmitt S, Haas SA, Hild M, Katsuyama T, Paro R (2007) Comparing active and repressed expression states of genes controlled by the Polycomb/Trithorax group proteins. Proc Natl Acad Sci U S A 104 (42):16615–16620
248. Holohan EE, Kwong C, Adryan B, Bartkuhn M, Herold M, Renkawitz R, Russell S, White R (2007) CTCF genomic binding sites in *Drosophila* and the organisation of the bithorax complex. PLoS Genet 3 (7):e112
249. Negre N, Hennetin J, Sun LV, Lavrov S, Bellis M, White KP, Cavalli G (2006) Chromosomal distribution of PcG proteins during *Drosophila* development. PLoS Biol 4 (6):e170
250. Tolhuis B, de Wit E, Muijrers I, Teunissen H, Talhout W, van Steensel B, van Lohuizen M (2006) Genome-wide profiling of PRC1 and PRC2 Polycomb chromatin binding in *Drosophila melanogaster*. Nat Genet 38 (6):694–699
251. Kahn TG, Schwartz YB, Dellino GI, Pirrotta V (2006) Polycomb complexes and the propagation of the methylation mark at the *Drosophila Ubx* gene. J Biol Chem 281 (39):29064–29075
252. Lanzuolo C, Roure V, Dekker J, Bantignies F, Orlando V (2007) Polycomb response elements mediate the formation of chromosome higher-order structures in the bithorax complex. Nat Cell Biol 9 (10):1167–1174
253. Tiwari VK, Cope L, McGarvey KM, Ohm JE, Baylin SB (2008) A novel 6C assay uncovers Polycomb-mediated higher order chromatin conformations. Genome Res 18 (7):1171–1179
254. Molto E, Fernandez A, Montoliu L (2009) Boundaries in vertebrate genomes: different solutions to adequately insulate gene expression domains. Brief Funct Genomic Proteomic 8 (4):283–296
255. Wallace JA, Felsenfeld G (2007) We gather together: insulators and genome organization. Curr Opin Genet Dev 17 (5):400–407
256. West AG, Gaszner M, Felsenfeld G (2002) Insulators: many functions, many mechanisms. Genes Dev 16 (3):271–288
257. Bell AC, West AG, Felsenfeld G (1999) The protein CTCF is required for the enhancer blocking activity of vertebrate insulators. Cell 98 (3):387–396
258. Huang S, Li X, Yusufzai TM, Qiu Y, Felsenfeld G (2007) USF1 recruits histone modification complexes and is critical for maintenance of a chromatin barrier. Mol Cell Biol 27 (22):7991–8002
259. Gurudatta BV, Corces VG (2009) Chromatin insulators: lessons from the fly. Brief Funct Genomic Proteomic 8 (4):276–282
260. Recillas-Targa F, Pikaart MJ, Burgess-Beusse B, Bell AC, Litt MD, West AG, Gaszner M, Felsenfeld G (2002) Position-effect protection and enhancer blocking by the chicken beta-globin insulator are separable activities. Proc Natl Acad Sci U S A 99 (10):6883–6888
261. West AG, Huang S, Gaszner M, Litt MD, Felsenfeld G (2004) Recruitment of histone modifications by USF proteins at a vertebrate barrier element. Mol Cell 16 (3):453–463
262. Gaszner M, Felsenfeld G (2006) Insulators: exploiting transcriptional and epigenetic mechanisms. Nat Rev Genet 7 (9):703–713

263. Phillips JE, Corces VG (2009) CTCF: master weaver of the genome. Cell 137 (7):1194–1211
264. Guelen L, Pagie L, Brasset E, Meuleman W, Faza MB, Talhout W, Eussen BH, de Klein A, Wessels L, de Laat W, van Steensel B (2008) Domain organization of human chromosomes revealed by mapping of nuclear lamina interactions. Nature 453 (7197):948–951
265. Park PJ (2009) ChIP-seq: advantages and challenges of a maturing technology. Nat Rev Genet 10 (10):669–680
266. Ren B, Robert F, Wyrick JJ, Aparicio O, Jennings EG, Simon I, Zeitlinger J, Schreiber J, Hannett N, Kanin E, Volkert TL, Wilson CJ, Bell SP, Young RA (2000) Genome-wide location and function of DNA binding proteins. Science 290 (5500):2306–2309
267. Dejardin J, Kingston RE (2009) Purification of proteins associated with specific genomic Loci. Cell 136 (1):175–186
268. Celniker SE, Dillon LA, Gerstein MB, Gunsalus KC, Henikoff S, Karpen GH, Kellis M, Lai EC, Lieb JD, MacAlpine DM, Micklem G, Piano F, Snyder M, Stein L, White KP, Waterston RH (2009) Unlocking the secrets of the genome. Nature 459 (7249):927–930
269. Ravasi T, Suzuki H, Cannistraci CV, Katayama S, Bajic VB, Tan K, Akalin A, Schmeier S, Kanamori-Katayama M, Bertin N, Carninci P, Daub CO, Forrest AR, Gough J, Grimmond S, Han JH, Hashimoto T, Hide W, Hofmann O, Kamburov A, Kaur M, Kawaji H, Kubosaki A, Lassmann T, van Nimwegen E, MacPherson CR, Ogawa C, Radovanovic A, Schwartz A, Teasdale RD, Tegner J, Lenhard B, Teichmann SA, Arakawa T, Ninomiya N, Murakami K, Tagami M, Fukuda S, Imamura K, Kai C, Ishihara R, Kitazume Y, Kawai J, Hume DA, Ideker T, Hayashizaki Y (2010) An atlas of combinatorial transcriptional regulation in mouse and man. Cell 140 (5):744–752
270. van Berkum NL, Lieberman-Aiden E, Williams L, Imakaev M, Gnirke A, Mirny LA, Dekker J, Lander ES (2010) Hi-C: a method to study the three-dimensional architecture of genomes. J Vis Exp (39):1869
271. Vassetzky Y, Gavrilov A, Eivazova E, Priozhkova I, Lipinski M, Razin S (2009) Chromosome conformation capture (from 3C to 5C) and its ChIP-based modification. Methods Mol Biol 567:171–188
272. Segal E, Widom J (2009b) From DNA sequence to transcriptional behaviour: a quantitative approach. Nat Rev Genet 10 (7):443–456
273. Strohner R, Wachsmuth M, Dachauer K, Mazurkiewicz J, Hochstatter J, Rippe K, Langst G (2005) A 'loop recapture' mechanism for ACF-dependent nucleosome remodeling. Nat Struct Mol Biol 12 (8):683–690
274. Agalioti T, Lomvardas S, Parekh B, Yie J, Maniatis T, Thanos D (2000) Ordered recruitment of chromatin modifying and general transcription factors to the IFN-beta promoter. Cell 103 (4):667–678
275. Chaban Y, Ezeokonkwo C, Chung WH, Zhang F, Kornberg RD, Maier-Davis B, Lorch Y, Asturias FJ (2008) Structure of a RSC-nucleosome complex and insights into chromatin remodeling. Nat Struct Mol Biol 15 (12):1272–1277

Chapter 12
Transcription Factor Effector Domains

Seth Frietze and Peggy J. Farnham

Abstract The last decade has seen an incredible breakthrough in technologies that allow histones, transcription factors (TFs), and RNA polymerases to be precisely mapped throughout the genome. From this research, it is clear that there is a complex interaction between the chromatin landscape and the general transcriptional machinery and that the dynamic control of this interface is central to gene regulation. However, the chromatin remodeling enzymes and general TFs cannot, on their own, recognize and stably bind to promoter or enhancer regions. Rather, they are recruited to cis regulatory regions through interaction with site-specific DNA binding TFs and/or proteins that recognize epigenetic marks such as methylated cytosines or specifically modified amino acids in histones. These "recruitment" factors are modular in structure, reflecting their ability to interact with the genome via one region of the protein and to simultaneously bind to other regulatory proteins via "effector" domains. In this chapter, we provide examples of common effector domains that can function in transcriptional regulation via their ability to (a) interact with the basal transcriptional machinery and general co-activators, (b) interact with other TFs to allow cooperative binding, and (c) directly or indirectly recruit histone and chromatin modifying enzymes.

12.1 Introduction

Transcriptional activation is a stepwise process that requires (a) creating and maintaining an open chromatin structure, (b) assembly of the preinitiation complex, and (c) transition to productive elongation (Fig. 12.1). Successful completion of each of these steps involves a diverse group of proteins, some of which function in a relatively promoter-specific manner whereas others regulate large sets of genes. Recent advances in molecular and computational biology allow histone and DNA modifications, TFs, and RNA polymerases to be precisely mapped throughout the genome,

P.J. Farnham (✉)
Department of Biochemistry and Molecular Biology, Norris Comprehensive Cancer Center, University of Southern California, 1450 Biggy Street, NRT 6503, Los Angeles, CA 90033, USA
e-mail: pfarnham@usc.edu

T.R. Hughes (ed.), *A Handbook of Transcription Factors*, Subcellular Biochemistry 52,
DOI 10.1007/978-90-481-9069-0_12,

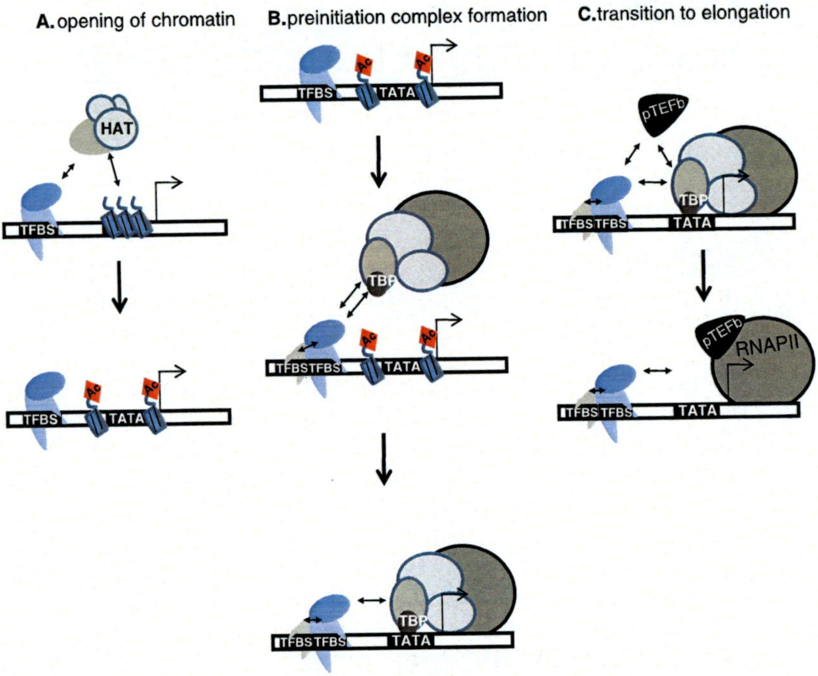

Fig. 12.1 Regulation of transcription. Shown is a schematic representing the three steps needed for productive transcription, including Step 1: the creation of open chromatin, which involves interactions between DNA-bound proteins and histone modifying enzymes (e.g. a HAT which can create an acetylated (Ac) histone); Step 2: assembly of the preinitiation complex, which can involve interactions between different DNA binding proteins and between DNA-bound proteins and general factors (such as TBP which binds to the TATA box); and Step 3: transition to productive elongation, which involves interaction between DNA-bound proteins and enzymes such as the pTEFb kinase. Although in this schematic the TFs are shown binding to transcription factor binding sites (TFBS) proximal to the transcription start site (indicated by the *bent arrow*), many transcriptional regulators can also bind to sites quite far from the core promoter regions

relative to active or silent promoters (see [1–3] for reviews). From this research, it is becoming clear that there is a complex interaction between the chromatin landscape and the transcriptional machinery and that the dynamic relationship of this interface is central to biological control over gene expression [4]. It is now recognized that regulatory factors can exert their influence on transcriptional activation either via co-localization with other proteins that are bound at or near core promoter regions or they can be recruited to distal enhancer regions and interact with promoter-bound proteins via looping mechanisms. However, generally speaking, the chromatin remodeling enzymes and the general transcription factors involved in initiation and elongation cannot, on their own, recognize and stably bind to the promoter or enhancer regions.

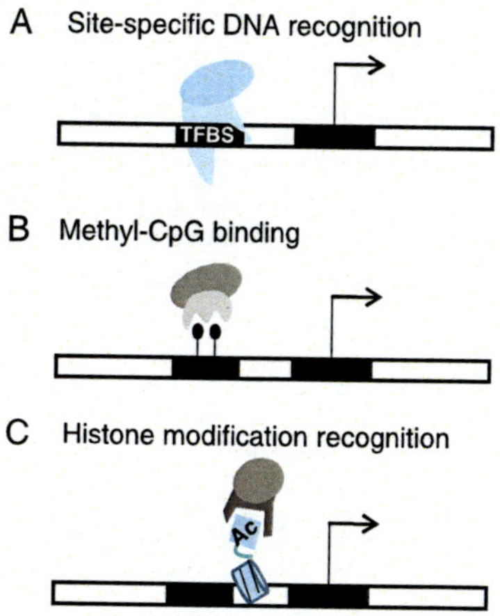

Fig. 12.2 Modular structure of effector domain-containing proteins. Effector domains can be recruited to specific genomic regions via **a** DNA binding domains that recognize short DNA sequence motifs (TFBS), **b** recognition of a methylated cytosine (shown as a *black ball*), and **c** recognition of a modified amino acid of a histone (e.g. an acetylated histone, shown as Ac)

One way in which chromatin remodeling enzymes and general transcription factors are recruited to cis-regulatory regions is through interaction with site-specific DNA binding TFs (Fig. 12.2a). The three largest classes of site-specific DNA binding proteins in mammals contact the genome via conserved DNA binding domains called zinc fingers, homeodomains, and helix–loop–helix domains [5] (Chapter 3 of this volume provides a catalog of eukaryotic DNA binding domains, and Chapters 4 and 5 specifically review C2H2 zinc fingers and homeodomains). Each of these classes of site-specific DNA binding factors contains many different proteins; for example, in humans there are over 650 zinc finger proteins, ~250 homeodomain proteins, and ~80 helix-loop-helix proteins [5]. Within each class, individual TFs can bind to and regulate hundreds to thousands of different genes. Site-specific TFs are modular in their structure reflecting their ability to bind to DNA via their DNA binding domains and simultaneously bind to other transcriptional regulatory proteins via so-called effector domains. The modular nature of site-specific TFs has been repeatedly demonstrated using in vitro and in vivo reporter assays. In these experiments, effector domains are separated from their natural DNA binding domains and then engineered to be part of a fusion protein having a heterologous DNA binding domain. Numerous studies have shown that simply bringing such effector domains to promoter regions can modulate transcription [6–8].

Another way in which chromatin remodeling enzymes and general transcription factors can be brought to the genome is via effector domains that reside in proteins that can recognize epigenomic marks. Similar to recognition of a short nucleotide motif by a DNA binding protein, other proteins can distinguish distinctively modified DNA and histone protein "motifs". For example, methylated cytosine in the 5′-CpG-3′ dinucleotide sequence is specifically recognized by members of a family of proteins containing a conserved methyl-CpG binding domain (MBD). MBD-containing proteins, which include MeCP2, MBD1, MBD2 and MBD4, bind specifically to methyl-CpG motifs located throughout the genome [9]; see Fig. 12.2b. MBD-containing proteins function by recruiting various co-regulators to methyl-CpG sites. For example, MeCP2 simultaneously binds promoter regions containing methyl-CpG motifs and the Sin3-containing histone deacetylase complex via a transcriptional repression domain (TRD), resulting in histone deacetylation and transcriptional silencing [10, 11]. Likewise, MBD1 and MBD2 copurify with distinct cellular complexes which link DNA methylation with chromatin modification and transcriptional repression. Similarly, posttranslational modifications of the amino termini of core histones are correlated to transcriptional states and are recognized by relevant chromatin-associated proteins (Fig. 12.2c). Several different histone modifications have been identified, including acetylation, phosphorylation, and methylation, and specific protein domains have evolved to recognize several of these different modifications. For example, different methylation states of histone H3 at lysine 4 can be recognized by tudor, chromo, and plant homeodomains (PHD), by malignant brain tumor (MBT) domains, and by WD40 repeat domains (many of these domains are structurally related and are collectively referred to as the "royal family" [12], reviewed [13, 14]). Other examples of this family include the chromodomain of HP1, which interacts with lower (mono- and di-) methylation states of lysine 9 of histone H3 but preferentially binds to the trimethylated state [15, 16] and the tudor domain of 53BP1, which can discriminate between the di- and tri-methyl state of H4K20, preferring the dimethyl form [17, 18]. Acetylated lysine is also recognized by specific protein modules called the bromodomain [19], which is found in many chromatin-associated proteins and in nearly all known nuclear histone acetyltransferases (HATs). Of course, epigenetic marks such as DNA methylation and histone modifications are located at specific genomic regions (which can vary in different cell types), indicating that DNA methylases and histone modifying enzymes must be recruited to the genome by sequence-specific mechanisms such as site-specific TFs or RNAs. For example, KRAB-ZNFs can recruit the KAP1/SETDB1 histone methylating complex and long non-coding RNAs can recruit the PRC2 histone methylation complex [20–23].

The focus of this chapter is on the effector domains that are brought to specific sites of the genome by DNA binding proteins, methyl-CpG binding proteins, or histone binding proteins. (The interaction of TFs with chromatin more generally is discussed in Chapter 11). We provide examples of common effector domains that can function in transcriptional regulation via their ability to influence each of the steps outlined in Fig. 12.1. Specifically, we discuss effector domains that can: (a) interact with the basal transcriptional machinery and general co-activators, (b) interact with other TFs to allow cooperative binding, and (c) directly or indirectly recruit histone and chromatin modifying enzymes. It is important to understand that a given effector domain does not have a one-to-one interaction with only one type of regulatory partner. Rather, some effector domains can interact with the general transcriptional machinery, with various co-activator complexes, and with chromatin remodeling proteins. To provide a specific example, nuclear receptors (NRs) are very specialized ligand-dependent TFs that regulate cellular gene expression programs in response to a variety of small molecules, including endocrine hormones, fatty acids, and lipid metabolites (discussed in detail in Chapter 6). NR transactivation domains (also referred to as Activating Function, or AF, domains) have the capacity to drastically alter transcriptional activities in a context-dependent manner by recruiting many different types of multi-protein co-regulatory complexes, often referred to as co-activators and co-repressors. For instance, the AF-1 and AF-2 terminal regions of the human glucocorticoid receptor alpha (hGRα) can link the receptor with different complexes depending on cellular signals. In the presence of glucocorticoids, the AF domains of hGRα interact with transcription-activating factors including basal TFs (e.g. RNA polymerase II, TATA-binding protein (TBP) and a host of TBP-associated proteins (TAFIIs)), coactivators such as p300/CBP and members of the p160/SRC family, site-specific factors including AP-1 and NFκB, and chromatin modulators such as the SWI/SNF and SAGA (Spt-Ada-Gcn5 acetyltransferase) complexes (reviewed in [24]). On the other hand, when bound to different gene regulatory regions, the same AF domains can, in response to glucocorticoid signals, recruit transcription repression complexes including corepressors, histone deacetylases (HDACs) and chromatin remodelers to down regulate transcriptional activity [25, 26]; see Fig. 12.3. Therefore, in addition to directing the TF to a specific genomic target, the precise nucleotide sequence of a regulatory factor

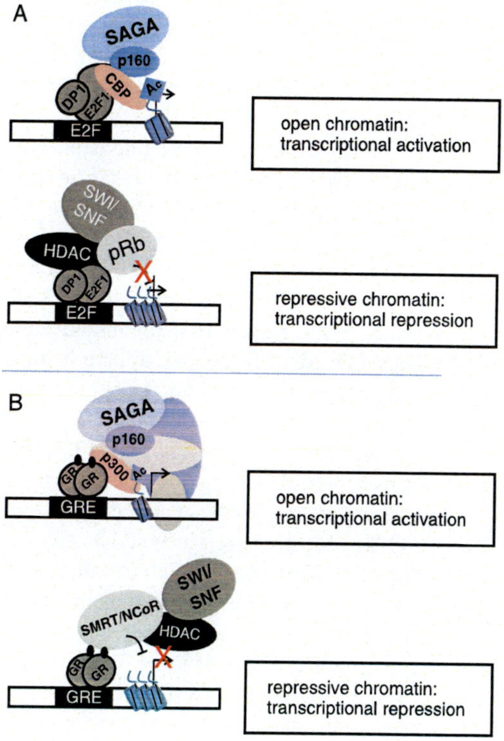

Fig. 12.3 Effector domains interact with different types of proteins to confer transcriptional regulation. The effector domains of both ubiquitously expressed factors such as E2F1 (**panel a**) and cell type-specific factors such as the glucocorticoid receptor (**panel b**) can interact with many different proteins, resulting in either transcriptional activation or transcriptional repression (see text for references and descriptions of the various proteins; see also [118] for a review on context-dependent transcriptional regulation)

binding site may also specify the mode of transcriptional regulation by directing the assembly of distinct regulatory complexes. The ability of a given DNA binding site to differentially affect hGRα activity has recently been investigated with much attention. Structural studies indicate that interactions between the hGRα DBD and different DNA elements can allosterically modulate interdomain interactions and thereby expose different surfaces for the recruitment of specific coregulator molecules [27].

12.2 Effector Domains Can Interact with the Basal Transcription Machinery

Sequence-specific transcriptional activators play an important role in transcription initiation by mediating the interaction of components of the transcriptional machinery with the DNA (see [28, 29] for a review of the eukaryotic basal transcriptional machinery). The domains that stimulate transcriptional activation through contacts with general TFs are called transactivation domains (TADs). Specifically, TADs interact with components of the preinitiation complex (PIC) to enhance recruitment and stabilization of the general factors at target promoters. The TADs from many regulatory TFs, such as E2F1, have been shown to make direct contacts with general TFs, including TATA-binding protein (TBP), TBP-associated factors (TAFs), TFIIA, TFIIB, and TFIIH from sites located both near to and far from core promoters [30–39]. TADs can also interact with and recruit components of the mediator protein complex, a multi-protein complex involved in activating a large number of genes [29].

Eukaryotic transactivation domains are typically classified with respect to their amino acid composition. TADs can be rich in acidic amino acid residues (e.g. E2F1 and p53), in glutamine residues (e.g. Oct1, Oct2, Sp1) or in proline residues (e.g. AP-2 and CTF/NF1). Each of these classes of transactivation domains has been shown to interact with various components of the basal transcriptional machinery, such as TFIIB and certain of the TBP-associated factors (TAFs) [40–42]. For example, the glutamine-rich transactivation domain of the site-specific DNA binding factor Sp1 can interact directly with a specific subunit of TFIID (TAFII 130) and point mutations within the transactivation domain inhibit binding of TFIID and reduce activation of transcription [32, 33, 43]. The discovery that the glutamine-rich domains of Sp1 interact with TAFII 130, whereas several acidic and proline-rich transactivation domains do not interact with TAFII 130, provides support for the association of specific transactivation domains with specific general coactivators [33, 43, 44]. However, it should be noted that not all glutamine-rich domains interact with TAFII 130, highlighting the limitations in understanding TAD function that arise by grouping activation domains by their most common amino acids [43, 44]. Along these lines, mutational analysis of the glutamine-rich TAD of SP1 and the acidic-rich TADs of p53 and RelA revealed that the ability of these TADs to stimulate transcription is more sensitive to mutation of bulky hydrophobic amino acids than to the mutation of the glutamine or acidic amino acids that broadly define them

[32, 45, 46]. Thus, the pattern of bulky hydrophobic residues may be more important than the more obvious features used to distinguish the classes of different activation domains.

More appropriately, eukaryotic TADs have been functionally grouped into those that stimulate initiation versus those that stimulate elongation, based on the different contacts they make with general transcription factors [47]. Prevailing models suggest that many activators act primarily at the level of transcription initiation [48]. However, contact between a TAD and the general transcriptional machinery can stimulate transcription not only by stabilizing the preinitiation complex (PIC), but can also facilitate promoter clearance and enhance the rate of elongation [45, 47, 49–55]. For example, in addition to stimulating transcriptional initiation, the activation domain of c-Myc also promotes transcription elongation through the recruitment of the RNA polymerase II Ser2 C-terminal domain (CTD) kinase called P-TEFb (positive transcription elongation factor b, which is composed of CycT1 and Cdk9) [55]. The c-Myc activation domain interacts directly with CycT1. Interestingly, the c-Myc transactivation domain can also increase mRNA cap maturation, polysome loading, and the rate of translation, processes that result from c-Myc-mediated phosphorylation of the RNA polymerase II CTD [56].

Structural studies of transactivation domains have revealed that many TADs are largely unstructured in solution [57]. For example, NMR studies have shown a lack of structure in the N-terminal region of p53 containing its acidic TAD [58, 59]. Further analysis revealed that specific motifs in the TAD fold into an α-helix upon binding to either the transcription initiation complex or to the p53 transcriptional attenuator Mdm2. Such studies propose that subdomains within the TADs become conformationally constrained upon interaction with a target protein [60–62]. Additionally, much evidence supports a structural and functional mechanism for the AF of hormone nuclear receptors that involves induced folding into an α-helical structure in response to protein–protein interactions and exposure to certain solutes [63–66]. These findings suggest that the target (i.e. a general transcription factor) is a template for the shaping of an unstructured TAD, allowing TADs to interact with numerous different components of the general transcriptional machinery. This mechanism creates a situation in which there is not a restricted relationship between certain general factors and specific types of TADs. Rather, TADs have evolved "flexible" ways to contact multiple components of the general machinery to activate transcription.

12.3 Effector Domains Can Interact with Other Site-Specific TFs

Another important type of effector domain that specifies TF functions is one that mediates direct interaction with other site-specific factors. Cooperative interactions between unrelated TFs expand the possibilities for extending DNA sequence recognition, perhaps allowing binding of a site-specific factor to a sequence not quite matching the preferred consensus motif. Additionally, the physical association of

TFs at enhancers or promoters not only stabilizes weak protein–DNA interactions of one factor to the genome but also allows combinatorial regulation, an important mechanism that enables integration of different signaling pathways [67].

One type of protein–protein interaction between site-specific DNA binding factors is the obligate hetero-or homodimer. bZIP, bHLH, and certain nuclear hormone receptors are examples of TFs that form dimers at their target genes. In such cases, the protein–protein interactions generally form in solution with the dimeric complex binding to DNA as a preassembled unit. The members of the E2F family of TFs, which are involved in the regulation of the cell cycle and many other cellular processes, also function via heterodimerization. These factors possess a centrally-located DNA binding domain immediately followed by a dimerization domain, which allows interactions with an obligate dimerization partner (DP) protein that contains similar DNA binding and heterodimerization domains. Dimerization between DP and E2F is required for high-affinity, sequence-specific DNA binding [68]. Thus, the ability of E2F TFs to form dimers can determine the strength of the resultant protein–DNA interactions as well as confer an ability to regulate a variety of different target genes [69]. In contrast to the E2F/DP dimeric complexes that are mediated by similar domains in each partner, other heterodimeric TF complexes dimerize using two dissimilar domains. In such cases, a heterodimeric TF complex might then preferably recognize half-binding sites arranged in a head-to-tail configuration. Examples of such an arrangement have been shown to occur in vivo for heterodimers of the retinoic acid receptor with vitamin D3 receptors, peroxisome proliferator-activated receptors or thyroid hormone receptors (reviewed in [70]).

Other effector domains mediate the interaction of one site-specific factor with another site-specific factor only subsequent to a DNA binding event. This DNA-dependent mode of association suggests that the individual proteins are unable to interact in solution, perhaps due to a relatively low dimerization constant. It is thought that the binding of a site-specific factor to DNA may induce an allosteric change in the protein structure, which in turn increases its affinity for another site-specific factor. This has been shown to be the case for binding of hGRα to different DNA motifs that differ by as little as a single base pair and for the DNA–dependent interaction of specific thyroid hormone receptor isoforms with the retinoid X receptor at specific DNA motifs, each of which can differentially affect the conformation and activity of the factors in response to hormone [27, 71–75]. Therefore DNA can be a sequence-specific allosteric ligand that modifies the activity of a site-specific factor at certain target genes. An important example of DNA–mediated protein interaction comes from studies of the Oct4 and Sox2 proteins, which are critical TFs involved in regulating embryonic stem cell (ESC) self-renewal and pluripotency. Oct4 co-localizes with different sets of TFs at many genomic sites, including promoters and enhancers [76–79]. The Oct4 binding sites co-occupied by Sox2 correlate with the ESC-specific expression of the nearby genes. Oct4 and Sox2 have low affinity for each other in solution, yet this affinity is critical for the cooperative binding of Oct4 and Sox2 proteins to adjacent sites on DNA. Electrophoretic mobility shift assays indicate that the Sox2-Oct4 heterodimer forms more efficiently on specific composite elements than do the single proteins [80]. Thus, the effector domains in Oct4 and Sox2 that mediate this specific protein–protein interaction play crucial roles in ESC-specific transcriptional regulation.

12.4 Effector Domains Can Recruit Chromatin-Modifying Enzymes

In addition to general and site-specific TFs, there are other types of regulatory proteins recruited by TFs to target genes. Many of these so-called transcriptional co-regulators harbor enzymatic activities that assist in gene regulation through post-translational histone modification. Numerous different histone-modifying enzymes have been identified; in particular, HATs and histone methyltransferases (HMTs) are critically involved in setting up active chromatin regions. Protein–protein interactions between TFs and histone modifying enzymes appear to play a dominant role in eukaryotic gene regulation and may ultimately determine the transcriptional output of a given promoter. For example, histone acetylation is associated with open chromatin and gene activation whereas histone methylation can be associated with both activation (e.g. methylation of lysine 4 of histone H3) and repression (e.g. methylation of lysine 9 or lysine 27 of histone H3). Although some subunits of the basal transcriptional machinery encode HAT functions (e.g. TAF1), in many cases the histone modifying enzyme is a component of a large multi-protein complex (see [29, 81] for reviews).

Many different TFs co-purify with histone modifying enzymes, including ubiquitous factors such as E2F family members and cell type-specific nuclear receptors. E2F family members possess domains that mediate interactions with histone modifying complexes that confer either activation or repression. For example, E2F family members can interact directly with the histone acetyltransferases p300/CBP [82, 83] via their C terminal transactivation domain. The transcriptional coactivators p300 and CBP (CREB binding protein) are versatile transcriptional regulator proteins that are highly related in primary structure and have many overlapping functions (thus they are referred to as p300/CBP). p300/CBP is a promiscuous acetyltransferase in that it catalyzes the acetylation of lysines on all four core histones, as well as acetylating more than 70 non-histone proteins, including itself. p300/CBP proteins have multiple protein interaction domains as well as a bromodomain, which recognizes acetylated lysines, thus providing extra contacts to specific "active" regions of the genome. E2F family members also interact with repressive histone-modifying complexes. For example, E2F6 copurifies in a repression complex with euchromatic HMTases called GLP and G9a, both of which are implicated in methylation of lysine 9 of histone H3 [84]. E2F6 has also been shown to interact with polycomb group protein complexes that contain H3K27me3-specific histone methyltransferases [85, 86]. These studies suggest that E2F6 may function to silence E2F-responsive genes via formation of heterochromatin. However, other studies [87, 88] have shown that E2F6 can also repress transcription via mechanisms other than lysine 9 or lysine 27 methylation, indicating that E2F6 must also be involved in other types of repressive complexes. Other E2F family members (i.e. E2F1-5) can interact with repressive chromatin complexes through interaction of their transactivation domains with members of the retinoblastoma (Rb) tumor suppressor protein family. Rb proteins serve as a bridge between E2Fs and histone methyltransferases that target H4K20, DNA methyltransferases, histone deacetylases, and chromatin compaction complexes [89–96]. Thus, the same effector domain in an E2F protein can mediate both activation and repression; see Fig. 12.3.

Nuclear receptors can also recruit various sets of co-regulators and histone modifying enzymes to their DNA binding sites (also called hormone response elements or HREs) to modulate target gene transcription [97]. For example, liganded nuclear receptors recruit the p160/SRC family of proteins that, in turn, provide a scaffold for the recruitment of HATs, such as p300/CBP, HMTs, and histone arginine methyltransferases such as CARM1 and PRMT1. These enzymes covalently modify histone and non-histone proteins to permit changes in the chromatin architecture and to alter the assembly of transcriptional complexes (for reviews see [98, 99]). The p160/SRC proteins have been shown to interact directly with the AF2 activation domain of NRs via conserved LxxLL motifs (where L stands for leucine and x is any other amino acid) [97]. Other transcriptional regulatory proteins have also been shown to associate with p160/SRC intermediaries, including AP-1, Smad3, NF-κB, E2F1, Rb, and p53 [100–105], demonstrating the widespread use of a p160/SRC scaffold to build transcription complexes. Additionally, ATP-dependent chromatin remodeling complexes, including the SWI/SNF, ISWI, and WINAC complexes, are recruited to HREs through direct interactions with NRs in a hormone-dependent manner, where they play critical roles in regulating transcriptional activation through remodeling chromatin structure [106]. In addition to NRs, other sequence-specific activators, including AP-1, ELKF, C/EBPβ and c-Myc can interact with ATP-dependent chromatin remodeling complexes. Interestingly, histone acetylation has been shown to stabilize SWI/SNF binding to nucleosomes (several SWI/SNF subunits, including BRG1, BAF250, BAF60a, and BAF57, contain bromodomains which are known to bind acetylated histone tails). Thus, multiple interactions are likely involved in both the recruitment and stabilization of SWI/SNF to activator-bound target genes. Once these multi-subunit complexes are recruited by effector domains to the regulatory regions of target genes, nucleosomal rearrangement and further chromatin modifications such as histone acetylation occur, allowing the Mediator complex, the general TFs, and the RNAPII machinery access to the promoter region to activate transcription.

A subset of nuclear receptors, including Thyroid hormone Receptor (TR), Retinoic Acid Receptor (RAR), and the Vitamin D Receptor (VDR), can repress transcription in the absence of their ligands [107]. Repression mediated by NRs involves the direct association of specific corepressor complexes containing the NCoR and SMRT corepressors. Analogous to coactivator recruitment, corepressors interact with nuclear receptors via effector domains and assemble in large multi-protein complexes that possess distinct enzymatic activities. However, rather than facilitating an open chromatin structure, corepressors generate repressive chromatin through the actions of HMTs, HDACs, histone demethylases, and specific chromatin remodeling complexes, including NURD [97]. In general, the specific mechanisms that lead to chromatin compaction and transcriptional repression are not well understood. Interestingly, many of these repressive chromatin complexes are shared for a number of site-specific TFs involved in transcriptional repression. For example, the BTB/POZ (Broad complex, Tramtrack, Bric-a-brac/POxvirus and Zinc finger) effector domain, a highly conserved protein–protein interaction domain, has been shown to interact with NCoR and SMRT corepressors [108–111]. There are approximately 80 different human BTB/POZ-containing proteins, including PLZF, HIC-1, BCL-6, Kaiso, FAZF and LRF, suggesting that this effector domain may be widely utilized for transcriptional repression [112]. Similarly, the extremely large family of TFs containing the Krüppel associated box (KRAB) domain, of which there are over 300 different members, have been suggested to repress the transcription of specific genes via an interaction with the KAP1 corepressor protein [113]. In turn, the KAP1 corepressor functions as a scaffold to recruit heterochromatin protein 1 (HP1) isoforms, histone deacetylases, and SETDB1, a SET-domain histone methyltransferase that methylates histone H3 at lysine 9 [23, 114, 115]. This modification is associated with closed chromatin and therefore KRAB effector domains of KRAB-zinc finger proteins link the KAP1 corepressor complex to specific genomic sites and silence gene expression by forming a facultative heterochromatin environment [116]. Due to the extremely large number of KRAB- zinc finger proteins, the KRAB domain may turn out to be one of the most commonly used effector domains involved in repression.

12.5 Summary

The molecular framework involved in transcription initiation consists of a multitude of cellular factors. A deep understanding of transcriptional regulation requires a detailed knowledge of the structural lattice in which TFs and co-regulators build hierarchical protein assemblies that provide control and specificity to transcriptional programs. The laths that link transcriptional regulators to their ultimate genomic targets are composed of a series of protein–protein interactions that recruit and confine transcriptional proteins to an appropriate regulatory location. Thus, knowledge of protein domains that serve as the biological effectors to recruit chromatin-modifying and nucleotide-synthesizing enzymes is critical for understanding how a cell type-specific transcriptome is established. A comprehensive cataloging of effector domains encoded in the human genome is beyond the scope of this review. However, we have provided examples of common effector domains utilized in eukaryotic transcriptional regulation. We suggest that researchers query the pfam website to identify conserved domains in specific TFs (http://pfam.janelia.org/; see also [117]). Although there are as many ways to regulate transcription as there are genes, several unifying themes can be derived from the many years of study of transcriptional regulation. These include:

(1) Effector domains can mediate gene activation or repression by promoting the formation of active or repressed chromatin, by interacting with domains in

other factors to form "platforms" for recruitment of co-regulatory proteins, or by stimulating or inhibiting preinitiation complex formation or productive elongation (Fig. 12.1).

(2) Effector domains can be brought to DNA in multiple ways, including as a modular domain of a site-specific DNA binding factor or as a domain or interacting partner with a protein that binds to methylated DNAs or modified histones (Fig. 12.2).

(3) There is not a one-to-one relationship between an effector domain and a specific co-regulatory protein; rather, many effector domains can interact with the same general factor, coregulator, or histone modifying complex and a single effector domain can interact with multiple other proteins, including proteins involved in both activation and repression (Fig. 12.3).

References

1. Park PJ (2009) ChIP-seq: advantages and challenges of a maturing technology. Nature Rev Genet (10):669–680
2. Laird PW, Jaenisch R (1994) DNA methylation and cancer. Human Molecular Genetics 3:1487–1495
3. Farnham PJ (2009) Insights from genomic profiling of transcription factors. Nature Rev Genet (10):605–616
4. Birney E, Stamatoyannopoulos JA, Dutta A, Guigo R, Gingeras TR, Margulies EH et al. (2007) Identification and analysis of functional elements in 1% of the human genome by the ENCODE pilot project. Nature 447 (7146):799–816
5. Vaquerizas JM, Kummerfeld SK, Teichmann SA, Luscombe NM (2009) A census of human transcription factors: function, expression and evolution. Nat Reviews Genetics 10 (4): 252–263
6. Keegan L, Gill G, Ptashne M (1986) Separation of DNA binding from the transcription-activating function of a eukaryotic regulatory protein. Science 231 (4739):699–704
7. Lin YS, Carey MF, Ptashne M, Green MR (1988) GAL4 derivatives function alone and synergistically with mammalian activators in vitro. Cell 54 (5):659–664
8. Brent R, Ptashne M (1985) A eukaryotic transcriptional activator bearing the DNA specificity of a prokaryotic repressor. Cell 43 (3 Pt 2):729–736
9. Sasai N, Defossez PA (2009) Many paths to one goal? The proteins that recognize methylated DNA in eukaryotes. Int J Dev Biol 53 (2–3):323–334
10. Jones PL, Veenstra GJ, Wade PA, Vermaak D, Kass SU, Landsberger N et al. (1998) Methylated DNA and MeCP2 recruit histone deacetylases to repress transcription. Nat Genet 19:187–191
11. Nan X, Ng HH, Johnson CA, Laherty CD, Turner BM, Eisenman RN et al. (1998) Transcriptional repression by the methyl-CpG-binding protein MeCP2 involves a histone deacetylase complex. Nature 393 (6683):386–389
12. Maurer-Stroh S, Dickens NJ, Hughes-Davies L, Kouzarides T, Eisenhaber F, Ponting CP (2003) The Tudor domain 'Royal Family': Tudor, plant Agenet, Chromo, PWWP and MBT domains. Trends Biochem Sci 28 (2):69–74
13. Ruthenburg AJ, Allis CD, Wysocka J (2007) Methylation of lysine 4 on histone H3: intricacy of writing and reading a single epigenetic mark. Mol Cell 25 (1):15–30
14. Taverna SD, Li H, Ruthenburg AJ, Allis CD, Patel DJ (2007) How chromatin-binding modules interpret histone modifications: lessons from professional pocket pickers. Nat Struct Mol Biol 14 (11):1025–1040

15. Bannister AJ, Zegerman P, Partridge JF, Miska EA, Thomas JO, Allshire RC et al. (2001) Selective recognition of methylated lysine 9 on histone H3 by the HP1 chromo domain. Nature 410 (6824):120–124
16. Lachner M, O'Carroll D, Rea S, Mechtler K, Jenuwein T (2001) Methylation of histone H3 lysine 9 creates a binding site for HP1 proteins. Nature 410 (6824):116–120
17. Botuyan MV, Lee J, Ward IM, Kim JE, Thompson JR, Chen J et al. (2006) Structural basis for the methylation state-specific recognition of histone H4-K20 by 53BP1 and Crb2 in DNA repair. Cell 127 (7):1361–1373
18. Kim J, Daniel J, Espejo A, Lake A, Krishna M, Xia L et al. (2006) Tudor, MBT and chromo domains gauge the degree of lysine methylation. EMBO Rep 7 (4):397–403
19. Mujtaba S, Zeng L, Zhou MM (2007) Structure and acetyl-lysine recognition of the bromodomain. Oncogene 26 (37):5521–5527
20. Khalil AM, Guttman M, Huarte M, Garber M, Raj A, Rivea Morales D et al. (2009) Many human large intergenic noncoding RNAs associate with chromatin-modifying complexes and affect gene expression. Proc Natl Acad Sci U S A 106 (28):11667–11672
21. Frietze S, O'Geen H, Blahnik KR, Jin VX, Farnham PJ (2010) ZNF274 recruits the histone methyltransferase SETDB1 to the 3' ends of ZNF genes. PLoS One 5 (12):e15082
22. Friedman JR, Fredericks WJ, Jensen DE, Speicher DW, Huang X-P, Neilson EG et al. (1996) KAP-1, a novel corepressor for the highly conserved KRAB repression domain. Genes & Dev 10 (16):2067–2078
23. Sripathy SP, Stevens J, Schultz DC (2006) The KAP1 corepressor functions to coordinate the assembly of de novo HP1-demarcated microenvironments of heterochromatin required for KRAB zinc finger protein-mediated transcriptional repression. Mol Cell Biol 26 (22): 8623–8638
24. Lavery DN, McEwan IJ (2005) Structure and function of steroid receptor AF1 transactivation domains: induction of active conformations. Biochem J 391 (Pt 3):449–464
25. Falkner KC, Pinaire JA, Xiao GH, Geoghegan TE, Prough RA (2001) Regulation of the rat glutathione S-transferase A2 gene by glucocorticoids: involvement of both the glucocorticoid and pregnane X receptors. Mol Pharmacol 60 (3):611–619
26. Szapary D, Huang Y, Simons SS, Jr. (1999) Opposing effects of corepressor and coactivators in determining the dose-response curve of agonists, and residual agonist activity of antagonists, for glucocorticoid receptor-regulated gene expression. Mol Endocrinol 13 (12):2108–2121
27. Meijsing SH, Pufall MA, So AY, Bates DL, Chen L, Yamamoto KR (2009) DNA binding site sequence directs glucocorticoid receptor structure and activity. Science 324 (5925):407–410
28. Thomas MC, Chiang CM (2006) The general transcription machinery and general cofactors. Crit Rev Biochem Mol Biol 41 (3):105–178
29. Lee TI, Young RA (2000) Transcription of eukaryotic protein-coding genes. Annu Rev Genet 34:77–137
30. Cujec TP, Cho H, Maldonado E, Meyer J, Reinberg D, Peterlin BM (1997) The human immunodeficiency virus transactivator Tat interacts with the RNA polymerase II holoenzyme. Mol Cell Biol 17 (4):1817–1823
31. Fry CJ, Slansky JE, Farnham PJ (1997) Position-dependent transcriptional regulation of the murine dihydrofolate reductase promoter by the E2F transactivation domain. Mol Cell Biol 17 (4):1966–1976
32. Gill G, Pascal E, Tseng ZH, Tjian R (1994) A glutamine-rich hydrophobic patch in transcription factor Sp1 contacts the dTAFII110 component of the *Drosophila* TFIID complex and mediates transcriptional activation. Proc Natl Acad Sci USA 91 (1):192–196
33. Goodrich JA, Hoey T, Thut C, Admon A, Tjian R (1993) Drosophila TAFII40 interacts with both a VP16 activation domain and the basal transcription factor TFIIB. Cell 75 (3):519–530
34. Horikoshi M, Hai T, Lin YS, Green MR, Roeder RG (1988) Transcription factor ATF interacts with the TATA factor to facilitate establishment of a preinitiation complex. Cell 54 (7):1033–1042

35. Kashanchi F, Piras G, Radonovich MF, Duvall JF, Fattaey A, Chiang CM et al. (1994) Direct interaction of human TFIID with the HIV-1 transactivator tat. Nature 367 (6460):295–299
36. Lin Y-S, Ha I, Maldonado E, Reinberg D, Green MR (1991) Binding of general transcription factor TFIIB to an acidic activating region. Nature 353 (6344):569–571
37. Roberts SG, Choy B, Walker SS, Lin YS, Green MR (1995) A role for activator-mediated TFIIB recruitment in diverse aspects of transcriptional regulation. Curr Biol 5 (5):508–516
38. Stringer KF, Ingles CJ, Greenblatt J (1990) Direct and selective binding of an acidic transcriptional activation domain to the TATA-box factor TFIID. Nature 345 (6278):783–786
39. Zhu H, Joliot V, Prywes R (1994) Role of transcription factor TFIIF in serum response factor-activated transcription. J Biol Chem 269 (5):3489–3497
40. Kim TK, Roeder RG (1994) Proline-rich activator CTF1 targets the TFIIB assembly step during transcriptional activation. Proc Natl Acad Sci U S A 91 (10):4170–4174
41. Chiang C-M, Roeder RG (1995) Cloning of an intrinsic human TFIID subunit that interacts with multiple transcriptional activators. Science 267 (5197):531–536
42. Tanese N, Pugh BF, Tjian R (1991) Coactivators for a proline-rich activator purified from the multisubunit human TFIID complex. Genes Dev 5:2212–2224
43. Hoey T, Weinzierl RO, Gill G, Chen JL, Dynlacht BD, Tjian R (1993) Molecular cloning and functional analysis of Drosophila TAF110 reveal properties expected of coactivators. Cell 72 (2):247–260
44. Chen J-L, Attardi DL, Verrijzer CP, Yokomori K, Tjian R (1994) Assembly of recombinant TFIID reveals differential coactivator requirements for distinct transcriptional activators. Cell 79 (1):93–105
45. Blair WS, Bogerd HP, Madore SJ, Cullen BR (1994) Mutational analysis of the transcription activation domain of RelA: identification of a highly synergistic minimal acidic activation module. Mol Cell Biol 14 (11):7226–7234
46. Lin J, Chen J, Elenbaas B, Levine AJ (1994) Several hydrophobic amino acids in the p53 amino-terminal domain are required for transcriptional activation, binding to mdm-2 and the adenovirus 5 E1B 55-kD protein. Genes & Dev 8 (10):1235–1246
47. Blau J, Xiao H, McCracken S, O'Hare P, Greenblatt J, Bentley D (1996) Three functional classes of transcriptional activation domains. Mol Cell Biol 16 (5):2044–2055
48. Choy B, Green MR (1993) Eukaryotic activators function during multiple steps of preinitiation complex assembly. Nature 366 (6455):531–536
49. Krumm A, Hickey LB, Groudine M (1995) Promoter-proximal pausing of RNA polymerase II defines a general rate-limiting step after transcription initiation. Genes Dev 9 (5):559–572
50. Mahanta SK, Scholl T, Yang FC, Strominger JL (1997) Transactivation by CIITA, the type II bare lymphocyte syndrome-associated factor, requires participation of multiple regions of the TATA box binding protein. Proc Natl Acad Sci U S A 94 (12):6324–6329
51. Yankulov K, Blau J, Purton T, Roberts S, Bentley DL (1994) Transcriptional elongation by RNA polymerase II is stimulated by transactivators. Cell 77 (5):749–759
52. Fuda NJ, Ardehali MB, Lis JT (2009) Defining mechanisms that regulate RNA polymerase II transcription in vivo. Nature 461 (7261):186–192
53. Rahl PB, Lin CY, Seila AC, Flynn RA, McCuine S, Burge CB et al. (2010) c-Myc regulates transcriptional pause release. Cell 141 (3):432–445
54. Selth LA, Sigurdsson S, Svejstrup JQ (2010) Transcript Elongation by RNA Polymerase II. Annu Rev Biochem 79:271–293
55. Eberhardy SR, Farnham PJ (2002) Myc recruits P-TEFb to mediate the final step in the transcriptional activation of the cad promoter. J Biol Chem 277 (42):40156–40162
56. Cowling VH, Cole MD (2007) The Myc transactivation domain promotes global phosphorylation of the RNA polymerase II carboxy-terminal domain independently of direct DNA binding. Mol Cell Biol 27 (6):2059–2073
57. Minezaki Y, Homma K, Kinjo AR, Nishikawa K (2006) Human transcription factors contain a high fraction of intrinsically disordered regions essential for transcriptional regulation. J Mol Biol 359 (4):1137–1149

58. Ayed A, Mulder FA, Yi GS, Lu Y, Kay LE, Arrowsmith CH (2001) Latent and active p53 are identical in conformation. Nat Struct Biol 8 (9):756–760
59. Dawson R, Muller L, Dehner A, Klein C, Kessler H, Buchner J (2003) The N-terminal domain of p53 is natively unfolded. J Mol Biol 332 (5):1131–1141
60. Chi SW, Lee SH, Kim DH, Ahn MJ, Kim JS, Woo JY et al. (2005) Structural details on mdm2-p53 interaction. J Biol Chem 280 (46):38795–38802
61. Uesugi M, Verdine GL (1999) The alpha-helical FXXPhiPhi motif in p53: TAF interaction and discrimination by MDM2. Proc Natl Acad Sci U S A 96 (26):14801–14806
62. Garza AS, Ahmad N, Kumar R (2009) Role of intrinsically disordered protein regions/domains in transcriptional regulation. Life Sci 84 (7–8):189–193
63. Brzozowski AM, Pike AC, Dauter Z, Hubbard RE, Bonn T, Engstrom O et al. (1997) Molecular basis of agonism and antagonism in the oestrogen receptor. Nature 389 (6652):753–758
64. Kumar R, Betney R, Li J, Thompson EB, McEwan IJ (2004) Induced alpha-helix structure in AF1 of the androgen receptor upon binding transcription factor TFIIF. Biochemistry 43 (11):3008–3013
65. Pike AC, Brzozowski AM, Hubbard RE, Bonn T, Thorsell AG, Engstrom O et al. (1999) Structure of the ligand-binding domain of oestrogen receptor beta in the presence of a partial agonist and a full antagonist. EMBO J 18 (17):4608–4618
66. Kumar R, Litwack G (2009) Structural and functional relationships of the steroid hormone receptors' N-terminal transactivation domain. Steroids 74 (12):877–883
67. Panne D (2008) The enhanceosome. Curr Opin Structural Biol 18 (2):236–242
68. Huber HE, Edwards G, Goodhart PJ, Patrick DR, Huang PS, Ivey-Hoyle M et al. (1993) Transcription factor E2F binds as a heterodimer. Proc Natl Acad Sci USA 90 (8):3525–3529
69. Helin K, Wu C-L, Fattaey AR, Lees JA, Dynlacht BD, Ngwu C et al. (1993) Heterodimerization of the transcription factors E2F-1 and DP-1 leads to cooperative *trans*-activation. Genes Dev 7 (10):1850–1861
70. Mangelsdorf DJ, Evans RM (1995) The RXR heterodimers and orphan receptors. Cell 83 (6):841–850
71. Remenyi A, Lins K, Nissen LJ, Reinbold R, Scholer HR, Wilmanns M (2003) Crystal structure of a POU/HMG/DNA ternary complex suggests differential assembly of Oct4 and Sox2 on two enhancers. Genes Dev 17 (16):2048–2059
72. Remenyi A, Tomilin A, Scholer HR, Wilmanns M (2002) Differential activity by DNA-induced quarternary structures of POU transcription factors. Biochem Pharmacol 64 (5-6):979–984
73. Kitayner M, Rozenberg H, Rohs R, Suad O, Rabinovich D, Honig B et al. (2010) Diversity in DNA recognition by p53 revealed by crystal structures with Hoogsteen base pairs. Nat Struct Mol Biol 17 (4):423–429
74. Leng X, Blanco J, Tsai SY, Ozato K, O'Malley BW, Tsai MJ (1994) Mechanisms for synergistic activation of thyroid hormone receptor and retinoid X receptor on different response elements. J Biol Chem 269 (50):31436–31442
75. Reginato MJ, Zhang J, Lazar MA (1996) DNA-independent and DNA-dependent mechanisms regulate the differential heterodimerization of the isoforms of the thyroid hormone receptor with retinoid X receptor. J Biol Chem 271 (45):28199–28205
76. Squazzo SL, Komashko VM, O'Geen H, Krig S, Jin VX, Jang S-W et al. (2006) Suz12 silences large regions of the genome in a cell type-specific manner. Genome Research 16 (7):890–900
77. Jin VX, O'Geen H, Iyengar S, Green R, Farnham PJ (2007) Identification of an OCT4 and SRY regulatory module using integrated computational and experimental genomics approaches. Genome Research 17 (6):807–817
78. Mathur D, Danford TW, Boyer LA, Young RA, Gifford DK, Jaenisch R (2008) Analysis of the mouse embryonic stem cell regulatory networks obtained by ChIP-chip and ChIP-PET. Genome Biol 9 (8):R126

79. Chen X, Xu H, Yuan P, Fang F, Huss M, Vega VB et al. (2008) Integration of external signaling pathways with the core transcriptional network in embryonic stem cells. Cell 133 (6):1106–1117
80. Rodda DJ, Chew JL, Lim LH, Loh YH, Wang B, Ng HH et al. (2005) Transcriptional regulation of nanog by OCT4 and SOX2. J Biol Chem 280 (26):24731–24737
81. Kouzarides T (2007) Chromatin modifications and their function. Cell 128:693–705
82. Trouche D, Cook A, Kouzarides T (1996) The CBP co-activator stimulates E2F1/DP1 activity. Nucleic Acids Res 24 (21):4139–4145
83. Fry CJ, Pearson A, Malinowski E, Bartley SM, Greenblatt J, Farnham PJ (1999) Activation of the murine dihydrofolate reductase promoter by E2F1: A requirement for CBP recruitment. J Biol Chem 274 (22):15883–15891
84. Ogawa H, Ishiguro K, Gaubatz S, Livingston DM, Nakatani Y (2002) A complex with chromatin modifiers that occupies E2F- and Myc-responsive genes in G0 cells. Science 296 (5570):1132–1136
85. Trimarchi JM, Fairchild B, Wen J, Lees JA (2001) The E2F6 transcription factor is a component of the mammalian Bmi1-containing polycomb complex. Proc Natl Acad Sci U S A 95 (6):2850–2855
86. Attwooll C, Oddi S, Cartwright P, Prosperini E, Agger K, Steensgaard P et al. (2005) A novel repressive E2F6 complex containing the polycomb group protein, EPC1, that interacts with EZH2 in a proliferation-specific manner. J Biol Chem 280 (2):1199–1208
87. Xu X, Bieda M, Jin VX, Rabinovich A, Oberley MJ, Green R et al. (2007) A comprehensive ChIP-chip analysis of E2F1, E2F4, and E2F6 in normal and tumor cells reveals iterchangeable roles of E2F family members. Genome Res 17 (11):1550–1561
88. Oberley MJ, Inman D, Farnham PJ (2003) E2F6 negatively regulates BRCA1 in human cancer cells without methylation of histone H3 on lysine 9. J Biol Chem 278 (43):42466–42476
89. Vandel L, Nicolas E, Vaute O, Ferreira R, Ait-si-ali S, Trouche D (2001) Transcriptional repression by the retinoblastoma protein through the recruitment of a histone methyltransferase. Mol Cell Biol 21 (19):6484–6494
90. Gonzalo S, Garcia-Cao M, Fraga MF, Schotta G, Peters AH, Cotter SE et al. (2005) Role of the RB1 family in stabilizing histone methylation at constitutive heterochromatin. Nat Cell Biol 7 (4):420–428
91. Robertson KD, Ait-Si-Ali S, Yokochi T, Wade PA, Jones PL, Wolffe AP (2000) DNMT1 forms a complex with Rb, E2F1 and HDAC1 and represses transcription from E2F-responsive promoters. nature genetics 25 (3):338–342
92. Pradhan S, Kim G-D (2002) The retinoblastoma gene product interacts with maintenance human DNA (cytosine-5) methyltransferase and modulates its activity. EMBO J 21:779–788
93. Trojer P, Li G, Sims RJ, 3rd, Vaquero A, Kalakonda N, Boccuni P et al. (2007) L3MBTL1, a histone-methylation-dependent chromatin lock. Cell 129 (5):915–928
94. Trojer P, Reinberg D (2008) Beyond histone methyl-lysine binding: how malignant brain tumor (MBT) protein L3MBTL1 impacts chromatin structure. Cell Cycle 7 (5): 578–585
95. Trojer P, Zhang J, Yonezawa M, Schmidt A, Zheng H, Jenuwein T et al. (2009) Dynamic Histone H1 Isotype 4 Methylation and Demethylation by Histone Lysine Methyltransferase G9a/KMT1C and the Jumonji Domain-containing JMJD2/KDM4 Proteins. J Biol Chem 284 (13):8395–8405
96. Longworth MS, Dyson NJ (2010) pRb, a local chromatin organizer with global possibilities. Chromosoma 119 (1):1–11
97. Rosenfeld MG, Lunyak VV, Glass CK (2006) Sensors and signals: a coactivator/corepressor/epigenetic code for integrating signal-dependent programs of transcriptional response. Genes Dev 20 (11):1405–1428
98. Lee YH, Stallcup MR (2009) Minireview: protein arginine methylation of nonhistone proteins in transcriptional regulation. Mol Endocrinol 23 (4):425–433

99. Spange S, Wagner T, Heinzel T, Kramer OH (2009) Acetylation of non-histone proteins modulates cellular signalling at multiple levels. Int J Biochem Cell Biol 41 (1):185–198
100. Lee SK, Kim HJ, Na SY, Kim TS, Choi HS, Im SY et al. (1998) Steroid receptor coactivator-1 coactivates activating protein-1-mediated transactivations through interaction with the c-Jun and c-Fos subunits. J Biol Chem 273 (27):16651–16654
101. Li G, Heaton JH, Gelehrter TD (2006) Role of steroid receptor coactivators in glucocorticoid and transforming growth factor beta regulation of plasminogen activator inhibitor gene expression. Mol Endocrinol 20 (5):1025–1034
102. Gao Z, Chiao P, Zhang X, Lazar MA, Seto E, Young HA et al. (2005) Coactivators and corepressors of NF-kappaB in IkappaB alpha gene promoter. J Biol Chem 280 (22): 21091–21098
103. Louie MC, Zou JX, Rabinovich A, Chen HW (2004) ACTR/AIB1 functions as an E2F1 coactivator to promote breast cancer cell proliferation and antiestrogen resistance. Mol Cell Biol 24 (12):5157–5171
104. Batsche E, Desroches J, Bilodeau S, Gauthier Y, Drouin J (2005) Rb enhances p160/SRC coactivator-dependent activity of nuclear receptors and hormone responsiveness. J Biol Chem 280 (20):19746–19756
105. Lee SK, Kim HJ, Kim JW, Lee JW (1999) Steroid receptor coactivator-1 and its family members differentially regulate transactivation by the tumor suppressor protein p53. Mol Endocrinol 13 (11):1924–1933
106. Belandia B, Parker MG (2003) Nuclear receptors: a rendezvous for chromatin remodeling factors. Cell 114 (3):277–280
107. Privalsky ML (2004) The role of corepressors in transcriptional regulation by nuclear hormone receptors. Annu Rev Physiol 66:315–360
108. Melnick A, Carlile G, Ahmad KF, Kiang CL, Corcoran C, Bardwell V et al. (2002) Critical residues within the BTB domain of PLZF and Bcl-6 modulate interaction with corepressors. Mol Cell Biol 22 (6):1804–1818
109. Yoon HG, Chan DW, Reynolds AB, Qin J, Wong J (2003) N-CoR mediates DNA methylation-dependent repression through a methyl CpG binding protein Kaiso. Mol Cell 12 (3):723–734
110. Muto A, Hoshino H, Madisen L, Yanai N, Obinata M, Karasuyama H et al. (1998) Identification of Bach2 as a B-cell-specific partner for small maf proteins that negatively regulate the immunoglobulin heavy chain gene 3' enhancer. EMBO J 17 (19):5734–5743
111. Dhordain P, Albagli O, Lin RJ, Ansieau S, Quief S, Leutz A et al. (1997) Corepressor SMRT binds the BTB/POZ repressing domain of the LAZ3/BCL6 oncoprotein. Proc Natl Acad Sci U S A 94 (20):10762–10767
112. Kelly KF, Daniel JM (2006) POZ for effect–POZ-ZF transcription factors in cancer and development. Trends Cell Biol 16 (11):578–587
113. Urrutia R (2003) KRAB-containing zinc-finger repressor proteins. Genome Biol 4 (10):231
114. Schultz DC, Ayyanathan K, Negorev D, Maul GG, Rauscher FJ, 3rd (2002) SETDB1: a novel KAP-1-associated histone H3, lysine 9-specific methyltransferase that contributes to HP1-mediated silencing of euchromatic genes by KRAB zinc-finger proteins. Genes Dev 16 (8):919–932
115. Schultz DC, Friedman JR, Rauscher F Jr (2001) Targeting histone deacetylase complexes via KRAB-zinc finger proteins: the PhD and bromodomains of KAP-1 form a coopeative unit that recruits a novel isoform of the Mi-2alpha subunit of NuRD. Genes Dev 15 (4):428–443
116. Groner AC, Meylan S, Ciuffi A, Zangger N, Ambrosini G, Denervaud N et al. (2010) KRAB-zinc finger proteins and KAP1 can mediate long-range transcriptional repression through heterochromatin spreading. PLoS Genet 6 (3):e1000869
117. Finn RD, Mistry J, Tate J, Coggill P, Heger A, Pollington JE et al. (2010) The Pfam protein families database. Nucleic Acids Res 38 (Database issue):D211–222
118. Fry CJ, Farnham PJ (1999) Context-dependent transcriptional regulation. J Biol Chem 274 (42):29583–29586

Chapter 13
Large-Scale Nuclear Architecture and Transcriptional Control

Juan M. Vaquerizas, Asifa Akhtar, and Nicholas M. Luscombe

Abstract Transcriptional regulation is one the most basic mechanisms for controlling gene expression. Over the past few years, much research has been devoted to understanding the interplay between transcription factors, histone modifications and associated enzymes required to achieve this control. However, it is becoming increasingly apparent that the three-dimensional conformation of chromatin in the interphase nucleus also plays a critical role in regulating transcription. Chromatin localisation in the nucleus is highly organised, and early studies described strong interactions between chromatin and sub-nuclear components. Single-gene studies have shed light on how chromosomal architecture affects gene expression. Lately, this has been complemented by whole-genome studies that have determined the global chromatin conformation of living cells in interphase. These studies have greatly expanded our understanding of nuclear architecture and its interplay with different physiological processes. Despite these advances, however, most of the mechanisms used to impose the three-dimensional chromatin structure remain unknown. Here, we summarise the different levels of chromatin organisation in the nucleus and discuss current efforts into characterising the mechanisms that govern it.

13.1 Introduction

Transcriptional regulation is a fundamental cellular process by which gene expression is activated or repressed [1]. In eukaryotes, transcription is regulated at many different levels ranging from the recruitment of the core transcriptional machinery to promoters by transcription factors, to modifications in the chromatin structure by histone remodelling and modification enzymes.

Chromosomes are the largest unit of chromatin organisation in the cell. In eukaryotes, these comprise single DNA molecules that are up to several hundred

J.M. Vaquerizas (✉)
EMBL-European Bioinformatics Institute, CB10 1SD Cambridge, UK
e-mail: jvaquerizas@ebi.ac.uk

T.R. Hughes (ed.), *A Handbook of Transcription Factors*, Subcellular Biochemistry 52,
DOI 10.1007/978-90-481-9069-0_13,

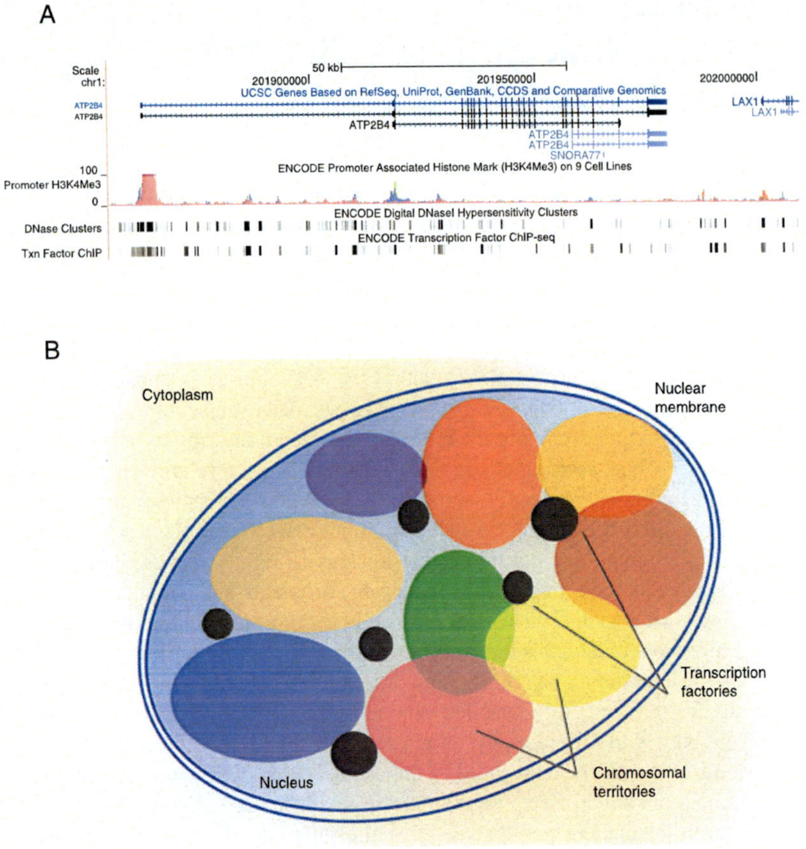

Fig. 13.1 Contrast between the linear representation of transcriptional regulation and a three-dimensional view of chromosomal structure. (**a**) UCSC Genome Browser representation of a 150 kb region of human chromosome 1 [75]. Tracks display gene annotation, histone modifications, DNase I hypersensitivity sites and transcription factor binding sites for the ATP2B4 gene. (**b**) Schematic representation of the 3D organisation of chromatin in an interphase nucleus. Chromatin is not organised linearly, but in the form of chromosomal territories inside the nucleus

mega-bases in length. Each chromosome encodes genes that must be transcribed, as well as regulatory elements that help determine when these events should occur. Given the structure of the DNA molecule, this information is encoded in an intrinsically linear manner, in which genes and regulatory sequences are interspersed between each other (Fig. 13.1a). Binding of sequence-specific transcription factors, as well as the presence or absence of chromatin marks such as DNA methylation or various histone modifications are crucial for the control of gene expression [2]. Owing to the lack of obvious compartments, it appeared as if transcription could take place anywhere within the nucleus; however, it is now becoming increasingly clear that the spatial localisation of regulatory elements and the gene in question plays an essential role in determining transcriptional activity (reviewed in [3–5]).

Early images already demonstrated that the nucleus is highly organised, with individual chromosomes occupying defined spatial territories [6, 7]. Recent publications utilising high-throughput techniques such as Hi-C and ChIA-PET have greatly renewed interest in the three-dimensional chromosomal structure [8–10]. Although the extent of chromosomal movement is still under debate, it is clear that there is substantial amount of chromatin re-organisation in response to changes in environmental signals or cellular state, and that this in turn leads to altered patterns of gene expression (reviewed in [11, 12]).

In this chapter, we introduce the latest advances in our understanding of how nuclear architecture and chromosomal organisation impact on transcription, and discuss some of the principal questions that remain unanswered. It is worth noting here that the field is progressing rapidly, and that there is not necessarily consensus for all the topics that we discuss.

13.2 Chromosomal Territories and Dynamics During Interphase

A common image of eukaryotic DNA is that of Giemsa-stained metaphase chromosomes, in which chromosomes are displayed as highly compacted rods, with a striped pattern of alternating light and dark bands representing gene-rich, early replicating and gene poor, mid-to-late replicating genomic regions [13]. Such images give the impression that chromosomes occupy the nucleus as discrete entities. Pioneering work by the Cremer brothers [14] and the development of in vivo imaging techniques such as fluorescence in-situ hybridisation (FISH) revealed that instead chromosomes are organised in a much more relaxed, but also well-defined conformations during interphase [15]. These regions, known as chromosomal territories, are irregularly shaped, compartmentalised structures of about 1–2 μm in diameter in which chromosomes are contained (for review see [3]; Fig. 13.1b). Advances in chromatin conformation capture techniques (eg, 3C, 4C, 5C; see Table 13.1) have allowed the spatial structure of a few loci to be described at high resolution [16, 17]. Most recently, availability of high-throughput-sequencing technologies now enable these approaches to be applied in a genome-wide fashion (eg, Hi-C and ChIA-PET), and studies have revealed that many seemingly distant genomic loci – when measured linearly along the chromosome – can be spatially close to each other within the nucleus [8–10, 18]. This observation is consistent with the existence of chromosomal territories since most spatially proximal loci belong to the same chromosome. Within each territory, chromatin is suggested to form of a fractal globule, which balances the requirements of efficient packing and ease of locally unfolding specific sections of chromatin [10].

Several studies have examined whether chromosomal territories display specific patterns of arrangement. These studies reported a correlation between gene density

Table 13.1 List of techniques to measure chromatin interactions

Analytical methods	Characteristics	Resolution
Single cell		
Chromosome banding	Characterisation of global chromosome structure using Giemsa staining	Whole chromosome
High-resolution/ 2D/3D DNA/ RNA FISH	Allowes to determine spacial localisation of DNA or RNA sequences by fluorescence in situ hybridisation using locus-specific probes	Variable (~kb – whole genome)
Immunostaining	Detects the location of specific proteins in the nucleus using fluorescent antibodies	Variable (single loci – whole nucleus)
Cell population		
ChIP (-seq, -chip)	Antibody-mediated chromatin precipitation followed by (high-throughput) sequencing or microarray hybridization	~bp (single locus – genome-wide)
DamID	Detects protein–DNA interactions using a *E. coli* DNA adenine methyltransferase fused to the protein of interest. Adenine-methylated DNA fragments are then isolated and quantified by qPCR, microarray or high-throughput sequencing	~bp (single locus – genome-wide)
Dnase I hypersensitivity	Allows the detection of open chromatin regions by digestion with Dnase I followed by ligation-mediated PCR amplification. Amplified producs are then sequenced of hibridised to a microarray	~bp (single locus – genome-wide)
3C	Measures physical interactions between loci in nuclear space by restriction enzyme-mediated digestion of fixed nuclei. This is followed by re-ligation in lax and dilute conditions to favour intramolecular ligation. Ligated pairs correspond to molecules that were in close proximity in the original nuclear space. These are detected through qPCR	~kb (single locus)
4C	Same as 3C, but allows the detection of interactions of a single locus against the rest of the genome by amplification of circularised 3C fragments using inverse PCR. The amplified library is then hybridised against a microarray or sequenced	~kb (single locus vs rest of the genome)
5C	Same as 3C but allows the detection of a large number of interactions between two sets of loci	~kb (multi-locus – genome-wide)
Hi-C/ChIA-PET	Same as 3C but allows detection of interactions for all loci against all the genome. This is achieved by selecting 3C fragments that are then determined using high-thoughput technologies	~0.1–1 Mb (depending on sequencing depth and genome size; multi-locus – genome-wide)

and nuclear localisation although the extent to which this occurs also depends on the cell type: small, gene-rich chromosomal territories often occupy more interior nuclear positions whereas gene-poor ones are located at the nuclear periphery [10]. This is exemplified by human chromosomes 18 and 19 (85 and 67 MB respectively): territories from the gene-rich chromosome 19 tend to localise at the nuclear interior, whereas those of the gene-poor chromosome 18 localise at the periphery [19].

The earliest evidence linking nuclear organisation and transcriptional regulation originated from microscopy studies showing that chromatin at the periphery tends to be compacted, and therefore silent (reviewed in [20]). Since then many studies have explored the relationship between the positioning of genes in these territories and their expression (reviewed in [12]). Within a territory, there are strong associations between expression and the spatial location of the gene with respect to the rest of its territory: active genes tend to be at the surface of a territory, whereas inactive ones tend to be buried in the interior (reviewed in [4, 11]). The Hi-C study observed that inter–loci interactions are highest between regions displaying the same type of expression activity [10]. However, active and inactive regions within a territory do interact also, and single-gene studies have shown that specific loci relocalise within the chromosomal territory depending on transcriptional requirements [21–23]. For example, the Hox B and D gene clusters relocate outwards from their respective territories upon activation; this in turn facilitates interactions of these loci with other chromosomal territories [24, 25]. Interestingly, open and closed chromatin conformations correlate highly with genomic regions that have early and late DNA replication times respectively [26].

In summary, in contrast to the stereotypical image of well-structured, discrete mitotic chromosomes, DNA in the interphase nucleus is highly organised. Individual chromosomes occupy well-defined territories, which bring together apparently distant loci. The spatial localisation of genes with respect to these territories has an important effect on gene expression, and studies of individual loci have reported physical movement between transcriptionally active and inactive locations within the nucleus. Therefore the position and interaction of genes with other chromosomal regions are likely to be important for their correct expression.

13.3 Co-localisation of Active Genes in Transcription Factories

Although the importance of chromosomal localisation for transcription is appreciated, the more detailed mechanisms underlying the regulation are less understood. In particular, it is not clear what drives specific loci to relocate from one place to another, and why genes occupying similar spatial positions tend to display correlated expression.

Genes with especially high levels of expression display a tendency to reside in specific locations with high concentrations of RNA polymerase II. Known as transcription factories, such nuclear regions are thought to produce much of the mRNA within a cell; however neither their identity, nor mode of action is understood [27]. Transcription factories were initially identified by Cook and colleagues who measured the incorporation of radiolabelled nucleotides into nascent transcripts [28]. Despite a prior expectation of random nuclear localisation, the modified nucleotide analogues were found in discrete foci across the nucleus. Transcription factories have been also visualised by immunofluorescence using antibodies against the elongating, Ser2-phosphorylated form of RNA polymerase II (Pol II) [29]. Although the number and size of these factories depend on the cell type, a typical eukaryotic nucleus is estimated to contain from a few hundreds to a few thousand factories, with an average of eight Pol II molecules in each [27].

The concept of transcription factories with an immobilised Pol II is attractive as it helps explain gene movement: since the number of factories is limited, chromosomal regions must position genes in the correct locations for expression. A consequence of the model is that Pol II molecules should remain largely fixed in space, with the DNA being threaded through in order to achieve transcription. However, several questions about transcription factories remain. For example, so far it has not been possible to purify the components of factories, so there is little understanding about what they contain and how they assemble. Further, the model fails to explain several features of the transcriptional process: (i) the existence of bursts of gene expression requires more Pol II units than that contained in an average factory [30–32]; (ii) simultaneous expression of divergently coded genes would not be allowed, as this would mean reeling the DNA fibre in opposite directions at the same time; and (iii) the presence of transcriptional activity at the nuclear periphery close to the nuclear pore, where at least in some cell types, transcription factories are usually not found [33].

An alternative explanation is the nuclear-speckle model, which is supported by the observation of splicing factor-enriched locations within the nucleus. By coupling transcription with mRNA-processing and export, expressed genes would then naturally congregate at specific nuclear locations [34, 35, 74]. Thus the accumulation of Pol II simply arises as a consequence, rather than a requirement of the model. There are difficulties here also however, since inhibition of transcription leads to decreased chromosomal mingling and relocation [25, 33, 36], suggesting that the transcriptional process itself influences nuclear compartmentalisation.

Both of the above models suggest that expressed genes aggregate in specific locations within the nucleus (Fig. 13.2). In either case, the mechanisms for controlling relocation are unknown and it will be interesting to see how the field develops in the near future.

13.4 Nuclear Co-localisation of Regulatory Elements

So far, we have focused on the co-localisation of the genes themselves. However, an important aspect of transcription is the involvement of cis-regulatory elements,

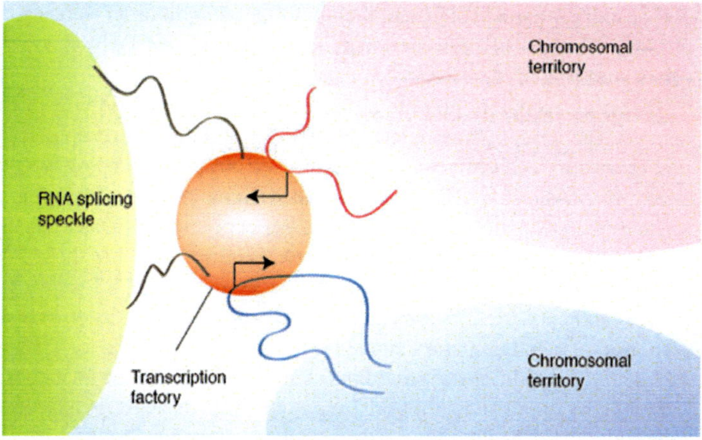

Fig. 13.2 Schematic representation of chromosomal territories, transcription factories, RNA speckles. Within each chromosomal territory genes are usually arranged according to their level of expression. Active genes are situated at the periphery of territories (*red* and *blue lines*) from which they can associate with a transcription factory (*light red*). The RNA products from these genes (*grey*) can then interact with the RNA-splicing machinery within nuclear speckles (*light green*)

typically transcription factor-binding sites. We will now discuss the relationship between spatial localisation and these elements.

Early studies of sequenced eukaryotic genomes assumed that most genes tend to be encoded randomly along chromosomes. Instead, it is now clear that co-expressed genes and those belonging to the same protein complex or pathways tend to be clustered. Well-characterised examples include the alpha- and beta-globin clusters, and the Hox genes [37, 38]. It is thought that by retaining these genes in clusters, common sets of regulatory enhancer elements could be used to ensure coordinated expression patterns [39–41]. This observation is supported by the observation in yeast that target genes of a given transcription factor are found in clusters on specific chromosomes [42].

An intriguing aspect of higher eukaryotic genomes is that regulatory elements such as transcription factor binding sites are often located many kilobases away – sometimes even on different chromosomes – from the target gene. This contrasts with most microbial genomes, for which binding sites usually reside within the promoter region directly upstream of genes. Genome-scale surveys of multiple transcription factors have reported such distal binding, including the oestrogen receptor, GATA1 and the Gli family of regulators [9, 43–46].

Long–range physical interactions between distal elements and their respective targets are thought to be a primary mechanism for transcriptional control [37]. There are known classes of transcription factors and co-factors that bend DNA, and so bring together distant loci: for instance, the SP1 and HMG families achieve this either by binding and distorting the major or minor DNA grooves or by binding two or more separated sites and then multimerising in order to bring them together [47–49]. These structures can form loops of protruding DNA that help co-localise genes with shared regulatory elements [16, 37].

Olfactory receptors provide an excellent example of looping in transcriptional control. The mouse genome encodes about 1,300 distinct receptors; however from this repertoire, each olfactory cell expresses only one receptor type from a single allele. Curiously, the "H" enhancer that controls olfactory receptor expression is located on a separate chromosome. Recent experiments measuring the position of the H element and the expressed receptor gene showed that these loci co-localise within the nucleus [50]. In another example, from $CD4^+$ T cells, the interferon-gamma and cytokine loci were shown to co-localise in order to activate quickly in response to an inflammatory signal [51].

Finally, in addition to providing control, it is proposed that these long–range interactions confer transcriptional memory [52]. In yeast, it has been shown that looping and the localisation of genes to the nuclear pore enables rapid re-induction following short periods of repression. When repression is prolonged however, the loop is lost and re-induction becomes much slower.

13.5 Possible Mechanisms for Chromatin Re-arrangement

In the interphase nucleus, chromatin moves according to the model of constrained diffusion: in other words, movement is much slower than that observed for DNA in solution [12]. This is likely to be caused by the interactions between chromatin and other nuclear elements including the different chromosomal territories themselves. However, in contrast to the slow movement at a chromosomal level, individual loci taking part in long-range intra- and inter-chromosomal interactions appear to move much more quickly. Unfortunately, little is known about how such fast transitions are achieved for long–range chromatin interactions, and how they change in response to cellular requirements.

Chromosomal movements are due at least in part to the transcriptional process, as inhibition of RNA polymerase II activity leads to reduced intermingling between chromosomes [25, 36]. However, it remains unclear whether transcription is itself responsible for chromatin re-organisation or whether chromatin has to be relocated to a transcriptionally active region in order to be transcribed. Interesting insight regarding this issue comes from the study of the mammalian inactive X chromosome. In females, one of the two copies of the X chromosome is randomly inactivated during development to compensate for a dose imbalance (reviewed in [53]). This is achieved through a silencing mechanism involving the expression of a non-coding RNA (Xist) that coats the inactive X chromosome and creates a barrier for the transcriptional machinery. Interestingly, a few X-linked genes escape inactivation and locate at the periphery of the inactive X chromosomal territory where they are expressed, whereas inactive genes localise at the interior of the chromosomal territory [54]. This suggests that relocation to a transcription factory is a pre-requisite for gene expression.

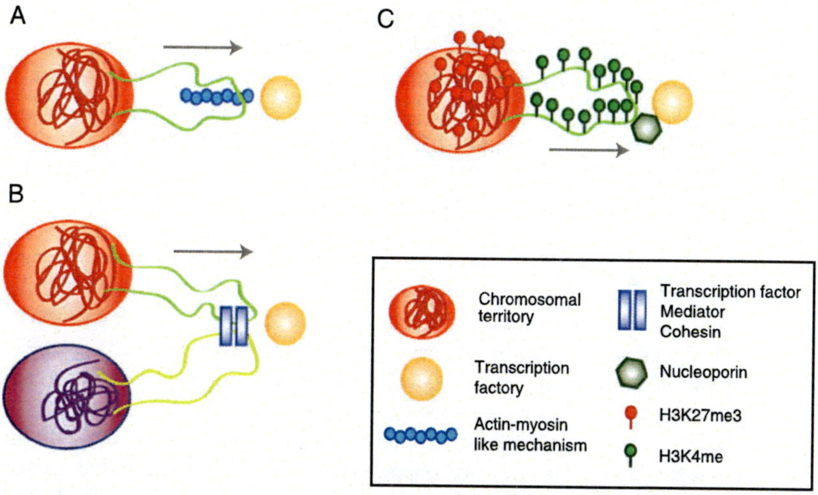

Fig. 13.3 Proposed mechanisms for forming chromatin architecture. (**a**) An actin-myosin like mechanism has been proposed for chromatin rearrangements [55]. (**b**) Transcription factors, such as Klf1, or components of the mediator or cohesin complex are involved in mediating the association of target genes with active transcription factories [56, 57]. (**c**) Chromatin modifications and nucleoporins might also modulate chromatin remodelling and association with transcription factories [58, 66]. (*Box*) Cellular components associated with chromosomal architecture

There are also further effectors beyond the polymerase. Recently, chromatin unfolding mediated by the transactivation domain of VP16 was shown to be sufficient to produce defined unidirectional chromosomal movements oriented perpendicular to the nuclear envelope, even under Pol II-inhibited conditions [55]. Moreover, long-range chromosomal movement was affected when cells were treated with inhibitors of nuclear actin-myosin. These results strongly suggest that there are motorised mechanisms driving chromosomal re-arrangements in the interphase nucleus (Fig. 13.3a).

If specific relocation mechanisms exist, what are the components involved in their functioning? Transcription factors (TFs) are one of the obvious candidates. This view is supported by the fact that several sequence-specific DNA-binding TFs are able to produce long-range chromatin-unfolding changes. These include well-studied members of this family of proteins, including p53 and the oestrogen receptor. The acidic activator domains of these TFs seem to be responsible for interacting, either directly or indirectly through other cofactors, with proteins that would mediate the re-localisation of the chromatin (Fig. 13.3b). These components, however, have not been discovered yet. Further support for this hypothesis results from a recent study of chromosomal interactions that revealed the role of the transcription factor Klf1 in mediating interactions between Klf1-regulated genes and specialised sets of transcription factories [56].

Most recently, work from the Taatjes, Dekker and Young laboratories shed some light on the components that determine nuclear architecture [57]. By systematically screening shRNA-mediated gene knockdowns, the authors identified a set of genes whose repression directly affects transcriptional regulation as measured by the loss of Oct-4 in embryonic stem cells. Surprisingly, the list of interfering genes contained a significant proportion of members of the mediator complex as well as several subunits of the cohesin complex. ChIP-seq revealed that mediator and cohesin bind and co-occupy enhancers and gene promoters of active genes. This co-localisation, which was shown to be involved in the formation of DNA loops, is cell-type specific, and hence might be important in determining chromosome architecture in particular cell types.

Another possibility for controlling chromosomal architecture is offered by chromatin modifications. Histone acetylation and methylation are known to play a fundamental role in regulating gene expression. Recent work by Shopland and colleagues showed specific patterns of higher-order folding for a ~4 MB fraction of mouse chromosome 14 (Fig. 13.3c). This organisation is accompanied by specific patterns of chromatin modifications such as H3K4me or H3K27me3 [58]. Some of these modifications have been linked to physical interactions with different sub-nuclear structures: for example, that of H3K27me3 with the nuclear lamina or H4K16ac with the nuclear pore (discussed in detail below). Such interactions explain how particular histone modifications help determine long–range chromatin interactions. In support of this, global changes in histone acetylation – mediated for example by inhibitors of histone deacetylases – result in a dramatic change of gene expression patterns and nuclear localisation [73]. However, establishing a causal link between chromatin marks and nuclear organisation is difficult and the mechanistic understanding of the hierarchical relationship between them in yet to be determined.

13.6 Nuclear Structural Proteins and Transcriptional Control

As noted above, a major source of chromatin organisation appears to be the interaction between chromatin and protein components of sub-nuclear structures [59, 60]. Such interactions were described almost 60 years ago and have been subjected to very intense research [61]. One of the best-studied mechanisms is the interaction of heterochromatin with the nuclear lamina, a mesh of proteins that coat the inner surface of the nuclear membrane. Genome-scale studies in fly and human examined their relationship with chromatin and transcriptional regulation using the DamID technique [62, 63]. The results demonstrated that these interactions occur across large, continuous, chromosomal regions that span up to 500 Mb in the human genome. These lamin-associated regions showed characteristic marks of repressive chromatin such as H3K27me3, and low levels of gene expression. These findings are consistent with earlier observations that DNA at the nuclear periphery tends to form compacted chromatin [20].

However, there is increasing evidence that the nuclear periphery can also be involved in gene activation. Initial work in yeast already showed that genes relocalise to the nuclear periphery, and in particular to the nuclear pore complex, upon induction [64], although similar experiments in humans failed to provide conclusive results [73]. Since the nuclear membranes of yeast and higher eukaryotes contain significantly different sets of proteins, it remained unclear whether the link between transcriptional activity and the nuclear pore is a general mechanism.

Independent studies by our own laboratories in flies provided early, compelling evidence for this link in higher eukaryotes. First, we identified physical interactions between members of the dosage compensation complex – which mediates large-scale transcriptional activity on the single male X chromosome – and subunits of the nuclear pore complex [65]. Most recently, using ChIP-chip we described interactions between two nucleoporins, Nup153 and Megator (the fly homologue of Tpr), with up to 25% of the fly genome, in the form of extended regions of high-density binding (Fig. 13.4) [66]. These interactions interspersed with the

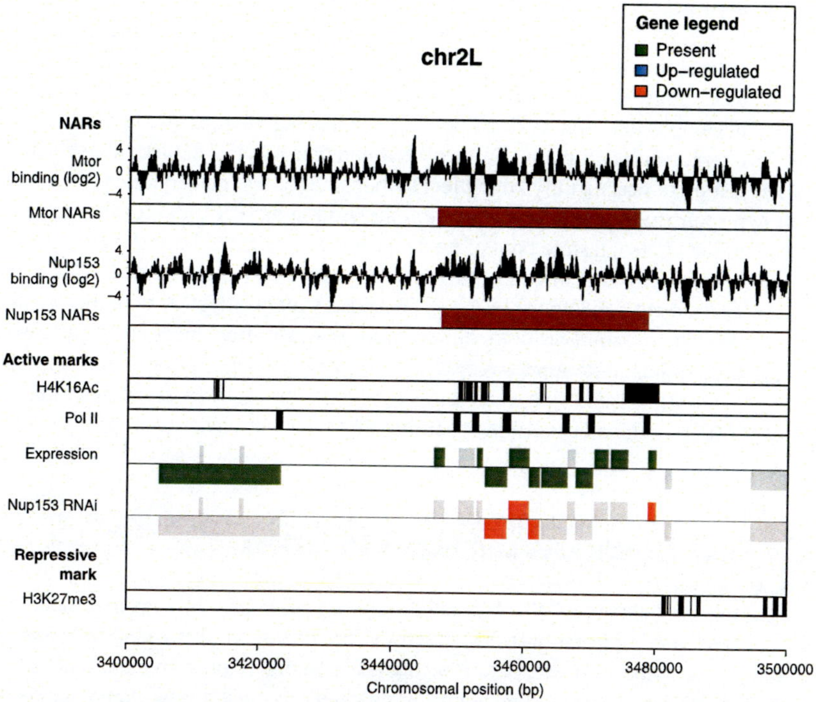

Fig. 13.4 Nucleoporin-associated regions define transcriptionally active regions in the genome. Genome-browser representation of a 100 kb section from *D. melanogaster* chromosome 2L. The first and third tracks depict binding profiles of the nucleoporins Nup153 and Mtor [66]. Nucleoporin associated regions (NARs; depicted in *dark red*) are enriched for transcribed genes compared with non-NARs (gene expression track; *green shading*). RNAi-mediated depletion of Nup153 results in a global down regulation of active genes in NARs (Nup153 RNAi track; *red shading*). Nucleoporin domains also align with markers of transcriptionally active chromatin such as RNA Pol II or H4K16ac, but exclude markers for inactive chromatin such as H3K27me3

lamin-bound regions and were significantly enriched in marks of active chromatin such as H4K16ac, MOF and RNA Pol II binding [67]. In support of this, the human orthologue Tpr has been recently associated with the formation of heterochromatin exclusion zones [68]. Gene-expression analysis revealed that genes within these nucleoporin-associated regions (NARs) tend to be expressed, and depletion of either nucleoporin caused a dramatic decrease in gene expression within these regions [66]. In agreement to this, another two studies showed an association of different nucleoporins with transcriptionally active genes involved in developmental processes and the cell cycle [69, 70]. Overall, these results suggested a strong implication of proteins of the nuclear pore complex in regulating gene expression and established nucleoporins as a major class of transcriptional regulators.

However, since the surface area provided by nuclear pores at the membrane is very small compared with the amount of chromatin in the nucleus, it was difficult to reconcile how so much binding could be achieved. Given that some nucleoporins are known to reside both at the nuclear pore and within the nucleoplasm [71], a possible answer is that the pool of soluble nucleoplasmic nucleoporin provides some of this binding. By performing three-dimensional FISH experiments to visualise the localisation of nucleoporin-associated loci, we showed that both nucleoplasmic and nuclear pore-associated populations of Nup153 are likely to interact with chromatin. Further we found that Nup153-depletion results in delocalisation of peripherally bound loci demonstrating that nucleoporins are at least partly responsible for chromatin positioning. Additional evidence for the regulatory role of the nucleoplasmic pool of nucleoporins comes from the studies of soluble nucleoporins involved in controlling expression for cell-cycle related or developmental genes mentioned above [69, 70]. Our observations, together with these reports, that describe intranuclear puffs of nucleoporin localisation under ecdysone treatment or heat shock conditions, suggest a strong link between nucleoporins and the establishment of transcription factories in a variation of the classical gene-gating model [60, 72]. The mechanisms governing these interactions remain to be found.

13.7 Conclusions and Future Directions

Our understanding of chromatin architecture and its implications have expanded dramatically in the last few years. On the one hand, this has been possible due to the detailed characterisation of nuclear dynamics for individual loci at very high resolution; on the other hand, these observations are now complemented by the development of high-throughput techniques such as Hi-C. These studies have confirmed that the spatial organisation of chromatin in the interphase nucleus has a profound effect on transcription.

Models have been proposed to explain how chromatin positioning might be coupled to changes in expression levels. However, none of these models can accommodate the full range of observations that been made in different organisms. Owing to the potential fundamental importance of chromatin architecture to gene

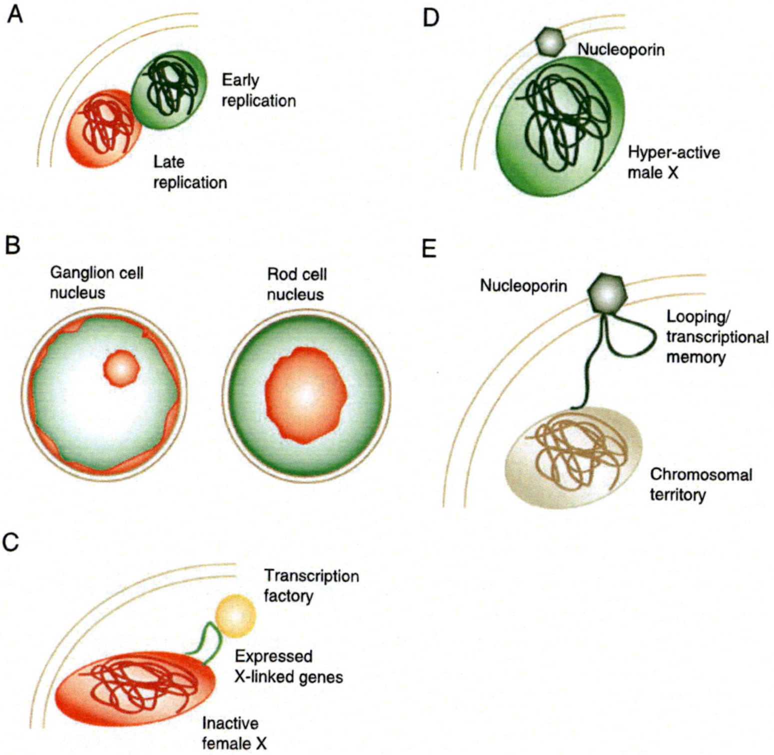

Fig. 13.5 Biological processes associated with chromatin remodelling. Many different biological processes have been associated with changes in chromosomal architecture. (**a**) Replication timing is one of them, in which early and late replication regions cluster together in the interphase nucleus in separate territories [26]. (**b**) Chromatin conformation in rod cells of nocturnal animals [76]. Retinal rod cells in these organisms have evolved a special organisation in which heterochromatic regions cluster at the middle of the nucleus instead of the periphery. This arrangement allows the nucleus to act as a lens that concentrates low levels of light. (**c**) Mammalian X-chromosome inactivation [77]. Expressed X-linked genes are evicted from the inactive X-chromosomal territory and re-positioned to a transcription factory. The product of the Xist gene is then be used to coat the inactive X chromosome. (**d**) Dosage compensation in fly. The single male X chromosome is up-regulated twofold to match the transcriptional output of the two active female X chromosomes. The single male X chromosome localises close to the nuclear periphery. Interestingly, we found an enrichment of interactions between the male X chromosome and subunits of the nuclear pore [66]. Depletion of nucleoporins leads to a relocation of peripheral regions to the interior of the nucleus. Furthermore, depletion of nucleoporins impairs the recruitment of members of the dosage compensation complex. (**e**) DNA looping at the nuclear periphery in yeast has been shown to contribute to transcriptional memory [52]. Active genes are localised at the nuclear periphery upon activation. Short-term changes in cellular conditions cause transcriptional repression but genes remain positioned at the nuclear periphery. When conditions revert to the original state, gene expression resumes promptly. However, if environmental changes are maintained for a longer period, genes are removed towards the interior of the nucleus and transcriptional activation is much slower. Full expression levels are only achieved after genes are relocated again to the nuclear periphery

expression – including short- and long-term memory to the modulation of immune responses – further insights into these mechanisms will be essential to understand how such cellular responses are controlled (Fig. 13.5).

Several key steps are still necessary to achieve this. The first step will be the full identification and characterisation of the components involved in establishing and modifying chromatin conformation. Once these components have been determined, a combination of biochemical, genetic and genomics experiments will provide mechanistic insights into how chromatin conformation impacts gene expression and its relationship to other components of the transcriptional regulatory system. Another inevitable step will be the development of high-throughput imaging techniques that allow examination of changes in chromatin conformations in vivo. Such developments will allow us eventually to understand the dynamics of nuclear architecture and its physiological function.

References

1. Jacob F, Monod J (1961) Genetic regulatory mechanisms in the synthesis of proteins. J Mol Biol 3:318–356
2. Lemon B, Tjian R (2000) Orchestrated response: a symphony of transcription factors for gene control. Genes Dev 14(20):2551–2569
3. Cremer T, Cremer C (2001) Chromosome territories, nuclear architecture and gene regulation in mammalian cells. Nat Rev Genet 2(4):292–301
4. Lanctot C, et al. (2007) Dynamic genome architecture in the nuclear space: regulation of gene expression in three dimensions. Nat Rev Genet 8(2):104–115
5. Misteli T (2007) Beyond the sequence: cellular organization of genome function. Cell 128(4):787–800
6. Dietzel S, et al. (1998) Separate and variably shaped chromosome arm domains are disclosed by chromosome arm painting in human cell nuclei. Chromosome Res 6(1):25–33
7. Zink D, et al. (1998) Structure and dynamics of human interphase chromosome territories in vivo. Hum Genet 102(2):241–251
8. Duan Z, et al. (2010) A three-dimensional model of the yeast genome. Nature 465(7296): 363–367
9. Fullwood MJ, et al. (2009) An oestrogen-receptor-alpha-bound human chromatin interactome. Nature 462(7269):58–64
10. Lieberman-Aiden E, et al. (2009) Comprehensive mapping of long–range interactions reveals folding principles of the human genome. Science 326(5950):289–293
11. Fraser P, Bickmore W (2007) Nuclear organization of the genome and the potential for gene regulation. Nature 447(7143):413–417
12. Hübner MR, Spector DL (2010) Chromatin dynamics. Annu Rev Biophys 39: 471–489
13. Holmquist GP (1992) Chromosome bands, their chromatin flavors, and their functional features. Am J Hum Genet 51(1):17–37
14. Zorn C, et al. (1979) Unscheduled DNA synthesis after partial UV irradiation of the cell nucleus. Distribution in interphase and metaphase. Exp Cell Res 124(1):111–119
15. Speicher MR, Carter NP (2005) The new cytogenetics: blurring the boundaries with molecular biology. Nat Rev Genet 6(10):782–792
16. Dekker J (2008) Gene regulation in the third dimension. Science 319(5871):1793–1794
17. Naumova N, Dekker J (2010) Integrating one-dimensional and three-dimensional maps of genomes. J Cell Sci 123(Pt 12):1979–1988

18. Rodley CDM, et al. (2009) Global identification of yeast chromosome interactions using Genome conformation capture. Fungal Genet Biol 46(11):879–886
19. Croft JA, et al. (1999) Differences in the localization and morphology of chromosomes in the human nucleus. J Cell Biol 145(6):1119–1131
20. Akhtar A, Gasser SM (2007) The nuclear envelope and transcriptional control. Nat Rev Genet 8(7):507–517
21. Branco MR, Pombo A (2006) Intermingling of chromosome territories in interphase suggests role in translocations and transcription-dependent associations. PLoS Biol 4(5):e138
22. Simonis M, et al. (2006) Nuclear organization of active and inactive chromatin domains uncovered by chromosome conformation capture-on-chip (4C). Nat Genet 38(11):1348–1354
23. Volpi EV, et al. (2000) Large-scale chromatin organization of the major histocompatibility complex and other regions of human chromosome 6 and its response to interferon in interphase nuclei. J Cell Sci 113(Pt 9):1565–1576
24. Chambeyron S, Bickmore WA (2004) Chromatin decondensation and nuclear reorganization of the HoxB locus upon induction of transcription. Genes Dev 18(10):1119–1130
25. Morey C, et al. (2007) Nuclear reorganisation and chromatin decondensation are conserved, but distinct, mechanisms linked to Hox gene activation. Development 134(5):909–919
26. Ryba T, et al. (2010) Evolutionarily conserved replication timing profiles predict long-range chromatin interactions and distinguish closely related cell types. Genome Res 20(6): 761–770
27. Sutherland H, Bickmore WA (2009) Transcription factories: gene expression in unions? Nat Rev Genet 10(7):457–466
28. Iborra FJ, et al. (1996) Active RNA polymerases are localized within discrete transcription "factories" in human nuclei. J Cell Sci 109(Pt 6):1427–1436
29. Phatnani HP, Greenleaf AL (2006) Phosphorylation and functions of the RNA polymerase II CTD. Genes Dev 20(21):2922–2936
30. Chubb JR, et al. (2006) Transcriptional pulsing of a developmental gene. Curr Biol 16(10):1018–1025
31. Darzacq X, et al. (2007) In vivo dynamics of RNA polymerase II transcription. Nat Struct Mol Biol 14(9):796–806
32. Jackson DA, et al. (1998) Numbers and organization of RNA polymerases, nascent transcripts, and transcription units in HeLa nuclei. Mol Biol Cell 9(6):1523–1536
33. Ragoczy T, et al. (2006) The locus control region is required for association of the murine beta-globin locus with engaged transcription factories during erythroid maturation. Genes Dev 20(11):1447–1457
34. Brown JM, et al. (2006) Coregulated human globin genes are frequently in spatial proximity when active. J Cell Biol 172(2):177–187
35. Osborne CS, et al. (2004) Active genes dynamically colocalize to shared sites of ongoing transcription. Nat Genet 36(10):1065–1071
36. Mahy NL, Perry PE, Bickmore WA (2002) Gene density and transcription influence the localization of chromatin outside of chromosome territories detectable by FISH. J Cell Biol 159(5):753–763
37. de Laat W, Grosveld F (2003) Spatial organization of gene expression: the active chromatin hub. Chromosome Res 11(5):447–459
38. Sproul D, Gilbert N, Bickmore WA (2005) The role of chromatin structure in regulating the expression of clustered genes. Nat Rev Genet 6(10):775–781
39. Hurst LD, Pál C, Lercher MJ (2004) The evolutionary dynamics of eukaryotic gene order. Nat Rev Genet 5(4):299–310
40. Spitz F, Gonzalez F, Duboule D (2003) A global control region defines a chromosomal regulatory landscape containing the HoxD cluster. Cell 113(3):405–417
41. Stamatoyannopoulos G (2005) Control of globin gene expression during development and erythroid differentiation. Exp Hematol 33(3):259–271

42. Janga SC, Collado-Vides J, Babu MM (2008) Transcriptional regulation constrains the organization of genes on eukaryotic chromosomes. Proc Natl Acad Sci USA 105(41):15761–15766
43. Fujiwara T, et al. (2009) Discovering hematopoietic mechanisms through genome-wide analysis of GATA factor chromatin occupancy. Mol Cell 36(4):667–681
44. Hallikas O, et al. (2006) Genome-wide prediction of mammalian enhancers based on analysis of transcription-factor binding affinity. Cell 124(1):47–59
45. Lin C, et al. (2007) Whole-genome cartography of estrogen receptor alpha binding sites. PLoS Genet 3(6):e87
46. Liu Y, et al. (2008) The genome landscape of ERalpha- and ERbeta-binding DNA regions. Proc Natl Acad Sci USA 105(7):2604–2609
47. Love JJ, et al. (1995) Structural basis for DNA bending by the architectural transcription factor LEF-1. Nature 376(6543):791–795
48. Sjøttem E, Andersen C, Johansen T (1997) Structural and functional analyses of DNA bending induced by Sp1 family transcription factors. J Mol Biol 267(3):490–504
49. Su W, et al. (1991) DNA looping between sites for transcriptional activation: self-association of DNA-bound Sp1. Genes Dev 5(5):820–826
50. Lomvardas S, et al. (2006) Interchromosomal interactions and olfactory receptor choice. Cell 126(2):403–413
51. Spilianakis CG, et al. (2005) Interchromosomal associations between alternatively expressed loci. Nature 435(7042):637–645
52. Tan-Wong SM, Wijayatilake HD, Proudfoot NJ (2009) Gene loops function to maintain transcriptional memory through interaction with the nuclear pore complex. Genes Dev 23(22):2610–2624
53. Straub T, Becker PB (2007) Dosage compensation: the beginning and end of generalization. Nat Rev Genet 8(1):47–57
54. Dietzel S, et al. (1999) The 3D positioning of ANT2 and ANT3 genes within female X chromosome territories correlates with gene activity. Exp Cell Res 252(2):363–375
55. Chuang C, et al. (2006) Long-range directional movement of an interphase chromosome site. Curr Biol 16(8):825–831
56. Schoenfelder S, et al. (2010) Preferential associations between co-regulated genes reveal a transcriptional interactome in erythroid cells. Nat Genet 42(1):53–61
57. Kagey MH, et al. (2010) Mediator and cohesin connect gene expression and chromatin architecture. Nature 467(7314):430–435
58. Shopland LS, et al. (2003) Clustering of multiple specific genes and gene-rich R-bands around SC-35 domains: evidence for local euchromatic neighborhoods. J Cell Biol 162(6): 981–990
59. Kind J, van Steensel B (2010) Genome–nuclear lamina interactions and gene regulation. Curr Opin Cell Biol 22(3):320–325
60. Strambio-De-Castillia C, Niepel M, Rout MP (2010) The nuclear pore complex: bridging nuclear transport and gene regulation. Nat Rev Mol Cell Biol 11(7):490–501
61. Brenner S (1953) The chromatic nuclear membrane. Exp Cell Res 5(1):257–260
62. Guelen L, et al. (2008) Domain organization of human chromosomes revealed by mapping of nuclear lamina interactions. Nature 453(7197):948–951
63. Pickersgill H, et al. (2006) Characterization of the *Drosophila melanogaster* genome at the nuclear lamina. Nat Genet 38(9):1005–1014
64. Casolari JM, et al. (2004) Genome-wide localization of the nuclear transport machinery couples transcriptional status and nuclear organization. Cell 117(4):427–439
65. Mendjan S, et al. (2006) Nuclear pore components are involved in the transcriptional regulation of dosage compensation in *Drosophila*. Mol Cell 21(6):811–823
66. Vaquerizas JM, et al. (2010) Nuclear pore proteins nup153 and megator define transcriptionally active regions in the *Drosophila* genome. PLoS Genet 6(2):e1000846
67. Kind J, et al. (2008) Genome-wide analysis reveals MOF as a key regulator of dosage compensation and gene expression in *Drosophila*. Cell 133(5):813–828

68. Krull S, et al. (2010) Protein Tpr is required for establishing nuclear pore-associated zones of heterochromatin exclusion. EMBO J 29(10):1659–1673
69. Capelson M, et al. (2010) Chromatin-bound nuclear pore components regulate gene expression in higher eukaryotes. Cell 140(3):372–383
70. Kalverda B, et al. (2010) Nucleoporins directly stimulate expression of developmental and cell-cycle genes inside the nucleoplasm. Cell 140(3):360–371
71. Rabut G, Doye V, Ellenberg J (2004) Mapping the dynamic organization of the nuclear pore complex inside single living cells. Nat Cell Biol 6(11):1114–1121
72. Blobel G (1985) Gene gating: a hypothesis. Proc Natl Acad Sci USA 82(24):8527–8529
73. Brown CR, et al. (2008a) Global histone acetylation induces functional genomic reorganization at mammalian nuclear pore complexes. Genes Dev 22(5):627–639
74. Brown JM, et al. (2008b) Association between active genes occurs at nuclear speckles and is modulated by chromatin environment. J Cell Biol 182(6):1083–1097
75. Karolchik D, et al. (2008) The UCSC Genome Browser Database: 2008 update. Nucleic Acids Res 36(Database issue):D773–9
76. Solovei I, et al. (2009) Nuclear architecture of rod photoreceptor cells adapts to vision in mammalian evolution. Cell 137(2):356–368
77. Chaumeil J, et al. (2006) A novel role for Xist RNA in the formation of a repressive nuclear compartment into which genes are recruited when silenced. Genes Dev 20(16):2223–2237

Index

T.R. Hughes (ed.), *A Handbook of Transcription Factors*, Subcellular Biochemistry 52, DOI 10.1007/978-90-481-9069-0, © Springer Science+Business Media B.V. 2011